Cities and Nature

Series Editors

Peter Newman, Sustainability Policy Institute, Curtin University, Perth, WA, Australia

Cheryl Desha, School of Engineering and Built Environment, Griffith University, Nathan, QLD, Australia

Alessandro Sanches-Pereira, Instituto 17, São Paulo, São Paulo, Brazil

Cities and Nature fosters high-quality multi-disciplinary research addressing the interface between cities and the natural environment. It provides a valuable source of relevant knowledge for researchers, planners and policy-makers. The series welcomes empirically based, cutting-edge and theoretical research in urban geography, urban planning, environmental planning, urban ecology, regional science and economics. It publishes peer-reviewed edited and authored volumes on topics dealing with the urban and the environment nexus, including: spatial dynamics of urban built areas, urban and peri-urban agriculture, urban greening and green infrastructure, environmental planning, urban forests, urban ecology, regional dynamics and landscape fragmentation.

More information about this series at http://www.springer.com/series/10068

Francesca Di Pietro · Amélie Robert
Editors

Urban Wastelands

A Form of Urban Nature?

Springer

Editors
Francesca Di Pietro
University of Tours
Tours, France

Amélie Robert
University of Tours
Tours, France

ISSN 2520-8306 ISSN 2520-8314 (electronic)
Cities and Nature
ISBN 978-3-030-74881-4 ISBN 978-3-030-74882-1 (eBook)
https://doi.org/10.1007/978-3-030-74882-1

This Springer imprint is published by the registered company Springer Nature Switzerland AG
The registered company address is: Gewerbestrasse 11, 6330 Cham, Switzerland

Urban Wastelands as a Form of Urban Nature: An Introduction

What are urban wastelands? How to define them? Wastelands are particular spaces inside and highly associated with the city. "Wastelands," "brownfields," "vacant lots," "vacant land," "terrains vagues"... Under different names and definitions, common characteristics emerge: these spaces have no official use, they are temporary, and they are located in urban areas. Highly variable in shape and size, in their place on the urban gradient, in their longevity and history, in their uses by residents, urban wastelands are generally present in all cities, whatever their size. Urban wastelands are former residential, industrial, or service areas involved in urban renewal or located in shrinking cities, or former cultivated fields encompassed in the city by urban sprawl. In all cases, the existence of urban wastelands is closely linked to the evolution of cities: they are an integral part of them.

Growing attention is nowadays paid to nature in cities. The literature on this topic is abundant, and the benefits of attendance and proximity to nature for the well-being and mental and physical health of city dwellers are widely studied. The public authorities' answer to this social demand for urban nature is the system of public green spaces. Scarce in quantity and normative in quality, this answer is today under discussion as the nature used by residents encompasses multiple forms: woodland, street trees and planted squares, cemeteries, private gardens, community and allotment gardens, outdoor sports centers... and wastelands.

Over the last 20 years, research work on urban wastelands has been constantly increasing and concerns several scientific fields, such as urban planning, geography, social sciences, and more recently urban ecology and soil science. The disciplinary field that has best dealt with urban wastelands is urban planning, from the perspective of building development. More recently, research has focused on urban wastelands in their present state, as informal spaces of nature, mainly in the social sciences and urban ecology. However, there is a lack of a book dealing specifically with urban wastelands as semi-natural spaces. The research studies presented in this book highlight the multiple values of urban wastelands: values *per se*, as urban nature, and comparative value, in the sense of an alternative to public green spaces. Moreover, we observe that in certain public green spaces, some piece of less regulated nature is sometimes left. These portions of "wild" vegetation are often called "wasteland" by green space designers and users. This choice, driven by both ecological and aesthetic concerns, shows that in our highly artificial cities there is emerging a preference for a freer informal nature—at different levels according to societies— this untamed nature whose attributes are pointed out by the philosopher Virginie Maris: its exteriority, its otherness, and its autonomy (Maris 2018).

Gradually, urban wastelands are evolving from the "mistake" that they once were into not only an alternative, but a model of nature.

Urban wastelands being spontaneously greened, they constitute on the one hand novel ecosystems, or emerging ecosystems (Hobbs et al. 2006), and on the other hand informal green spaces in the city. To what extent can we consider these spaces as a form of nature in cities, albeit temporary? And how can we protect these urban biodiversity hotspots and their informal uses? To fully answer these questions, it is essential to consider the ecological, social, and urban planning points of view. This book thus examines these three dimensions of green urban wastelands, by considering their ecology, the practices and perceptions of their users and other inhabitants, and the way they are included in urban planning.

Urban Wastelands as Biodiversity Hotspots

Recent research in urban ecology is paying growing attention to urban wastelands and considers them as valuable places for urban biodiversity. Due to their spontaneous vegetation, urban wastelands occupy a special place in the range of semi-natural spaces in cities: their management, if any, is generally rare and irregular. Urban wastelands are thus the least managed green space in the city. The weak management of these areas makes them the best indicators of the diversity and history of the urban landscape, both being potentially included in the soil seed bank (Johnson et al. 2017).

Over a period of years, such sites can develop habitat structures unique to urban areas (Kowarik 2013), providing valuable retreats as well as substitute or stepping-stone habitats for animal and plant species. As a result, plant and animal communities develop as a function of the characteristics of the wasteland itself (its surface area, period of abandonment, type of soil, microclimate, kind of previous use, etc.) and of the area surrounding it: contiguity with other vegetated areas or other wastelands, adjacent land use, height of adjacent buildings, etc. (Bonthoux et al. 2014).

Due to their highly variable longevity and local and landscape characteristics, urban wastelands may potentially host a high biodiversity. As a result of the little intense management, they may, however, also provide habitats for alien, sometimes invasive species, arriving there in a spontaneous manner or having been introduced into surrounding spaces. This is stressed by Machon (Chapter "Urban Wastelands Can Be Amazing Reservoirs of Biodiversity for Cities"), who underlines some factors and conditions of this ecological diversity and highlights that their diversity is optimal when their size exceeds a few thousand square meters, when their age is between 15 and 20 years, and when they are connected to other biodiversity reservoirs by vegetated linear infrastructures.

Urban wastelands are novel ecosystems, which bring together species unlikely to be found in other "historical" environments, as the result of human action. Gallagher (Chapter "Natural Transformation of Post-industrial Lands: Liberty State Park in Jersey City, NJ (USA)") shows an interesting example of a novel ecosystem with an urban wasteland on metal-contaminated soils (a former railyard that has developed into a mix of urban forest shrubland and meadow, with its legacy of metal contaminants), where the plant communities are a novel mix of old and new world species and do not follow the traditional paths of succession. The contaminant transfer throughout the food web, examined at several trophic levels, appears to be minimal; exotic species better tolerate high levels of metals in the soil, and even provide ecosystem services, as bioremediation.

Local and landscape factors contribute to urban wasteland biodiversity, as confirmed by McKinney (Chapter "Strategies for Increasing Biodiversity Conservation in Cities Using Wastelands: Review and Case Study"), who underlines the biodiversity factors of urban wastelands in comparison with other semi-natural spaces. In order to increase public exposure to biodiversity, he suggests strategies to include urban wastelands in urban planning, on the basis of a local project carried out with many stakeholder groups. This project led to the acquisition of many parcels of vacant land with spatial habitat diversity, and to their management in maintaining different stages of ecological succession and increasing connectivity.

To what extent are wasteland plant communities linked to the urban landscape? Wasteland plant communities are influenced by spatial proximity, and especially urban similarity: increasingly different urban characteristics result in increasingly dissimilar plant communities, as stressed by Brun & Di Pietro (Chapter "Urban Wastelands' Contribution to Ecological Connectivity"), who show that plant communities are more similar in low urbanization contexts and more diversified in highly urbanized areas.

Muratet and co-authors (Chapter "Wasteland, a Refuge for Biodiversity, for Humanity") also question the migration of plants, showing that the species migrate between large habitat patches, which thus form a functional network, whereas smaller sites are isolated in the urban matrix. They show that urban wastelands are a refuge for non-urbanophile species of plant and insects. Moreover, linking plant communities with human practices in wastelands used as temporary housing, through an original study of objects found in urban wastelands, they highlight wastelands as a refuge for a plurality of human and non-human species.

Nature and People: Uses and Perceptions of Urban Wastelands

This form of urban nature questions the residents' perceptions. Sometimes, these bushy, disordered spaces are not assimilated to nature; sometimes the residents' view of them is negative: they are seen as a failure of urban planning. On the contrary, other residents may see and experience them as resources, and/or as a space of freedom in the city, as a piece of urban nature. If so, wastelands are informal green spaces in the city.

Urban wastelands are sometimes used as places for walking, for sporting activities or simply for passing through, playgrounds for children, areas of isolation for teenagers… These uses have evolved in recent decades, in relation to changes in the use of public spaces by city dwellers: the use of urban wastelands as a playground for children, for example, has become marginal during recent decades (Rivière 2016). Moreover, the use of these spaces, especially when they are in former industrial areas, is not free from threats to human health (Bambra et al. 2014). Meanwhile, urban wastelands are often important for residents who use wastelands regularly in an informal manner.

What kinds of resources do wastelands represent for residents, and how do local authorities interact with them? Two types of uses can be distinguished: vital uses, as food and housing (Chapters "Dwelling in an Urban Wasteland: Struggles for Resources" and "Long-Standing Wastelands"), and recreational activities (Chapters "Let It Grow? Social Representations of Nature on Contaminated Brownfields," "Getting to Know Urban Wasteland—A Look at Vacant Lands as Urban Green Space in Japan," "Wastelands at Port-City Interfaces. The Search for Water Spaces to Evade the Constant Hustle and Bustle of City Life").

Through two ethnographic surveys, Mattoug (Chapter "Dwelling in an Urban Wasteland: Struggles for Resources") highlights that green urban wastelands can be important resources for residents and informal users: gardens, farms, and settlements without a legal status, as well as traffic for gray economies. These temporary uses transform urban wastelands into self-organizing niches not only for recreational needs, but also for subsistence. At the same time, local authorities develop innovative management plans for these areas and are willing to delegate urban open space management to NGOs.

Muçi and Dorso (Chapter "Long-Standing Wastelands") question a paradox: how to explain the long-term existence of inner-city wastelands in cities characterized by densification, sharp land pressure, and large-scale urban projects? The lack of official management is in no way an abandonment, and their work highlights the informal uses and political arrangements. One of the factors that helps to explain the longtime scale here relates, in part, to the place and role of vegetation: the practical and protective aspects of nature perform important services.

Tendero and Bazart (Chapter "Let It Grow? Social Representations of Nature on Contaminated Brownfields") focus on a kind of wasteland, contaminated brownfields, to question their representations by city dwellers. Residents link different forms of nature to contaminated brownfields, from the wild garden to the more domesticated forms such as vegetable gardens and crop fields. A link is suggested between the lack of perception of contaminated urban wastelands and the lack of identification of urban wastelands with nature, although the main orientation wished for the regeneration of urban wastelands is that of public green spaces.

Kim and Rupprecht (Chapter "Getting to Know Urban Wasteland—A Look at Vacant Lands as Urban Green Space in Japan") review the potential of vacant lands as urban green spaces. They have investigated the perception of urban vacant lands and show that residents with higher exposure to traditional green spaces in their daily lives were more aware of the existence of vacant lands. As a result, vacant lands may have a high tendency to be perceived by residents as an intimate, local space, suggesting usability. The authors highlight that vacant lands can serve an alternative or supplementary role in cities limited in opportunities to create new urban green spaces.

Mazy (Chapter "Wastelands at Port-City Interfaces. The Search for Water Spaces to Evade the Constant Hustle and Bustle of City Life") explores the points of view of users of city-port interfaces, mostly residents of these neighborhoods, grounded in a study of territorial representations. These representations are clearly different from the conflictual environment of decision-making. The image of a peaceful, breathing space, capable of stimulating the senses, combined with moments of relaxation or retreat, dominates the comments of residents. However, some citizens beg for more "controlled" nature to transform an area that is seen as dirty, gray, or polluted.

Planning Nature With Urban Wastelands

This part of the book raises issues of power in urban wastelands, of relationships between users and public authorities, and of integration of informal spaces as urban wastelands in the formal urban fabric, in a perspective that progressively broadens the spatial and temporal field of view.

In shrinking cities, the quantity of urban wastelands is increasing sharply. Conversely, in a context of urban growth, wastelands, waiting to be used, are generally scheduled to be built upon. The injunction to densify cities, in order to limit urban sprawl, leads urban planners to systematically recommend the urbanization of these "empty places." Faced with the uncertainties of urbanization in a context of crisis, this vision of urban wastelands has been renewed by the emergence of temporary urbanism, or "tactical" urbanism (Webb 2018).

With the growing attractiveness of urban nature, some wastelands can be transformed into urban parks. But the regeneration of wastelands, incorporating new environmental amenities, can lead to gentrification, as shown by Carrière and Farthing (Chapter "The Regeneration of Urban Riverbanks: A Dilemma Between Environmental and Social Issues"). They highlight that the recycling of wastelands in urban riverbanks and waterfronts has become a policy priority in metropolises as well as in medium-sized cities. However, the implementation of such planning strategies reinforces urban socio-spatial segregation and reveals significant conflicts within local planning policies, particularly in relation to "the right to the city," following Henri Lefebvre, for all their inhabitants.

Reclamation of urban wastelands, for urbanization or transformation into recreational or agricultural areas, even for a temporary project, often needs soil decontamination, and represents important land and planning issues. Petit-Berghem and co-authors (Chapter "Renaturation and Ecosystem Services of Contaminated Urban Wastelands in France") highlight the importance of considering the stakeholders, experts as well as users, in reclamation projects involving soil contamination, and the necessity to gather feedback from *in situ* experiments carried out in projects of bioremediation of formerly polluted wastelands.

How to include green urban wastelands in urban planning other than through urban densification? Mathey and Roessler (Chapter "Approaches to Developing Urban Wastelands as Elements of Green Infrastructure") outline the potential of urban wastelands to supplement urban green infrastructure, urban green spaces being desired by residents, who often prefer their retention to the re-development of wasteland sites. They consider how different "designs" of urban wastelands are perceived and used by residents. Moreover, based on findings regarding biodiversity, ecosystem services, and the perception/acceptance of vegetation-covered wastelands, various planning and development approaches are presented.

Based on the analysis of three semi-natural spaces located at the edges of a dense city (Brussels, Belgium), within or without five projects of green network, Vanbutsele (Chapter "From Isolated Wastelands to Informal Open Spaces Connected to a Metropolitan Park System") shows that the more a site is integrated in a park system, the lower is the expectation to build upon it, and vice versa. She concludes that in order to preserve urban wastelands as informal open spaces, connecting them with metropolitan green continuities could be essential.

This part of the book also broadens the time scale and questions the specific origins of wastelands (Chapter "Becoming Urban Wastelands") and their opposition to official green spaces (Chapter "Unscripted Spaces. Urban Green Space and *Terrains Vagues* in Historical Perspective. Antwerp (Belgium) c. 1900").

Verdelli and co-authors (Chapter "Becoming Urban Wastelands") deal with the processes leading to the creation of urban wastelands in an expanding city (Chennai, India). Areas formerly used as part of an irrigation system linked to agricultural uses, large open water reservoirs and a multitude of waterbodies, the last remnants of the "traditional" water management system in the villages of Tamil Nadu, are gradually being surrounded by urbanization and abandoned. The authors show how such community-owned lands managed by the village authorities have been perceived since the colonial period as uncultivated lands.

Tritsmans (Chapter "Unscripted Spaces. Urban Green Space and *Terrains Vagues* in Historical Perspective. Antwerp (Belgium) c. 1900") puts urban wastelands in a historical perspective too, comparing urban green spaces and *terrains vagues* in late nineteenth and early twentieth centuries (in Antwerp, Belgium). His research compares the perspective of citizens with the official policy on urban public green spaces and reveals the importance of these "unscripted spaces" for the urban population, as opposed to the negative connotation that is generally assigned to them and in contrast with the official policy about open spaces at the turn of the century. His chapter points out that we cannot fully understand cities without considering the important role of *terrains vagues.*

The critique of this dualistic thinking is also led by Beau in his concluding chapter. For a long time, in the framework of the dualistic thinking of the relationship between humans and nature which permeated the modern vision, wastelands have been geographically and intellectually kept on the margins of society. However, the deconstruction of the nature-culture dualism opened up a new theoretical landscape that, combined with the development of urban ecology in a context characterized by the scale of human presence on earth, tends to place wastelands at the center of attention. In postmodern ecological thinking, wastelands are places where new ways of living in an unstable world are being developed.

In the wake of the concept of novel ecosystems, this concluding chapter, just like Chapters "Natural Transformation of Post-industrial Lands: Liberty State Park in Jersey City, NJ (USA)" and "Unscripted Spaces. Urban Green Space and Terrains Vagues in Historical Perspective. Antwerp (Belgium) c. 1900," shows that wastelands challenge our representations of places and forms of life in the Anthropocene.

Francesca Di Pietro
Amélie Robert

References

Bambra C, Robertson S, Kasim A, et al (2014) Healthy land ? An examination of the area-level association between brownfield land and morbidity and mortality in England. Environ Plan A 46:433–454

Bonthoux S, Brun M, Di Pietro F, et al (2014) How can wastelands promote biodiversity in cities? A review. Landsc Urban Plan 132:79–88. https://doi.org/10.1016/j.landurbplan.2014.08.010

Hobbs RJ, Arico S, Aronson J, et al (2006) Novel ecosystems: theoretical and management aspects of the new ecological world order. Glob Ecol Biogeogr 15:1–7. https://doi.org/10.1111/j.1466-822x.2006.00212.x

Johnson AL, Borowy D, Swan CM (2017) Land use history and seed dispersal drive divergent plant community assembly patterns in urban vacant lots. J Appl Ecol 1–10. https://doi.org/10.1111/1365-2664.12958

Kowarik I (2013) Cities and wilderness. A new perspective. Environ Pollut 159:1974–1983. https://doi.org/10.1002/2014GB005021

Maris V (2018) La Part sauvage du monde. Penser la nature dans l'Anthropocène [The Wild Part of the World. Thinking Nature in the Anthropocene]. Seuil, 272 p

Rivière C (2016) « Les temps ont changé ». Le déclin de la présence des enfants dans les espaces publics au prisme des souvenirs des parents d'aujourd'hui ["Times have changed." The decline of children's presence in public spaces through the prism of the memories of today's parents]. Les Ann la Rech urbaine 111:1–17

Webb D (2018) Tactical urbanism: Delineating a critical praxis. Plan Theory Pract 19:58–73. https://doi.org/10.1080/14649357.2017.1406130

Contents

Editors and Contributors

About the Editors

Francesca Di Pietro Associate Professor of Environmental Geography at the University of Tours (France), in the Planning and Environment Department, works on plant biodiversity in anthropogenic environments, at first in agricultural landscapes, later in urban ones. Her research focuses on plant communities of urban habitats such as public green spaces, urban wastelands, roadsides, and community and individual gardens. Since 2011 she has worked on urban wastelands, coordinating work on their plant communities, informal uses, and how they are perceived.

Amélie Robert is currently a contract researcher and part-time lecturer at the University of Tours. Her research focuses on landscape dynamics, the identification of their drivers and ecosystem services, from a geohistorical perspective. More generally, she questions the interrelationships between societies and their environments in different contexts. She pays special attention to novel ecosystems, particularly in urban contexts, having contributed to several research programs in this field.

Contributors

Geeva Chandana Balasubramanian Department of Architecture and Interior Design, SRM University, Chennai, India

Mathilde Baude LBLGC-INRAE, Université d'Orléans, Orléans, France

Cécile Bazart CEE-M, Montpellier, France

Rémi Beau Institut d'Ecologie et des Sciences de l'environnement de Paris (IEES Paris), Sorbonne Université, CNRS, UMR 7618 Paris, France

Marion Brun UMR CITERES-CNRS, Université de Tours, Tours, France

Marine Canavese Max Weber Center, UMR 5283, University Lyon 2, Lyon, France

Jean-Paul Carrière UMR CITERES-CNRS, Université de Tours, Tours, France

Francesca Di Pietro UMR CITERES-CNRS, Université de Tours, Tours, France

Franck Dorso UPEC – Lab'Urba – IFEA, Paris, France

Stuart Farthing Urban Planning, University of the West of England, Bristol, UK

Colin Fontaine CESCO (Centre d'Ecologie et des Sciences de la Conservation), Paris, France

Frank J. Gallagher Department of Landscape Architecture, Rutgers - The State University, New Brunswick, NJ, USA

Rukkumany R. Harishankar Department of Architecture, SAP School of Architecture and Planning, Anna University, Chennai, India

Minseo Kim Department of Academic Research and Industry-Government Collaboration, Planning and Promotion Division, Hiroshima University, Hiroshima, Japan

Nathalie Machon CESCO (Centre d'Ecologie et des Sciences de la Conservation), MNHN, CNRS, Sorbonne-University, Paris, France

Juliane Mathey Leibniz Institute of Ecological Urban and Regional Development (IOER), Dresden, Germany

Cécile Mattoug UMR Géographie-Cités (EHGO), Paris I Panthéon-Sorbonne/ADEME, Paris, France

Kristel Mazy University of Mons, Mons, Belgium

Michael L. McKinney University of Tennessee, Knoxville, USA

Stela Muçi APPUII – Lab'Urba, Paris, France

Audrey Muratet UMR LIVE-CNRS, Université de Strasbourg, Strasbourg, France

Myr Muratet Paris, France

Marie Pellaton Paris, France

Yves Petit-Berghem Landscape Project Research Laboratory (Larep), National School Superior of Landscape, Versailles, France

Amélie Robert UMR CITERES-CNRS, Université de Tours, Tours, France

Elisabeth Rémy INRAE-AgroParisTech, UMR SAD-APT, Team "Proximity", Paris, France

Stefanie Rößler Leibniz Institute of Ecological Urban and Regional Development (IOER), Dresden, Germany

Christoph D. D. Rupprecht FEAST Project, Research Institute for Humanity and Nature, Kyoto, Japan

Assaf Shwartz Faculty of Architecture and Town Planning, Technion - Israel Institute of Technology, Haïfa, Israël

Marjorie Tendero ESSCA School of Management, SMART-LERECO, Angers, France

Bart Tritsmans History Department, University of Antwerp, Antwerp, Belgium

Séréna Vanbutsele Haute école d'ingénierie et d'architecture, Fribourg, Switzerland

Laura Verdelli UMR CITERES-CNRS, Université de Tours, Tours, France

Part I
Urban Wastelands as Biodiversity Hotspots

Urban Wastelands Can Be Amazing Reservoirs of Biodiversity for Cities

Nathalie Machon

Abstract Urban wastelands are areas previously exploited, then abandoned, and colonized by spontaneous vegetation. These areas are characterized by the abandonment of management, which leads to the coexistence of the flora once cultivated or horticultural flora with spontaneous flora. There follows a succession of plant species benefiting from much less intensive disturbance regimes than in other urban areas. To what extent are these wastelands reservoirs of biodiversity? Their diversity is optimal when their size exceeds a few thousand square meters, when their age is between 15 and 20 years, when they are connected to other biodiversity reservoirs by vegetated linear infrastructures, and when the pressure of human activities remains low. In this case, the preservation of these wastelands represents a significant benefit for the preservation of a city's biodiversity, and for the benefit, among other services, of the quality of life of its citizens.

Keywords Urban ecology · Ecosystem services · Ecological networks

1 Introduction

In cities, humans have organized the space according to their needs and activities. The ecological characteristics of cities are quite particular due to the intense and variable activities people engage in and to the concentration of all kinds of constructions they build. These characteristics may vary according to the density of human populations, the geographical location of the city (city center, periurban districts, etc.), and the types of activities that are carried out.

The city is not properly set up to shelter flora and fauna. However, a certain number of animal and plant species spontaneously colonize it, live and reproduce within it, and even, in some cases, proliferate (Schmidt et al. 2014). Next to the very mineral human structures, more natural spaces, i.e., at least covered with vegetation: parks, gardens, wastelands, stadiums, cemeteries, embankments of transport infrastructures,

N. Machon (✉)
CESCO (Centre d'Ecologie et des Sciences de la Conservation), MNHN, CNRS, Sorbonne-University, Paris, France
e-mail: nathalie.machon@mnhn.fr

F. Di Pietro and A. Robert (eds.), *Urban Wastelands*, Cities and Nature,
https://doi.org/10.1007/978-3-030-74882-1_1

etc., shelter most of this urban biodiversity (Shwartz et al. 2013). Urban species are, therefore, often confined within small areas, isolated from each other by buildings (Pauleit and Duhme 2000; Sukopp 2004; Bierwagen 2007; Fahrig 2017).

A growing literature clearly shows that the quality of life of urban dwellers and even their health are closely linked to the quality of biodiversity in their district (Threlfall and Kendal 2018; Aerts et al. 2018). Nature in the urban environment provides many ecosystem services such as temperature regulation, depollution of water, air, and soil, and attenuation of noise pollution (Bolund and Hunhammar 1999). The possible pleasant landscaping given by green spaces encourages residents to use them for walking or playing sports. As a result, they have psychological and physical benefits for urban dwellers when they live in rich neighborhoods in green spaces. Studies show that in greener neighborhoods, urbanites are less prone to allergies, cardiovascular diseases, etc. (Tzoulas et al. 2007). Urban biodiversity can also have cultural and educational virtues. It can provide an opportunity to raise awareness of environmental issues among a wide audience, starting with children (Threlfall and Kendal 2018).

Among all the types of urban greenspaces, wastelands have particular characteristics which could have a great influence on biodiversity and ecosystem services provided to city dwellers. According to Kowarik (2013), in wastelands, in particular, high numbers of non-native species mix with native ones and thus constitute novel types of urban ecosystems in which self-regulation occurs in terms of soil formation, species immigration and extinction, and biotic interactions. They give birth to novel types of wildness which might participate in the set of services provided by urban biodiversity, strengthen links between nature and urban residents and also have a positive impact on the pursuit of conservation beyond cities.

In this chapter, I describe biodiversity in wastelands, centering my description on French city wastelands (Muratet et al. 2017), which are mostly the same as those of other European cities (Kowarik 2013), but which can show some variations from one region to another depending on the climate, soil, etc. Afterwards, I show how biodiversity in such spaces is likely to produce services that are essential to a better quality of life for city dwellers.

2 Wastelands, Special Places in Ecological Transition Where Many Animal and Plant Species Live

Among all urban natural spaces, wastelands are particular: they have been previously exploited, then abandoned, and colonized by spontaneous vegetation (Muratet et al. 2007a, b). These include industrial or military wastelands, abandoned railway wastelands, wastelands containing old buildings or agricultural wastelands created during urban sprawl.

They are therefore characterized by a transitional state, where animal and plant communities linked to past activities coexist with spontaneous species that have arrived since management ceased. The transitional state is due to the fact that, naturally, ecosystems evolve, passing through different stages that depend on their initial state (Zipperer et al. 2000; Gallagher et al. 2011). If the ground is initially bare (Fig. 1), with a virgin substrate such as stripped soil or mineral structures (concrete slabs, for example), the first species (bacteria, fungi, plants) constituting the pioneer communities become established. These are gradually replaced by other more robust and competitive species and the wasteland naturally tends to evolve towards a woody type environment. Indeed, plant succession generally leads to the closure of environments, i.e., the progressive colonization of herbaceous environments by shrubs and tree species.

As successive plant species settle, animals colonize these areas because they can find habitats and food resources. Soil fauna of open environments and/or wooded areas can live in wastelands (Jerzak et al. 2011). Animals succeed in settling depending on the stage of vegetation they find, their ability to colonize from another site of the city, and the competition that is established between them (Godefroid et al. 2007; Shochat et al. 2010). Depending also on the disturbance caused by human activities, the turnover of species is more or less rapid.

Fig. 1 Wasteland at an early stage: bare ground. © N. Machon

3 Which Species Can Be Found in Urban Wastelands?

The major terrestrial taxonomic groups are almost all represented in urban wastelands (Eyre et al. 2003; Lundholm and Richardson 2010; Menke et al. 2011; Holzer 2011; Eckert et al. 2017; Hall et al. 2017; Muratet et al. 2007a, b). Plants and herbivorous and carnivorous animals can be found in them. Depending on the history of the sites, i.e. the past and present activities that are developed on them, they may host a mix of wild species with cultivated or domestic ones. Former orchards or agricultural plots, for example, can host rich and original plant communities (Muratet et al. 2007a, b). In her Ph.D. thesis, Muratet (2006) explored the biodiversity of all kinds of green spaces of the French Hauts-de-Seine department. Among them, wastelands after cultivation, embankments of transport infrastructures, and nitrophilic wastelands were among the richest in plant species.

Wild plant species from wastelands are quite numerous. In general, species are very common. Many are pioneer or ruderal species, from natural habitats whose conditions are similar to urban ones (Ignatieva et al. 2000; Lundholm and Marlin 2006). They are well adapted to the disturbances linked to the activities of city dwellers, such as trampling, mowing or regular uprooting.

The species composing this habitat depend on past activities. Muratet et al. (2017) described the vegetation of wastelands as follows:

Wastelands are made up of a set of environments that are distinguished from each other by environmental factors such as the nature of the soil, proximity to water, exposure to the sun, etc. The variability of these factors will allow the development of communities of particular species.

High numbers of non-native plant species often prevail on urban wasteland and may constitute novel types of urban woodlands (Kowarik 2005). Areas of both traditional wilderness and novel wildness show a functioning of ecosystems without deliberate human interventions.

The environment of wastelands and their associated plants is, therefore, a combination of environmental and anthropogenic factors. Within the same wasteland, depending on its size, location, and uses, six different types of vegetation can coexist (Rebele 2013; Muratet et al. 2017):

- vegetation of mineralized interstices
- vegetation of pioneer communities
- proper vegetation of wastelands
- grasslands and meadows
- thickets and woods
- riverbanks and wetland vegetation.

The next sections present a description of each type of vegetation according to Muratet et al. (2017). The taxonomic reference for the species was the French Reference TAXREF v8.0 (Gargominy et al.2014).

3.1 *Vegetation of Mineralized Interstices*

On the ground of some wastelands, there may still be more or less extensive areas of impermeable elements, remains of old buildings or old slabs: concrete slabs, asphalt surfaces, paving stones, gravel, sand, remains of railways, and ballast.

In the interstices of these mineral structures, a rather discreet flora develops that is adapted to the specificity of the environment: small quantities of soil and water available in the cracks of the ground, sometimes heavily trampled and highly exposed to the sun. The forms that adaptations take are diverse: they can be prostrate like *Amaranthus deflexus*, or *Sagina procumbens*, they can develop water reserve tissues like *Sedum sp.*, rigid tissues like *Catapodium rigidum*, be pubescent like *Dittrichia graveolens* or *Cerastium semidecandrum*. These biological devices that limit exposure of the plants to the sun and thus prevent their drying out or reduce their evapotranspiration have allowed these plants to colonize, maintain themselves and reproduce in mineral crevices. Surviving and reproducing in these difficult conditions leads plants like *Draba verna* or *Saxifraga tridactylites* to flower very early in the season and quickly complete their life cycle. Their existence is ephemeral during the year.

In this type of environment, the species are relatively sparse and thus interact little among themselves and with animal species. Species richness is often low.

3.2 *The Vegetation of Pioneer Communities*

The pioneer vegetation is the first to grow following disturbances of the ground, which can be major, such as mechanical action of bulldozers, construction of embankments, digging of trenches, and supply of soil or rubble by dump trucks. These disturbances are often designed to modify the site deeply in order to make it difficult to access. They can also be the consequences of human uses for camping, car breakage or waste disposal and incineration in a small part of the wasteland or over its entire area.

As in agricultural fields that have just been plowed, pioneer vegetation quickly settles these remodeled soils. Annual plants such as *Amaranthaceae*, *Brassicaceae,* and *Papaveraceae*, quickly produce offspring. They generally do not have specific mechanisms to adapt to harsh conditions such as high humidity, extreme pH or exposure to hot sun, and so are unable to compete favorably with other species. Thus, they are rapidly replaced by perennials that become established more slowly and constitute the true vegetation of wastelands following the pioneer vegetation (Muratet et al. 2017). In the case of new disturbances, this pioneer vegetation may rapidly recolonize the site thanks to the seed bank it constitutes in the soil.

3.3 Proper Wasteland Vegetation

The vegetation that succeeds in pioneer environments is considered proper wasteland vegetation because it is only found in spaces that had a particular use before they were abandoned, thus constituting the specificity of wastelands. Over time, pioneer vegetation is enriched with perennial species, including high biennial plants such as *Actium, Verbascum,* and *Oenothera* sp., which form leaf rosettes in the first year and spectacular inflorescences in the second. The moisture and nutrient content of the soil influence the range of species encountered in this habitat. Those growing on rather dry, stony or sandy soils develop underground reserve organs (roots, rhizomes) that provide the plant with the necessary resources during dry seasons. This is the case for common fennel, carrot, and parsnip (*Foeniculum vulgare, Daucus carota,* and *Pastinaca sativa*). Wastelands with human excrement and animal feces are rich in nitrates, favoring nitrophilous species such as *Urtica dioica, Erigeron, Sysimbrium* or *Lamium* sp. This flora can be maintained over long periods if moderate human disturbances persist. Otherwise, it evolves towards bush and then woody communities.

It is usually in this type of vegetation that exotic or invasive species develop: *Senecio inaequidens* (Fig. 2), *Solidago canadensis, Reynoutria japonica, Scolymus hispanicus, Lepidium virginicum,* introduced in European countries either voluntarily for the decoration of parks and gardens (Kendal et al. 2012) or accidentally under the

Fig. 2 *Senecio inaequidens*, an invasive species very present in wastelands from the 1990s. © N. Machon

wheels of cars, or hanging from clothing (Von der Lippe and Kowaric 2007). They escape from their introduction sites and invade wastelands more or less intensively, adapt to their new environment, and take advantage of disturbances to develop.

3.4 *Grasslands and Meadows*

When wastelands result from the transformation of old parks and gardens with lawns, vegetation can evolve as grasslands and meadows where *Poaceae* are dominant. This vegetation mainly comprises plants such as *Arrhentherum elatius, Holcus lanatus,* and *Calamagrostis epigeios*, and more sparsely *Lathyrus pratensis*, *Tragopogon pratensis,* and *Trifolium pratense.*

The species of grasslands and meadows, unlike the pioneers, are very interactive. They coexist easily and form dense communities that largely cover the soil. They use high amounts of resources to develop vegetatively, i.e., they favor the growth of their roots, stems, and leaves. Rapid seed production and long-term installation are equally important issues for them.

Meadows and grasslands are favored areas for a multitude of interactions between plants and animals that feed and develop in them. These fauna-flora interactions can alter the plants through consumption of flowers, leaf stems or other organs, and weaken them, for example, through the removal of part of the elaborated sap. However, these interactions are most often beneficial to plant communities because they can contribute to their reproduction by pollination (Pellissier et al. 2012). Mammals, birds, and ants help disperse fruits or seeds, thus allowing the plants to spread among sites (Howe and Smallwood 1982). Plants also benefit from symbiosis with bacteria or fungi, thus improving their extraction of nutrients from the soil.

This vegetation most often grows on the edges of wastelands or in areas on the fringes of cities. It is sometimes crushed or mowed, maintained by regular and light human management, for example, to prevent it from colonizing adjacent spaces.

3.5 *Thickets and Woodlands*

In the absence of intervention or management, bushes and woodlands inevitably follow the vegetation of wastelands and meadows. Shrubs with fleshy fruits such as brambles (*Rubus fruticosus*), roses (*Rosa canina*), elderberries (*Sambucus sp.*), and dogwoods (*Cornus sp.*) colonize the sites in a few years. They feed birds and small mammals and also serve as lodges and refuges (Aronson et al. 2014).

The heart of the bushes is generally dominated by a single plant species, sometimes so dense that it is impenetrable, like hawthorn (*Crataegus monogyna*), buddleia (*Buddleia davidii*), and willow (*Salix sp.*). The natural evolution is towards afforestation (Fig. 3). Wastelands that have matured to the woodland stage are rare in urban areas since their slow maturation (several decades) is not very compatible with their

Fig. 3 Trees bushes and herbaceous species in wastelands: sharing the space depending on the activities © N. Machon

life span. Thus, they do not remain abandoned for long, except for example in some old industrialized regions where they have had time to develop undisturbed since the 1970s (Dettmar 2005).

Trees can then exceed 5 m in height. They are mostly *Robinia pseudoaccacia, Ailantus altissima* and *Populus nigra and Acer sp.* At their base, herbaceous vegetation appreciating the shade and cool of the undergrowth develops, such as *Anthriscus sylvestris* and *Brachipodium sylvaticum.*

3.6 Wetland Vegetation

When wastelands border watercourses, they host species adapted to wetlands, such as *Reynoutria japonica*, *Urtica dioica*, and *Equisetum arvense*, associated with specifically hygrophilic species, which only grow in very humid soils *Epilobium hirsutum*, *Lepidium latifolium* or trees like *Alnus glutinosus* and *Salix alba* (Fig. 4).

3.7 Invasive Species

Among the plants that live in wastelands, some species come from remote areas and have significant proliferation capacities. They are called invasive alien species. Many of them, coming from hot countries, find in cities conditions favorable to their

Fig. 4 Wasteland of riverbanks: here, along the Yvette river (91, France) © N. Machon

proliferation. Roads give them access to the hearts of cities. Wastelands, at least in their early stages, provide them with privileged reception areas (Maurel et al. 2010; Godefroid and Ricotta 2018).

3.8 Animals

Wastelands are places where many small mammals, birds, insects, etc., live. The most visible are the birds. Several dozen species populate them. According to Malher et al. 2010, in Paris, common linnets (*Linaria cannabina*), melodious warblers (*Hippolais polyglotta*), common whitethroats (*Sylvia communis*), marsh warblers (*Acrocephalus palustris*), European stonechats (*Saxicola rubicola*), and little ringed plovers (*Charadrius dubius*) are among the most common species living in Parisian wastelands. Some exotic species, such as beetle pigeons (*Columba livia*) or ring-necked parakeets (*Psittacula krameri*), can also be observed (Clergeau and Machon 2014).

There are only about ten species of mammals likely to live on wasteland. Hedgehogs (*Erinaceus europaeus*), moles (*Talpa europaea)*, martens (*Martes foina*), squirrels (*Sciurus vulgaris*), rats (*Ratus sp.*) and mice (*Mus musculus*), rabbits (*Oryctolagus cuniculus*), foxes (*Vulpes vulpes*) and some species of bats (for example, the common pipistrelle (*Pipistrellus pipistrellus*)) are about the only mammals found in these areas (Clergeau and Machon 2014).

Domestic cats or feral cats (*Felis silvestris catus*) can be extremely numerous in wastelands. They have a strong impact on the regulation of wilderness biodiversity. Studies show that these animals are fearsome predators, killing billions of birds, lizards, and small mammals (Clergeau and Machon 2014).

The warm urban environment and the high availability of food resources also attract many small invertebrates (Gardiner et al. 2013). Beetles, for example, can be observed in many wastelands (Eyre et al. 2003). Studies have shown that the more near-natural spaces are connected to each other by hedges or other vegetated structures, the more they contain these small detritus feeders (Vergnes et al. 2012). In general, wastelands appear to be a refuge for generalist urban insects. Butterflies often desert city centers but can be found in large numbers in unmanaged areas where floral resources are quite abundant. Other pollinating species are also found in wastelands, especially solitary bees, which are responsible for most of the pollination of entomophilous plants in cities (Everaars et al. 2018; Twerd and Banaszak-Cibicka 2019), when they are not too much in competition with honey bees from the many hives placed in certain districts (Cane and Tepedino 2016).

4 Characteristics of Wastelands that Influence Their Biodiversity

Unlike more rural areas, but like other urban green spaces, urban wastelands are often limited in size and surrounded by buildings, streets, walls, and structures that are not appropriate for a number of animal and plant species because they do not provide food or shelter. In fact, the isolation of these spaces, within an inhospitable urban matrix, prevents most organisms from migrating from or to other areas (Bierwagen 2007). The consequence of this isolation is a loss of viability of plant or animal populations through genetic impoverishment and inbreeding.

A particular characteristic of urban environments is the climate that wastelands, among others, are subject to. Conditions are warmer and drier than in the surrounding areas of the cities. This phenomenon called "urban heat island" is due to human activities concentrated in cities, some of which produce heat, such as factories, vehicle engines, building heating, air conditioning, hot water circulating in pipelines, etc. It is also the result of the artificialization of urban soils that absorb solar radiation and then return it as heat notably during the night. Finally, animal and plant organisms living in wastelands are subject to air, water, and soil pollution to which they are more or less sensitive. This pollution is the result of emissions of different gases from road traffic, industrial activities and district heating in the neighborhood, or former polluting activities. Soil pollution is particularly frequent in industrial or military wastelands and also in ancient agricultural areas where pesticides have been used on a massive scale over long periods of time. Other sources of pollution in cities are artificial light and noise, which cause serious damage to sensitive species. In wastelands, artificial light and noise may be less intense than in other urban green spaces, because they are abandoned, and therefore, less equipped and less frequented.

Nevertheless, depending on the intensity of the defensive measures that characterize urban wastelands, it is possible for species to find the tranquility that is not found in other types of areas, which makes them privileged areas for biodiversity. Even in cases where marginal activities, which are more or less legal, can be organized, the number of visitors often remains low compared to other types of urban green spaces. When activities intensify and when the human population that frequents the wasteland increases significantly (for example, the establishment of a slum or a garden for the production of vegetables), the impact on biodiversity becomes strong, but the site can no longer be considered to meet the definition of wasteland.

As seen above, the characteristics of the soil are factors that select flora and thus fauna of wastelands (Diaz et al. 1998; Knapp et al. 2008).

Invasive species, by occupying large areas, tend to impose themselves as monospecific stands, to the detriment of many other species. Buddleias (*Buddleia davidii*) and knotweed from Japan (*Reynoutria japonica*) can invade some wastelands, impoverish their natural heritage, and degrade the services they could provide (Maurel et al. 2010).

The possible connections with other wastelands or other green spaces also influence the richness of the plant and animal communities. In particular, the presence of hedgerows, greenways or alignment trees in the neighborhood of wastelands can favor the movement and thus the life of certain organisms among them (Vergnes et al. 2012).

5 The Conditions that Make Urban Wastelands Reservoirs of Biodiversity for Cities

The most important factor governing the quality of biodiversity is the space allocated to it. As a result, the larger the surface area of the wastelands, the more varied the species they will be able to host (Muratet et al. 2007a, b). In the wastelands studied by Muratet in the Hauts-de-Seine (Muratet 2006), plant species richness was correlated with their size. The largest ones could host up to 80 species of plants. The age of the wastelands is also a factor that affects their richness. Ecological successions often lead to the closure of the environment (Rall and Haase 2011). Intermediate age areas composed of several strata (herbaceous, shrubby, and tree) will be the most favorable to local biodiversity.

The positive impact of wastelands on city biodiversity depends (1) on their size, (2) on low turnover to allow them to reach a good age of maturation (Rall and Haase 2011), (3) a good soil, (4) low disturbance regimes, and (5) their capacity to export pollen and seeds to other plant populations and also to allow migration of animals (Howe and Smallwood 1982; Bozonnet and Belarbi Allard 2006).

Wastelands should not be surrounded by walls, buildings or streets impassable to many species, but benefit from green structures for the dissemination of species in their neighborhood (Vergnes et al. 2012).

6 Ecosystem Services Provided by Wastelands

The abandonment of any form of management and preservation away from any human activity leads wastelands to be multi-strata and species-rich ecosystems, richer than most other highly managed green spaces of cities, and thus much more heterogeneous in their structure and composition. This characteristic is an essential asset in terms of the provision of services because it is well known that the most complex ecosystems are those with the most ecological interest (Harrison et al. 2014). Furthermore, due to their potential size and cumulative surface area, wastelands can play a major role in the provision of ecosystem services to urban dwellers (Dearborn and Kark 2010). Actually, depending on their number and size, they are likely to significantly enhance the presence of vegetation in cities, and thus services like support and regulation services: temperature regulation, air, and water depollution control or wild biodiversity hosting.

With the time of abandonment, many trees can grow in wastelands. They are most often descendants of trees growing in the surrounding area, including alignment trees planted along streets. While there are many, they act like real urban woodlands: they sequester some of the greenhouse gases, including carbon dioxide, and produce oxygen through photosynthesis. They purify a fraction of the air, reducing particulate and gaseous pollution. Trees in particular absorb significant amounts of pollutants. They play a significant role in the carbon cycle and have a significant impact on neighborhood temperatures, especially in the case of severe heat waves (Mathey et al. 2014).

Along urban rivers, by absorbing the mineral elements necessary for their growth, plants of wastelands trap a significant part of the substances from human activities that transit to the watercourse by runoff or infiltration, and certainly do so better than other kinds of greenspaces with less complex vegetation.

In addition to these purely utilitarian arguments, preserving the nature of wasteland means preserving part of the biodiversity of certain regions. Indeed, since urban territories are becoming increasingly large and urban sprawl sometimes affects areas rich in fauna and flora, preserving biodiversity in wastelands is potentially a way to preserve certain wild species (Kong et al. 2010), which is not the case of managed greens paces like parks or gardens.

Similarly, the preservation of areas still hospitable for a wide range of species, such as wastelands, allows them to act as stepping-stone corridors for fauna and flora, to enable the species to cross cities from periurban zones (Jim 2004).

7 Conclusion

In conclusion, wastelands are spaces with considerable opportunities for preserving biodiversity in cities (Bonthoux et al. 2014). Their positive influence implies a good quality of biodiversity and low turnovers within them. Their management and activities should be light and localized in space and time (Politi Bertoncini et al. 2012). These conditions depend strongly on political will because there is a need for coordinated and successive freezing of certain large sites over 10–20 years between construction projects, to the detriment of short-sighted economic interests. The urban matrix must be arranged to allow population flows thanks to a good quality of green networks (Kong et al. 2010). Even if spontaneous vegetation on urban wastelands often consists of very robust plants that do not require special care and thus represent a cost-effective variant of urban greenery, management policies may also produce novel and interesting urban ecosystems by, for example, reintroducing native species in wasteland vegetation (Fischer et al. 2013).

Thanks to a network of functional wastelands, the global biodiversity of cities and the ecosystems they form should provide essential and amazing services that improve the lives of urban dwellers.

Acknowledgements This text was largely inspired by the work of Audrey Muratet (https://www.audreymuratet.com/fr/). The English text was corrected by Dominique Macabies (https://dominique.macabies.fr/). I would like to thank Juliane Mathey, Amélie Robert and Francesca Di Pietro for their attentive reviews and their precious and wise advice on how to improve the text.

References

Aerts R, Honnay O, Van Nieuwenhuyse A (2018) Biodiversity and human health: mechanisms and evidence of the positive health effects of diversity in nature and green spaces. Br Med Bull 127(1):5–22. https://doi.org/10.1093/bmb/ldy021

Aronson MF, La Sorte FA, Nilon CH, Katti M, Goddard MA et al (2014) A global analysis of the impacts of urbanization on bird and plant diversity reveals key anthropogenic drivers. Proc R Soc Lond B Biol Sci 281. https://doi.org/10.1098/rspb.2013.3330

Bierwagen BG (2007) Connectivity in urbanizing landscapes: the importance of habitat configuration, urban area size, and dispersal. Urban Ecosyst 10(1):29–42

Bolund P, Hunhammar S (1999) Ecosystem services in urban areas. Ecol Econ. https://doi.org/10.1016/S0921-8009(99)00013-0

Bonthoux S, Brun M, Di Pietro F, Greulich S, Bouché-Pillon S (2014) How can wastelands promote biodiversity in cities? A review. Landsc Urban Plann 132:79–88

Bozonnet ER, Belarbi Allard F (2006) Modelling air flows around buildings in urban environment. Paper presented at the international workshop on energy performance and environmental quality of buildings. hal.archives-ouvertes.fr/hal-00312385

Cane VJ, Tepedino JH (2016) Gauging the effect of honey bee pollen collection on native bee communities. Conserv Lett. https://doi.org/10.1111/conl.12263

Clergeau P, Machon N (2014) Où se cache la biodiversité en ville? [Where is biodiversity hiding in the city?]. Editions Quae

Dearborn DC, Kark S (2010) Motivations for conserving urban biodiversity. Conserv Biol. https://doi.org/10.1111/j.1523-1739.2009.01328.x

Dettmar J (2005) Forests for shrinking cities? The project "Industrial Forest of the Ruhr". In: Kowarik I, Körner S (eds) Wild urban woodlands. New perspectives for urban forestry. Springer, Heidelberg, New York, Berlin, pp 263–276

Diaz S, Cabido M, Casanoves F (1998) Plant functional traits and environmental filters at a regional scale. J Veg Sci 9(1):113–122. https://doi.org/10.2307/3237229

Eckert S, Möller M, Buchholz S (2017) Grasshopper diversity of urban wastelands is primarily boosted by habitat factors. Insect Conserv Diver. https://doi.org/10.1111/icad.12221

Everaars J, Settele J, Dormann CF (2018) Fragmentation of nest and foraging habitat affects time budgets of solitary bees, their fitness and pollination services, depending on traits: results from an individual-based model. PlosOne. https://doi.org/10.1371/journal.pone.0188269

Eyre MD, Luff ML, Woodward JC (2003) Beetles (Coleoptera) on brownfieldsites in England: an important conservation resource? J Insect Conserv 7:223–231. https://doi.org/10.1023/B:JICO.0000021020.66549.1e

Fahrig L (2017) Ecological responses to habitat fragmentation. Per Se Annu Rev Ecol Evol Syst 48(1):1–23. https://doi.org/10.1146/annurev-ecolsys-110316-022612

Fischer LK, von der Lippe M, Rillig MC, Kowarik I (2013) Creating novel urban grasslands by reintroducing native species in wasteland vegetation. Biol Conserv 159:119–126

Gallagher FJ, Pechmann I, Holzapfel C, Grabosky J (2011) Altered vegetative assemblage trajectories within an urban brownfield. Environ Pollut 159:1159–1166. https://doi.org/10.1016/j.envpol.2011.02.007

Gargominy O, Tercerie S, Régnier C, Ramage T, Schoelinck C, Dupont P et al (2014) TAXREF v8.0, référentiel taxonomique pour la France: Méthodologie, mise en oeuvre et diffusion. Rapport SPN Vol. 42

Gardiner MM, Burkman CE, Prajzner SP (2013) The value of urban vacant land to support arthropod biodiversity and ecosystem services. Environ Entomol 42:1123–1136. https://doi.org/10.1603/EN12275

Godefroid S, Ricotta C (2018) Alien plant species do have a clear preference for different land uses within urban environments. Urban Ecosyst 21(6):1189–1198. https://doi.org/10.1007/s11252-018-0792-4

Godefroid S, Monbaliu D, Koedam N (2007) The role of soil and microclimatic variables in the distribution patterns of urban wasteland flora in Brussels, Belgium. Landsc Urban Plan 80(1–2):45–55

Hall DM, Camilo GR, Tonietto RK, Ollerton J, Ahrné K, Arduser M, Ascher JS, Baldock KCR, Fowler R, Frankie G, Goulson D, Gunnarsson B, Hanley ME, Jackson JI, Langellotto G, Lovenstein D, Minor ES, Philpot SM, Potts SG, Sirohi MH, Spevak EM, Stone GN, Threlfall CG (2017) The city as a refuge for insect pollinators. Conserv Biol 31:24–29

Harrison PA, Berry PMS, G, Haslett JR, Blicharsk M, Bucur M, Dunford R, Egoh B, Garcia-Llorent M, Geamănă N, Geertsema W, Lommelen E, Meiresonne L, Turkelboom F (2014) Linkages between biodiversity attributes and ecosystem services: a systematic review. Ecosyst Serv. https://doi.org/10.1016/j.ecoser.2014.05.006

Holzer KA (2011) Amphibian use of constructed and remnant wetlands in an urban landscape. Urban Ecosyst. https://doi.org/10.1007/s11252-014-0373-0

Howe HF, Smallwood J (1982) Ecology of seed dispersal. Annu Rev Ecol Syst 13(1):201–228

Ignatieva M, Meurk CD, Newell C (2000) Urban biotopes: the typical and unique habitats of city environments and their natural analogues. Paper presented at the urban biodiversity and ecology as a basis for holistic planning and design. Proceedings of a workshop held at Lincoln University, pp 46–53

Jerzak L, Böhner J, Kavanagh B (2011) Urban fauna. Studies of animal biology, ecology and conservation in European cities. UTP, Bydgoszcz, pp 227–236

Jim C (2004) Green-space preservation and allocation for sustainable greening of compact cities. Cities 21(4):311–320

Kendal D, Williams KJ, Williams NS (2012) Plant traits link people's plant preferences to the composition of their gardens. Landsc Urban Plan 105(1–2):34–42

Knapp S, Kuehn I, Wittig R, Ozinga W, Poschlod P et al (2008) Urbanization causes shifts of species' trait state frequencies. Preslia 80:375–388

Kong F, Yin H, Nakagoshi N, Zong Y (2010) Urban green space network development for biodiversity conservation: identification based on graph theory and gravity modeling. Landsc Urban Plan 95(1–2):16–27

Kowarik I (2005) Novel urban ecosystems, biodiversity, and conservation. Environmental Pollution. https://doi.org/10.1016/j.envpol.2011.02.022

Kowarik I (2013) Cities and wilderness. A new perspective. Int J Wilderness 19(3):32–36

Lundholm JT, Marlin A (2006) Habitat origins and microhabitat preferences of urban plant species. Urban Ecosyst 9(3):139–159. https://doi.org/10.1007/s11252-006-8587-4

Lundholm JT, Richardson PJ (2010) Habitat analogues for reconciliation ecology in urban and industrial environments. J Appl Ecol 47:966–975. https://doi.org/10.1111/j.1365-2664.2010.01857.x

Malher F, Lesaffre G, Zucca M, Coatmeur J (2010) Oiseaux nicheurs de Paris. Un atlas urbain [Nesting birds of Paris. An urban atlas]. Delachaux et Niestlé

Mathey J, Rößler S, Banse J, Lehmann I (2014) How can urban brown-fields contribute to climate adaptation and human well-being in cities? In: Korn H, Stadler J, Bonn A, Bockmühl K, Macgregor N (eds) Proceedings of the European conference "Climate Change and Nature Conservation in Europe—an ecological, policy and economic perspective", Bonn, pp 114–115

Maurel N, Salmon S, Ponge JF, Machon N, Moret J, Muratet A (2010) Does the invasive species Reynoutria japonica have an impact on soil and flora in urban wastelands? Biol Invasions 12:1709–1719. https://doi.org/10.1007/s10530-009-9583-4

Menke SB, Guénard B, Sexton JO et al. (2011) Urban areas may serve as habitat and corridors for dry-adapted, heat tolerant species; an example from ants. Urban Ecosyst. https://doi.org/10.1007/s11252-010-0150-7

Muratet A, Machon N, Jiguet F, Moret J, Porcher E (2007a) The role of urban structures in the distribution of wasteland flora in the greater Paris area, France. Ecosystems 10(4):661

Muratet A, Machon N, Jiguet F, Moret J, Porcher E (2007b) The role of urban structures in the distribution of wasteland flora in the greater Paris area, France. Ecosystems 10:4

Muratet A, Muratet M, Pellaton M (2017) Flore des friches urbaines [Flora of urban wastelands]. Editions Xavier Barral

Muratet A (2006) Diversité végétale en milieu urbain: l'exemple des Hauts-de-Seine [Plant diversity in urban areas: the example of the Hauts-de-Seine area]. PhD thesis, Pierre et Marie Curie University, Paris

Pauleit S, Duhme F (2000) Assessing the environmental performance of land cover types for urban planning. Landsc Urban Plan 52(1):109–111

Pellissier V, Muratet A, Verfaillie F, Machon N (2012) Pollination success of Lotus corniculatus (L.) in an urban context. Acta Oecologica 39:94–100

Politi Bertoncini A, Machon N, Pavoine S, Muratet A (2012) Local gardening practices shape urban lawn floristic communities. Landsc Urban Plan 105(1–2):53–61

Rall EL, Haase D (2011) Creative intervention in a dynamic city: a sustainability assessment of an interim use strategy for brownfields in Leipzig, Germany. Landsc Urban Plan 100:189–201. https://doi.org/10.1016/j.landurbplan.2010.12.004

Rebele F (2013) Differential succession towards woodland along a nutrient gradient. Appl Veg Sci 16:365–378

Schmidt KJ, Poppendieck H, Jensen K (2014) Effects of urban structure on plant species richness in a large European city. Urban Ecosyst 17(2):427–444

Shochat E, Lerman SB, Anderies JM, Warren PS, Faeth SH, Nilon CH (2010) Invasion, competition, and biodiversity loss in urban ecosystems. Bioscience 60(3):199–208. https://doi.org/10.1525/bio.2010.60.3.6

Shwartz A, Muratet A, Simon L, Julliard R (2013) Local and management variables outweigh landscape effects in enhancing the diversity of different taxa in a big metropolis. Biol Conserv 157:285–292

Sukopp H (2004) Human-caused impact on preserved vegetation. Landsc Urban Plan 68(4):347–355

Threlfall CG, Kendal D (2018) The distinct ecological and social roles that wild spaces play in urban ecosystems. Urban for Urban Green 29:348–356. https://doi.org/10.1016/j.ufug.2017.05.012

Twerd L, Banaszak-Cibicka W (2019) Wastelands: their attractiveness and importance for preserving the diversity of wild bees in urban areas. J Insect Conserv

Tzoulas K, Korpela K, Yli-Pelkonen V et al (2007) Promoting ecosystem and human health in urban areas using Green Infrastructure: a literature review. Landsc Urban Plan. https://doi.org/10.1016/j.landurbplan.2007.02.001

Vergnes A, Le Viol I, Clergeau P (2012) Green corridors in urban landscapes affect the arthropod communities of domestic gardens. Biol Conserv. https://doi.org/10.1016/j.biocon.2011.11.002

Von der Lippe M, Kowarik I (2007) Long-distance dispersal of plants by vehicles as a driver of plant invasions. Conserv Biol 21(4):986–996

Zipperer WC, Wu J, Pouyat RV, Pickett ST (2000) The application of ecological principles to urban and urbanizing landscapes. Ecol Appl 10(3):6

Nathalie Machon is a Professor of Urban Ecology at the National Museum of Natural History, in Paris, France. She tries to understand which urban structures form the urban green network, i.e., the structures which shelter animal and plant species and allow them to migrate through the matrix formed by the buildings and other artificial surfaces. Among others, she studies the ecological role of wastelands in the urban green network.

Natural Transformation of Post-industrial Lands: Liberty State Park in Jersey City, NJ (USA)

Frank J. Gallagher

Abstract Abandoned post-industrial or wasteland landscapes present an interesting paradox. Do the ecological benefits of these naturally assembled systems present an "attractive nuisance" due to the potential for contaminate transfer or do they provide valuable ecosystem services and functions? This case study presents data from a former railyard that since its abandonment in the 1960s has developed into a mix of urban forest shrubland and meadow. The heterogeneous structure and the legacy metal contaminants have been characterized. The diversity of the naturally assembled plant community is also well documented, as are its successional trajectories. The ecological risk associated with contaminant transfer has been examined at the primary, secondary, and tertiary trophic levels. In general, the findings indicate that while the plant communities are a novel mix of old and new world species and do not follow traditional trajectories, they are providing significant ecosystem functions. In addition, and most significantly, the risk associated with contaminant transfer throughout the food web appears to be minimal. After 30 years of investigations, researchers have concluded that this wasteland provides valuable ecosystem services.

Keywords Post-industrial landscape · Soil metals · Diversity · Productivity · Ecological risk

1 Background and History of the Site

The extent of human impact on the ecological systems within which we function has been documented for well over 100 years (Marsh 1882). While we recognize, or perhaps because we recognize and bemoan the degradation we have caused, we tend to maintain a strict sense of appropriate typologies, and legacy plant and animal assemblages that "should" exist. In addition, post-industrial sites or wastelands have often been associated with contamination that results in both ecological and human-related health risks. Therefore, the naturally assembled floral and faunal communities

F. J. Gallagher (✉)
Department of Landscape Architecture, Rutgers - The State University, New Brunswick, NJ, USA
e-mail: fgallagh@sebs.rutgers.edu

F. Di Pietro and A. Robert (eds.), *Urban Wastelands*, Cities and Nature,
https://doi.org/10.1007/978-3-030-74882-1_2

associated with post-industrial landscapes have been associated with the risk that industrial contamination brings. This is especially true in the United States as the "Love Canal," a case in Niagara Falls, New York, where a 70-acre landfill used by the Hooker Chemical Company and later sold to the municipality for the construction of an elementary school caused an extreme cancer cluster. The resulting advocacy effort and litigation, which began in the late 1970s and continues to this day, help to foster a paradigm that all post-industrial sites present a human health risk. Unlike Europe and especially Germany which have embraced the ecological potential of post-industrial landscapes and celebrated them in places like Südgelände and Landschaftspark, the United States has yet to develop or let develop a Fourth Nature Park. Such an engrained negative perception has resulted in both the loss of productive ecological assemblages within urban areas and, as importantly, the loss of an urban ecological identity which is critical for understanding "Place." The case study presented here is one of the few in the United States where the site has been abandoned long enough and there has been adequate research to prove its ecological value and the actual health risk.

The land now known as Liberty State Park (LSP) can be viewed as a retrospective of the interaction between humans and the environment of the New York Harbor for at least the last 8,000 years. Prior to European colonization, Native People built a social structure based upon the area's abundant natural resources. Summer residents of Communipaw Cove (Fig. 1) lived on the fish and shellfish of the bay. Early colonial settlers seeking respite from the social oppression of European monarchies found a land rich in the resources needed to support their new life. Oyster Bay, as it was known to the Dutch, and later the English settlers, supplied abundant resources.

The Industrial Revolution created a need for transportation modes of increased capacity and a larger labor force. The harbor's protected access to the ocean and abundant fish and shellfish populations provided both. The labor force came from all over the world. While numbers vary considerably, approximately 37 million people immigrated to America within a span of 100 years. The Great Wave, which brought many Western Europeans to the United States between 1903 and 1920 as the Industrial Revolution spiked, saw at least 10 million immigrants pass through the doors of Ellis Island not 100 m from the shores of New Jersey. The Central Railroad of New Jersey designated a special track and train car to deliver these prospective Americans to their new homes.

Since the tidal salt marshes and mudflats of Communipaw Cove were too shallow for heavier vessels, the Central Railroad used millions of cubic yards of construction debris, refuse from New York City, and ballast from ocean-going vessels to get to deeper water and in the process created a 1,000-acre railyard. The fill occurred primarily in six stages between the years 1860 and 1916. Regardless of what was used as fill, the area was leveled with cinder, ash, and gravel to form a base for rails. For over 100 years, railroad, ferry service, and shipyard activities dominated the area. In 1967, with the bankruptcy of the Central Railroad of New Jersey, the lands went fallow. Gradually, various sections were purchased first by the City of Jersey City then the State of New Jersey. The construction of the park began in the early 1970s and has continued ever since. Extensive soil testing has revealed that

Fig. 1 Location of the study site. Almost all of the New York harbor coastline consists of fill material. The study site, now part of Liberty State Park, was once Communipaw Cove. It was gradually filled for use as a railyard between 1860 and 1920 (Hartman et al. 2017)

most areas were contaminated with hydrocarbon derivatives and metals. In several small areas where the concentrations of contaminants were considered severe, soil was removed. However, the vast majority of the area was considered historic fill and when developed was capped with clean material. Fortunately, funds were not available to develop the entire site, and 235 acres in the interior of the park remain undeveloped. The vegetative community is the result of natural colonization and has developed into one of the oldest if not the oldest urban wildland or wasteland in the United States.

2 Resulting Soil Conditions

In 2010, as part of the United States Department of Agriculture's urban soils classification initiative, the soils within the interior section of the park were officially classified. The resulting Lady Liberty soil series "consists of very deep, moderately well-drained soils with moderately low to moderately high saturated hydraulic conductivity. These soils are formed in (sic) a thick mantle of human transported material consisting of coal slag, dredged materials, and/or any geologic deposits ranging from till, outwash, alluvium, or coastal plain sediments usually from a local source. They occur on anthropogenic landforms in and near major urbanized areas of the Northeast." What this description understates is the extreme heterogeneity of the soil—the variation in any one soil character. Triplicate samples taken within one meter of each other can yield standard deviations of up to 100%. Despite this variation, there are large areas of consistent vegetation typologies which has led to the theory that facilitation or cooperation between plant species plays a much larger role in wastelands than in traditional settings (Krumins et al. 2015).

Metals are common constituents in many anthropogenic soils. Typically, cadmium (Cd), copper (Cu), zinc (Zn), lead (Pb), and others are present above established screening criteria (Dudka et al. 1996). Soil metals at LSP are typically found in concentrations considerably above ecological screening criteria (Table 1, Gallagher 2008). The variation in the concentration was often above 20%, with spikes commonly an order of magnitude greater than the mean or average concentration. However, since metals are often adsorbed by carbonates, organic matter, and ferric oxides, their impact on the associated biologic community varies considerably and is in general difficult to predict (Adriano 1986; Ross 1994).

To better understand the potential ecological risk associated with the soils of the site, two different indices were developed (Qian et al. 2017). The geochemical index compares the soil metal concentrations of the site to known background levels that result from natural processes associated with the breakdown of the parent geologic materials. Originally developed for sediment analysis (Ghrefat 2011; Müller 1969), it was then modified and used for soil contamination evaluation. The geoaccumulation index at the park demonstrated that all six metals being investigated were above background concentrations. Copper (Cu) and Lead (Pb) consistently yielded the highest concentrations relative to the background. Zinc (Zn) had the greatest

Table 1 Comparison of the mean arsenic (As), chromium (Cr), mercury (Hg), sodium (Na), lead (Pb), zinc (Zn), vanadium (Va) concentrations (μg g^{-1}), and the standard deviation (stdev) from the 24 sites that were studied in 2005. The majority of the samples exceeded both residential and ecological standards (Gallagher et al. 2008)

Metal	As	As stdev	Cr	Cr stdev	Cu	Cu stdev	Hg	Hg stdev
Min	13.33	2.08	10.90	3.29	53.65	2.75	0.00	0.00
5%	13.33	2.18	12.66	3.29	66.45	8.44	0.00	0.00
25%	17.03	3.88	19.28	4.85	93.52	18.99	0.06	0.03
Median	22.60	7.44	62.95	12.66	285.40	31.24	0.29	0.19
75%	69.97	27.89	109.35	23.63	435.50	55.19	3.15	0.95
95%	325.61	201.85	282.55	107.50	1320.17	371.58	182.33	4.51
Max	384.41	253.15	334.56	141.83	2200.38	486.07	400.00	6.87
Metal	Na	Na stdev	Pb	Pb stdev	Zn	Zn stdev	Va	Va stdev
Min	0.75	0.07	96.58	8.08	10.88	10.81	0.00	0.00
5%	0.82	0.08	125.11	17.08	10.88	13.31	0.00	0.00
25%	1.76	0.40	220.85	49.13	22.44	17.98	16.22	15.85
Median	3.52	0.61	445.43	103.24	235.88	76.16	56.40	18.82
75%	6.77	1.29	753.85	143.90	437.24	282.98	103.08	33.13
95%	11.68	3.33	3510.46	1085.96	1883.49	1746.08	269.95	142.14
Max	14	3.77	6673.22	1603.66	2326.77	2046.85	316.80	207.33

variation; almost absent at some locations, it was five times the normal concentration at others. Interestingly, arsenic (As) contamination was not widespread, while chromium (Cr) and mercury (Hg) were basically absent. The overall concentration of metals associated with the soils of LSP follows the order of Cu > Pb > Zn > As > Cr > Hg.

The ecological risk index is often used to assess the potential impact of the combined or synergistic character of contaminated sites. It is generally calculated as the summation of the toxic response factor for each species of metal (Hakanson 1980). In this case, the ecological index was reflective of the geochemical index, with several differences relating to the concentration and mobility of the various metal ions. While the metals at higher concentrations, Cu, Pb, and Zn remained in the same order of potential risk; the other three metals were reordered as follows: Hg > Zn > Cr (Qian et al. 2017).

In order to understand the combined impact of soil metals on the emerging plant community of the park, a total soil metal load (TML) map was developed (Gallagher et al. 2008). The TML was derived as the negative summation of the concentrations of As, Cu, Cr, Pb, and Zn after these are normalized using logarithmic transformation. The summation is then rank-ordered on a scale of 1–5 (Juang et al. 2001). Finally, that dataset was block-kriged, to account for the high standard deviation in the original datasets and exported as in ArcMap (Fig. 2). The result was a single variable TML

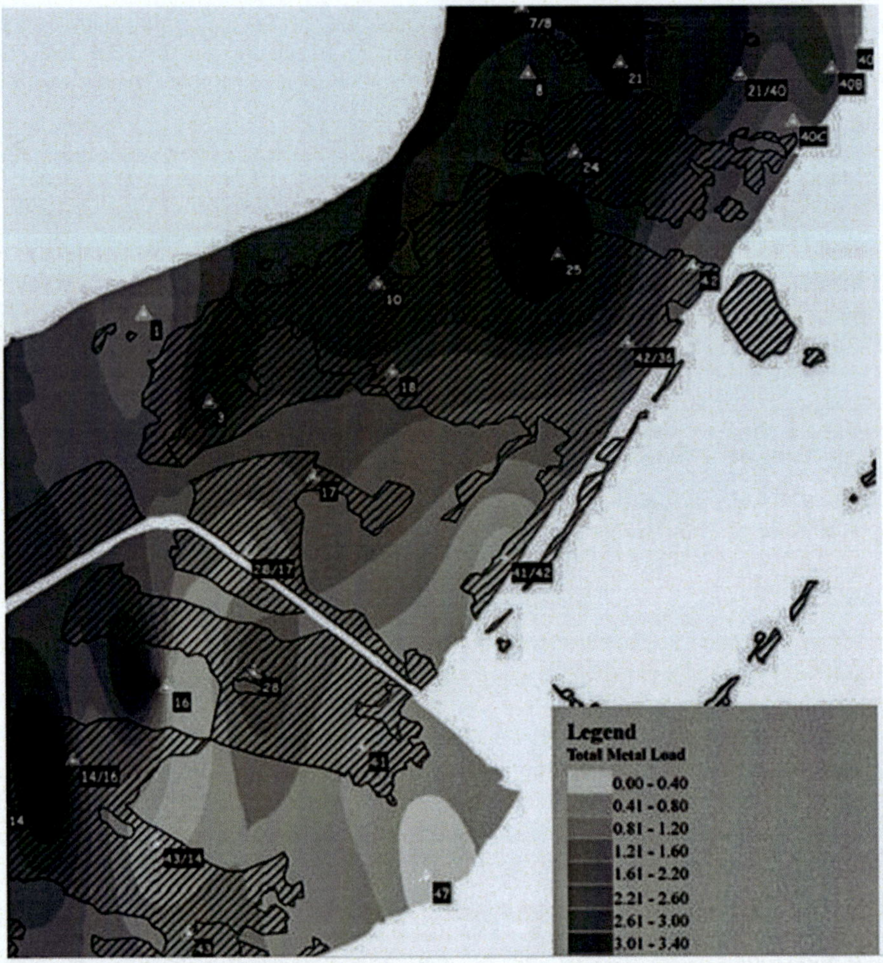

Fig. 2 Total soil metal load. The total soil metal load is the summation of the normalized concentrations of the five metals found in significantly high concentrations. It is a unitless metric which scales from 0 to 5. It is a comparative index for use within the study area (Gallagher et al. 2008)

map which could be used for comparison against biological and ecological characteristics of the community. It served as the foundation for most future ecological studies.

3 Species Composition and Distribution at LSP

Plant species composition at LSP, and in the railyard before the park was constructed, had been a subject of interest for well over 100 years. In 1878, Addison Brown published an article entitled "Plants Introduced with Ballast and on Made Land"

(Brown 1878). His focus was on the unique plants primarily from Western Europe that had been accidentally introduced primarily in the ballast of ocean-going ships. Making observations weekly from May through September, he identified 64 species, 24 of which were not in Gray's Manual of Botany, and not known to formally exist in this country. Little else was done during the intensively active railroad years, however, by the 1970s, as the land was being purchased by the State of New Jersey, an Environmental Impact statement was required. Known as the Texas Instrument Study (Texas Instrument 1976), it identified 113 species. It characterizes the vegetation of the site as that of very early succession dominated by non-native herbaceous species. *Artemisia vulgaris* had the greatest importance value as it occurred at every sampling site, while *Phragmites australis* had the second greatest importance value and occurred in dense stands. Interestingly, in many areas *Artemisia* has given way to later successional species, while Phragmites decreased by approximately 9 acres between the years 1996 and 2003 (Gallagher 2008). Tree communities were considered sparse with large trees entirely absent. A few small stands of *Populus deltoides* saplings existed at one location. Shrubs, which was *Rhus copallinum,* was dominant were also rare.

During the summers of 1995 and 1996, David McFarlane, then a graduate student at Rutgers, The State University of New Jersey, conducted a survey of the plants and animals of the site. The survey produced the first georeferenced vegetative communities map of the 235 undeveloped acres within the interior section of the park. McFarlane notes that the unique soils and microtopography conditions have produced several communities consisting of rare and unusual species, some of which are unknown in other parts of the state. This study documented at least 150 species within that section of the park. In 2003, as part of the Hudson-Raritan Estuary Environmental Restoration Study conducted by the Army Corps of Engineers (USACE), a point count vegetative survey of the interior section of the park was conducted. While the survey was carried out in the latter part of the summer and missed many early and ephemeral species, the USACE encountered considerable diversity and characterized the area as follows: "Although largely altered by human activity and colonized by non-native invasive plant species, the site supports a minimum of 184 plant species of 125 genera and provides suitable habitat for a variety of wildlife species" (USACE 2004).

An effort by the New York Metropolitan Floral Project documents the greatest number of species at 562, however, the entire park including the estuarine habitats was surveyed. A Torey Botanical field trip conducted in 2009 listed 189 species. Finally, the Fusion Ecology Lab from Rutgers has been examining species richness within the interior section of the park since the early 2000s and has documented at least 182 species. In summary, these data appear to suggest that the site underwent something akin to primary succession, which was dominated early by several invasive species, and then diversified, becoming what can clearly be described as a rich novel community, a mixture of native and non-native species.

Understanding that high concentrations of soil metals can be a considerable filter to the recruitment and establishment of plant species, we examined both the diversity and productivity values throughout the site in 2007. Diversity was examined in 0.1-ha plots representing all guilds within the site. Productivity was determined using true color aerial multispectral imagery and calculated as normalized difference vegetation index (NDVI) over the entire site. We had hypothesized that both diversity and productivity would decrease as the total soil metal load (TML) increased and that the relationship would be linear. That hypothesis proved incorrect as productivity, especially in the hardwood forest, and to a lesser extent diversity changed little within the vegetative guilds until the TML exceeded three (Gallagher 2008, Fig. 3). In other words, the plant communities within the site could tolerate the increasing concentration of soil metals until a critical threshold was reached. At that point, photosynthetic productivity and community diversity decreased rapidly.

When we examined how much of the site had TML values between three and five, areas that clearly showed an impact of soil contamination they accounted for approximately 20 acres or less than 1/10 of the entire site. Despite the 100 years of industrial use resulting in soil metal contamination above most ecological and residential soil screening criteria, the vast majority of the site had developed a fairly robust ecological community that demonstrated relatively high diversity and productivity values.

To understand the mechanistic properties associated with the loss of productivity above the TML critical threshold, Allison Salisbury carried out an experiment that focused on the co-dominant tree species *Betula populifolia* Marsh (Salisbury et al. 2017). Using a transect that established a TML gradient, she measured the gas exchange of representative leaves from the same tree over the course of several growing seasons. The results of the study demonstrate that most metabolic functions (net assimilation, light compensation points, quantum efficiencies, dark respiration rates, net assimilation, transportation, and stomatal conductance) were statistically

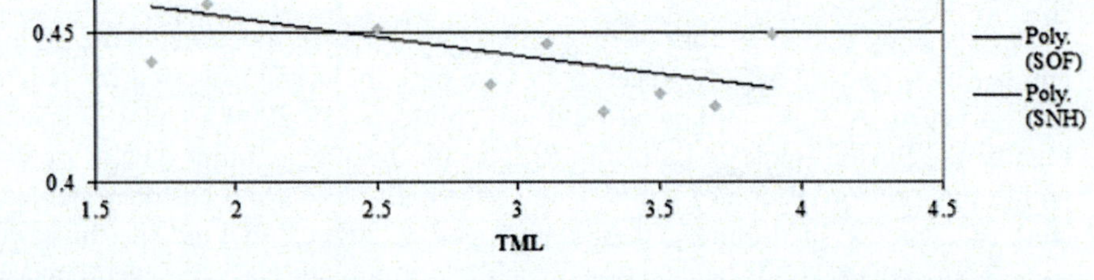

Fig. 3 Productivity within the Successional Northern Hardwood Forest (SNH) and Successional Old Field (SOF). Neither vegetative community exhibited a linear negative relationship, rather both had stronger polynomial relationships indicating a threshold relationship between total soil metal load (TML) and productivity as measured by normalized difference vegetation index (NDVI) (Gallagher 2008)

similar. However, leaf mass in trees located at sites above the critical TML threshold was significantly less. This study indicated that while the metabolic processes are similar, the resulting resource allocation is different, with more energy being used to maintenance in the areas with high soil metal concentrations.

To answer the question of whether or not the vegetative diversity of the site has plateaued or reached some sort of urban maximum, a comparative study was conducted of similar data collected in permanently established plots during 2008 and 2016. Interestingly, both species richness (species accumulation curves) and diversity (Shannon index) increased during that time frame (Fig. 4) (Salisbury et al. 2020). The increase in species diversity was highest in areas with lower TML and greatest in the Successional Northern Hardwood Forest.

Evenness, however, demonstrated no statistical difference during this period. The percentage of woody species increased in the grassland and forb dominated areas, while vine species increased in all habitat types. The increase of woody species was still associated with areas of higher soil metal loads, a trend that was originally identified in earlier studies (Gallagher et al. 2008). While this study did not prove why diversity values increased, as such increases could have been related to attenuation of soil metals, disturbance by Superstorm Sandy in 2012, or other factors, it did demonstrate that even after five decades of development post-industrial sites can be dynamic and exhibit strong and increasing diversity values. In addition, the communities within such sites are driven by both the unique soil properties as well as traditional competitive and facultative biotic filters.

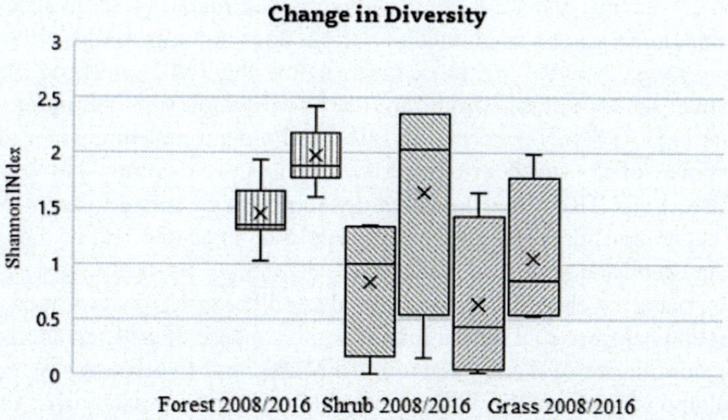

Fig. 4 Change in Shannon index of diversity. Error bars indicate one standard deviation. The three examined vegetative guilds all showed trends of increasing diversity between the years 2008 and 2016 (Salisbury et al. 2019)

4 Plant Community Trajectory

As mentioned earlier, the naturally assembled plant communities of LSP represent unique associations of both native and non-native species that can be considered the by-product of the cultural events that have taken place over the last 100 years. Understanding the future trajectory of such communities is a topic of considerable interest in urban ecology. Our research demonstrated that traditional successional scenarios, which have formed our concept of the norm since their introduction by Clements in 1916, may not apply to wasteland sites. In such sites, abiotic limiting factors can impair recruitment, resulting in communities that develop more slowly and can exhibit alternate stable states for long periods of time (Hobbs and Norton 2004). Conversely, certain guilds such as early successional forests may leapfrog typical succession sequences and develop along with grass communities, while other guilds such as shrubs may be totally absent for long periods of time.

We used historic true color aerial photographs from 1976, 1984, 1993, 1996, 2000, 2003, and 2009 and a 1969 gray scale photograph to determine the extent of the four recognized plant communities which included early hardwood forest, shrub, herbaceous (grasses and forbs), and wetland communities (Gallagher et al. 2011). We then divided the site into two distinct area typologies, above and below the TML critical threshold of three. In the 2009 map, the distribution of the pioneer forest assemblage dominated by *Betula populifolia* Marsh and two species of *Populus* spp., both of which are known to be metal tolerant, could be strongly correlated with soils above the TML threshold. Herbaceous species including both grasses and forbs were recruited early while the shrub guild, which is dominated by three species of *Rhus* spp. (Sumac), not known to be metal tolerant, does not appear until 1996 and never comprised more than 10% of those areas below the TML threshold. Successional development of herb/grass, shrub, and early hardwood tree communities in areas where the TML is below the critical TML threshold has developed in a pattern fairly representative of the northeastern region of the United States. However, in areas above the critical TML threshold, the development of metal tolerant pioneer tree species began considerably earlier than would be expected and at the time of this writing has persisted (Fig. 5).

In summary, the early successional hardwood community developed earlier than expected and has persisted longer than expected, a case of arrested succession. How long the dominance of *Betula populifolia* Marsh and two species of *Populus* will persist on the soils above the TML threshold is currently a question of some debate. Is the soil metal filter that fostered this unique community still a driving force after 60 years of development?

The attenuation of soil metals can occur via two processes. Leaching into the groundwater is always possible and it is known that groundwater at the site flows north at less than 1 m per year. We assumed that leaching was extensive in the early development of the site. However, we questioned whether the soil leaching had reached some type of equilibrium. To answer that question, we compared the analyzed total concentration of As, Cr, Cu, Pb, and Zn soil data collected from 1995,

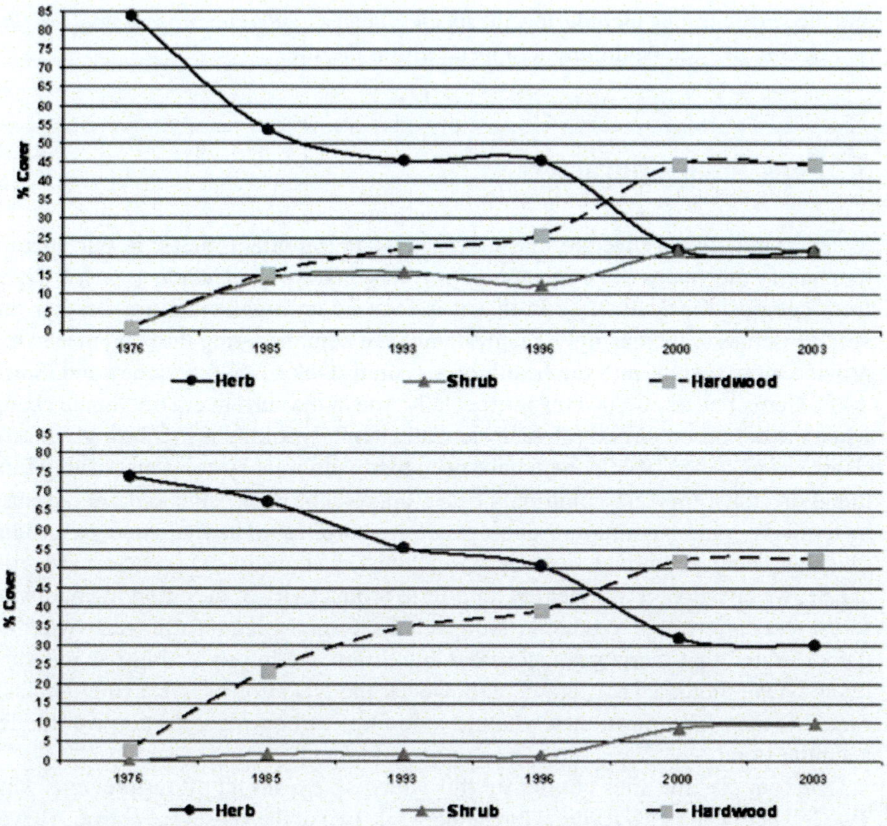

Fig. 5 Guild trajectories both above (top) and below (bottom) the critical total soil metal load (TML) threshold. Interestingly, the hardwood forest community establishes early in areas above the critical TML while the shrub community is delayed (Gallagher et al. 2011)

2005, and 2015. While there was no change in the concentration of any of these metals in the upper 30 cm of the soil between 1995 and 2005, and from 2005 to 2015, As and Cr concentrations increased in the soil C1 horizon (approximately 5- to 25-cm depth). We proposed that the observed increases in As and Cr resulted from downward migration through leaching from the upper 5 cm of soil and subsequent immobilization in the C1 horizon. However, as the C1 zone is well within the range of root growth, the impact on plant growth should have remained constant throughout the study period.

As mentioned earlier, the adsorption of metals to organic matter is a well-known phenomenon, which can reduce the concentration of liable metals. It is therefore possible that the continued accumulation of organic material resulting from vegetative growth could mitigate the soil metal filter and allow for the establishment of late successional tree species. The regional species within the remainder of the park and the surroundings include several species, such as *Quercus rubrum* and *Fraxinus americana*, that are beginning to appear within the site.

5 Functional Character

Recently, functional diversity has emerged as an important metric in our attempts to quantify the values of ecosystem performance (Diaz and Cabido 2001). It differs from compositional diversity in that it is focused on organismal traits rather than species richness. Increasingly, many studies are demonstrating that this metric may provide more insight into the health of a system (Dukes 2000; Petchey and Gaston 2007; Hooper et al. 2005). This is especially true in the current era of climate change when the definition of natural, historic, and altered typology is increasingly blurred (Hobbs et al. 2009). While there have been many different systems classifying functional diversity, they are commonly based upon niche dimension and are measured in terms of traits such as morphological structure, reproductive strategy, productivity, and interspecific relationships. To address the question of ecosystem function at LSP, we questioned whether the long-term productivity of the emerging hardwood forest was comparable to other less-impacted sites in the region. Our early work had demonstrated that there was a considerable difference in productivity as measured by an NDVI along a TML gradient, however, the NDVI measures a single point in time. We therefore questioned whether long-term carbon sequestration would yield a similar trend.

The four specific sites chosen for this study were a subset of our previous work. Two fell below the TML critical threshold while two of the sites were above. All were clearly dominated by *Betula populifolia* Marsh. The results of the study proved that our original hypothesis was not totally correct. There were considerable differences in the tree diameter at breast height and age relationship (Fig. 6), as it took trees at sites above the TML threshold longer to reach diameters similar to those in less-impacted

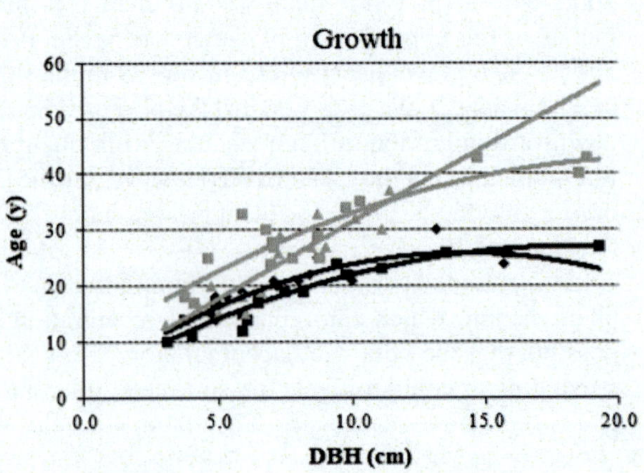

Fig. 6 Relationship between diameter at breast height (DBH) and age class in years. There is a statistically significant difference between the sites above the critical total soil metal load (TML) threshold (trend line in black) and sites below the TML threshold (trend lines in gray) which grew faster (Dahle et al. 2014)

sites. However, there was no difference in the above-ground biomass stored in trees of similar diameter regardless of location. Therefore, resource allocation between growth and maintenance must differ between sites, with trees at higher TML sites dedicating more energy to metabolic functions involved in maintenance rather than growth. The total carbon sequestration between sites was also considerably different. The two sites with relatively low TML yielded 40,017 and 50,362 kg/ha. One of the high TML sites yielded 48,504 kg/ha C, which is within a standard deviation of the 2 sites below the TML threshold. However, the second site above the TML threshold yielded 71,935 kg/ha C, the highest of the four study sites. This latter result is comparable to less impacted regional environments such as the New Jersey Pine Barrens which on average yield approximately 84,940 kg/ha in areas where trees have similar basal areas and stem density (Shafer et al. 2010).

Obviously, tree age and size combined with stocking densities are all factors in the potential for above-ground carbon sequestration. What is interesting about this study was that despite similar age and species compositions of the two sites above the TML threshold, stand density and hence C flux patterns differed significantly. We suggested that these differences were the result of edaphic and topographic differences in the site. The site with higher C sequestration rates was slightly lower in elevation and the topography allowed for organic matter to accumulate more easily. The increased organic matter in the soil not only mitigated some of the TML filters but also allowed for greater soil moisture, resulting in increased stocking densities.

6 Ecological Risk

While there is no doubt that soil metals are part of the legacy of the industrial revolution and remain one of the most persistent contamination issues of our time, the standards that guide planning and policy are based on extremely conservative procedures. Developed by the regulatory community responsible for ensuring the health and safety of the human community, these standards must take into account every precaution. In addition, conducting bio-assessments for every case where soil contamination is suspected is simply not feasible. I do not argue here against standards, or that the ecological risk has been grossly exaggerated. Soil metal impact on metabolic processes has been long defined (Tyler et al. 1989; Giller et al. 1998), and easily translatable standards are necessary for coherent policy. However, such standards do not consider variation in toxicity related to the actual availability or the liable concentration that results from the complex relationships between total soil metal concentration and soil chemistry, particularly pH, soil moisture, and organic matter. Most tests have also been conducted with soil material that has not been leached before the experiment, which results in higher soluble concentrations. In addition, most standards are written based on bioassays conducted with sensitive organisms which exhibit impairment at far lower concentrations than metal tolerant species. In urban wastelands, however, the sites are exposed to weather and leaching for long periods of time, and successful early migrants have to be metal tolerant by definition. Therefore, most current standards do not reflect the actual ecological risk associated with wastelands.

In addition to the empirical modeling of risk mentioned above, we also examined the uptake and translocation of six of the most well-known metals in the dominant colonizing species in plant, insect, and bird species at LSP. We tested (Gallagher, et al. 2008) for arsenic (As), chromium (Cr), copper (Cu), lead (Pb), zinc (Zn), vanadium (V), and sodium (Na) in ten of the dominant or more common species within the study site. We focused on As, Cr, Cu, Pb, and Zn as they are more commonly associated with wastelands and have been shown to be generally toxic at higher concentrations. We did not include mercury as soil concentrations of this element were minimal and its distribution was sporadic. Tests for the metals were conducted on root, stem, and leaf tissue of the various plants. In general, the concentration of the elements was an order of magnitude less in the root tissue than in the soil. The aerial sections of the plant also contained concentrations of metals that were greatly below the root concentrations (Table 2). In fact, with the exception of Pb and Zn, the tested elements were not or minimally detected in leaf tissue. Stem and leaf concentrations of Pb in *Betula populifolia* Marsh. were detected at an order of magnitude less than in the roots. Interestingly, the only element to translocate to the leaf tissue at relatively high concentrations was Zn. Zn in both *B. populifolia* Marsh. and *Populus tremuloides* Michx. accumulated in the leaf tissue at concentrations higher than the ambient soil concentrations. While not a hyper-accumulator, the accumulation of Zn in both of these species is well documented from several different sites (MacFarlane and Burchett 2000).

Clones of *Betula pendula* Roth. show a broad range of soil metal concentrations (Hillary and Wilkins 1987), demonstrating a genetic plasticity that undoubtedly infers a competitive advantage on metalliferous soils and is one of the reasons they are so successful on post-industrial landscapes. Our data suggest that many of the dominant species associated with wasteland sites are successful as they have inherent mechanisms that mitigate the effect of soils with high metal concentrations. Mechanisms such as ion exclusion or sequestration result in assemblages of plant species that innately mitigate the potential for metal translocation and hence reduce and possibly even eliminate the associated ecological risk. Naturally assembled wasteland plant communities can therefore be considered an effective phytostabilization management strategy.

Translocation, even at minimal concentrations, however, has been proven to pose a significant ecological and human health risk if bioaccumulation occurs in higher trophic level species. For example, Hg tends to accumulate in the fatty tissue of fish (Goyer 2000) and can impair neurological and cardiovascular functions in humans. To assess whether accumulation and/or biomagnification was occurring at the study site, several translocation studies were conducted. Interestingly, the concentrations of metals at higher trophic levels were always far less than those in any part of the plant tissue.

In our initial study, nestling house wrens (*Troglodytes aedon*) and nestling American robins (*Turdus migratorius*) were used as indicators of metal bioaccumulation (Hofer et al. 2010). Both species feed their young with resources found locally, so the

Table 2 Concentrations of the various metals in the soil (s) and in the tissues (r: root; st: stem; and l: leaf) of the common plant species at the study site

Species	As	mg/kg			Cr	mg/kg			Cu	mg/kg		
	s	r	s	1	s	r	st	1	s	r	st	1
Artemisia vulgaris mean (n = 4)	19.2	1.4	19.2	0	29.26	5.248	0.145	0.373	222.5	67.66	8.241	20.93
Betula populifolia mean (n = 8)	31.08	2.317	31.08	0.02	48.84	13.07	0.261	0.261	124.5	46.07	1.74	5.619
Onoclea sensibilis	35.07	2.906	35.07	0	53.42	13.36	0.51	0.718	67	47.97	4.166	4.621
Phragmites australis	>MD	>MD	>MD	0	>MD	>MD	0	0.272	>MD	>MD	4.169	5.408
Polygonum cuspidatum	12.45	2.561	12.45	0	43.39	12.71	0.06	0.512	48.29	40.34	6.511	4.289
Populus sp. mean	124.7	6.741	124.7	0.009	47.35	6.954	0.015	0.302	520.3	68.87	3.818	7.486
Rhus copallinium mean (n = 5)	29.8	0.929	29.8	0	48.14	5.357	0.039	0.317	208.5	54.61	2.053	4.254
Solidago virgaurea	>MD	>MD	>MD	0	>MD	>MD	0.122	0.471	>MD	>MD	10.08	9.305
	Pb	mg/kg			Zu	mg/kg			Va	mg/kg		
	s	r	st	1	s	r	st	1	s	r	st	1
Artemisia vulgaris mean (n = 4)	408.8	56.87	0.645	0	506.7	287.3	123.5	94.97	79.81	7.842	0	0
Betula populifolia mean (n = 8)	266.2	129.3	11.26	2.252	156.5	220.6	123.1	904.7	53.77	10.83	0	0
Onoclea sensibilis	168.2	47.21	0	>MD	38.69	37.53	25.46	2.685	72.32	21.78	>MD	>MD
Phragmites australis	>MD	>MD	0.566	>MD	>MD	>MD	30.18	2.088	>MD	>MD	>MD	>MD
Polygonum cuspidatum	156	42.56	1.219	>MD	96.29	131.3	59.06	98.4	18.11	22.54	>MD	>MD
Populus sp. mean	1446	106.5	3.91	0	632.8	373	104.8	1189	31	12.45	0	0
Rhus copallinium mean (n = 5)	435.2	54.62	1.194	0	216.9	258.7	15.07	17.83	28.01	7.36	0	0
Solidago virgaurea	>MD	>MD	1.282	>MD	>MD	>MD	395.6	69.46	>MD	>MD	>MD	>MD

translocation of metals would have to come from the local environment. Both feather and fecal samples were analyzed for six of the metals (As, Cr, Cu, Fe, Pb, and Zn) found in high concentrations in the soil. Interestingly, the concentrations of metals in the bird samples were significantly less than those found in the plants of the same site. However, concentrations of Pb, Fe, and Zn in wrens were statistically higher than those in robins, while Cu and Cr concentrations were lower. This differential uptake by bird species could indicate a species preference for selected metals or a difference in soil or food resource metal concentrations within the study site. Finally, the study did find increased metal concentrations in the examined bird species in the study site when compared to controls located within the Hutchinson Memorial Forest at Rutgers University, a site which does not exhibit soil metal enrichment.

While both species accumulated more metals in the fecal and feather samples, the house wren exhibited the greatest concentrations. In spite of the significantly increased metal load in the wrens, the birds did not exhibit any differences in size metrics or fledge rates during the breeding season, compared to nestlings from the control site. In fact, the fledging rate at LSP was higher than that of the control. There could be many reasons for the greater nesting success rates at the study site, the most obvious of which is the difference in predation pressure. LSP appears to have fewer nest predators than the Hutchinson Memorial Forest.

Litwhiler (2015) carried out an extensive study of the availability of food sources to migrating birds, in an examination of whether or not the study site was actually an "attractive nuisance," an area that attracts wildlife but then negatively impacts that wildlife or provides a vehicle for the transfer of the contaminant during migration. Overall trophic transfer was low with Pb and Cd yielding the highest relative concentrations in shrub fruits. Litwhiler's data indicated that many species forage within the brownfield during migration as the site provides a critical refuge in the urban context. Litwhiler concludes, as have the majority of studies at the site, that while trophic transfer of selected metals is possible the ecological risk associated with that transfer does not appear to impair function. The working hypothesis that is developing as a result of long-term ecological studies within naturally assembled urban wastelands is that colonizing species are successful as they can tolerate high soil metal loads. Over time, the above- and below-ground feedbacks and accumulating organic matter tend to decrease the soil metal filter, and functional novel plant communities develop. Such communities have unique mixtures of species that are not only functional but also tend to reduce the risk normally associated with high soil metals.

7 Changing Ecological Paradigms

Due to the rapid loss of biodiversity in general (Meyers 1996; Sax and Gaines 2003), plant community diversity has been a strong focus of ecological research for several decades. The literature includes many studies where greater diversity correlates with healthier or more resilient systems (Robinson 2006). Given this context, urban biodiversity and ways to support its enhancement, although controversial, have received much attention. In addition, while there is a strong preference for native diversity, as it has generally been accepted that legacy assemblages, especially those with long tenures, function more efficiently, many native species cannot tolerate the abiotic conditions of the urban environment (Pouyat et al. 2007). The acceptance then of novel plant communities with mixtures of old and new world species, a common phenomenon in the urban context, provides a viable alternative that can be functional, as our work at LSP has demonstrated. Such communities can provide provisioning, regulating, and cultural ecosystem services and as part of their supporting service they can reduce the risk of soil contaminant transfer.

Succession in such wasteland communities also needs to be reconsidered. The classic deterministic model (Clements 1916) states that vegetative assemblages tend to follow established patterns of development in which a predetermined state or end point is predictable. Within the dynamic urban context, however, such assumptions may not be practical. On the other hand, the stochastic model (Gleason 1939; Van der Maarel and Sykes 1993) allows for a more random process that anticipates filtering according to the abiotic character of a given environment, the regional species pool composition, and the order of arrival. Such a model appears much more applicable to wasteland sites.

In 1990, the Ecological Society of America adopted and published the following definition: "Ecological restoration is the process of intentionally altering a site to establish a defined, indigenous, historic ecosystem. The goal of this process is to emulate the structure, function, diversity and dynamics of the specified ecosystem" (McDonald et al. 2016). While this technical definition has been modified over time, species composition and diversity remain the primary metrics. In addition, most regulatory agencies throughout the United States base their judgment of successful restoration on plant community composition. Interestingly, many if not most restoration initiatives fail according to the definition (Lockwood and Pimm 1999) as they are unable to maintain the threshold level of legacy species. The assumption is that non-native or invasive species will decrease diversity and result in decreased functionality. While there are numerous studies that support this concept (Powell and Powell 1986; Canterbury et al. 2000), they either ignore the surrounding context and focus on a site or series of sites, or they are limited in the time during which observations are made. More recent ecological work, especially in urban wastelands such as post-industrial sites like LSP, demonstrates that the perception of biological impoverishment associated with such lands was clearly overstated. In fact, biologic diversity, even if uniquely represented as novel assemblages of both old and new world species, is common in urban areas (Aronson et al. 2014). Perhaps that is why the Science and Policy Working Group from the Ecological Society of America recently revised its definition to "the process of assisting the recovery of an ecosystem that has been damaged, degraded or destroyed" (Gann et al. 2019).

In conclusion, plant communities associated with wastelands develop through unique patterns of species composition/diversity/distribution, models of primary productivity, and patterns of carbon sequestration. Such patterns are driven by threshold tolerances to soil metals and develop along nontraditional guild trajectories. Most importantly, the ecological risk associated with the uptake and transfer of various soil metal contaminants appears not to follow traditional paths.

Acknowledgements Over the past several decades, there have been many researchers who have added to our understanding of the undeveloped portion of Liberty State Park. Most are cited here. They have all helped to shape a new understanding of urban ecology.

References

Adriano DC (1986) Trace elements in the terrestrial environment. Springer, New York

Aronson MFJ, La Sorte FA, Nilon CH, Katti M, Goddard MA, Lepczyk, CA, Warren PA, Williams NSG, Cilliers S, Clarkson B (2014) A global analysis of the impacts of urbanization on bird and plant diversity reveals key anthropogenic drivers. Proc R Soc B

Brown A (1878) Plants introduced with ballast on manmade land. Bull Torrey Bot Club 6(45):255–258

Canterbury GE, Martin TE, Petit DR, Petit LJ, Bradford DF (2000) Bird communities and habitat as ecological indicators of forest condition in regional monitoring. Conserv Biol 14:544–558

Clements FE (1916) Plant succession. Publication 242. Carnegie Institution Washington DC

Dahle GA, Gallagher FJ, Gershenson D, Schäfer KVR, Grabosky JC (2014) Allometric and mass relationships of *Betula populifolia* in a naturally assembled urban woodlot. Urban Ecosyst 17:1147–1160

Diaz S, Cabido M (2001) Vive la difference: plant functional diversity matters to ecosystem processes. Trends Ecol Evol 16:646–655

Dudka S, Ponce-Hernandez R, Tate G, Hutchinson TC (1996) Forms of Cu, Ni and Zn in soils of Sudbury, Ontario and the metal concentrations in plants. Water Air Soil Pollut 90:531–542

Dukes JS (2000) Biodiversity and invasibility in grassland microcosms. Oecologia 126:563–568

Gallagher FJ (2008) The role of soil metal contamination in the vegetative assemblage development of an urban brownfield. Rutgers The State University of New Jersey, School of Graduate Studies, ProQuest Dissertations Publishing, 3330871

Gallagher FJ, Pechmann I, Bogden J, Grabosky J, Weis P (2008) Soil metal concentrations and plant productivity in an urban brownfield. Environ Pollut 153:351–361

Gallagher FJ, Pechmann I, Holzapfel C, Grabosky J (2011) Altered vegetative assemblage trajectories within an urban brownfield. Environ Poll 159:1159–1166

Ghrefat HA (2011) Application of geoaccumulation index and enrichment factor for assessing metal contamination in the sediments of Kafrain Dam. Jordan Environ Monit Assess 178:95–109

Giller KE, Witter E, McGrath SP (1998) Toxicity of heavy metals to microoganisms and microbial processes in agricultural soils: a review. Soil Biol Biochem 30:1389–1414

Gleason HA (1939) The individualistic concept of plant association. Am Midl Nat, 21(1) 92–110

Goyer RA (2000) Toxicological effects of methylmercury. National Academy Press, Washington DC

Gann GD, McDonald T, Walder B, Aronson J, Nelson CR, Jonson J, Hallett JG, Eisenberg C, Guariguata MR, Liu J, Hua F, Echeverria C, Gonzales, EK, Shaw N, Decleer K, Dixon KW (2019) International principles and standards for the practice of ecological restoration. Second edition. Restor Ecol S1–S46

Hakanson L (1980) An ecological risk index for aquatic pollution control. A sedimentological approach. Water Res 14:975–1001

Hartman JM, Smith Z, Czarny T, Quispe J (2017) Jersey city environmental resource inventory. Appendix A-15

Hillary JD, Wilkins DA (1987) Zinc tolerance in Betula spp., effect of external concentration of zinc on growth and uptake. New Phytol 106:517–524

Hobbs JD, Norton DA (2004) Ecological filters, thresholds and gradients in resistance to ecological reassembly. In: Templeton VMRJ, Hobbs T, Nuttle T, Halle S (eds) Assembly rules and restoration ecology. Island Press, Washington, pp 72–95

Hobbs RJ, Higgs E, Harris JA (2009) Novel ecosystems: implications for conservation and restoration. Trends Ecol Evol 24(11):599–605

Hofer C, Gallagher FJ, Holzapfel C (2010) Metal accumulation and growth performance of nestlings of passerine bird species at an urban brownfield site. Environ Pollut 158:1207–1213

Hooper DU, Chapin III FS, Ewel JJ, Hector A, Inchausti P, Lavorel S, Lawton JH, Lodge DM, Loreau M, Naeem S, Schmid B. Setälä H, Symstad AJ Vandermeer J, Wardle DA (2005) Effect of biodiversity on ecosystem functioning: aconcensus of current knowledge. Ecolo Monogr 75(1):35–33

Juang KW, Lee DY, Ellsworth TR (2001) Using rank-order geostatistics for spatial interpolation of highly skewed data in heavy-metal contaminated site. J Environ Qual 30:894–903

Krumins JA, Goodey NM, Gallagher FJ (2015) Plant-soil interactions in metal contaminated soils. Soil Biol Biochem 80:224–231

Litwhiler ME (2015) Attraction and risk in urban bird habitats. New Jersey Institute of Technology Dissertations. 131. https://digitalcommons.njit.edu/dissertations/131

Lockwood JL, Pimm SL (1999) When does restoration succeed? In: Weiher E and Keddy PA (eds) Ecological assembly rules: perspective, advances and retreats. Cambridge University Press

MacFarlane GR, Burchett MD (2000) Cellular distribution of copper, lead and zinc in the grey mangrove Avicennia marina (Forsk) Vierh. Aquat Bot 68:45–59

Marsh GP (1882) The earth as modified by human action. A new ed. of Man and nature. New York Scribner's Sons

McDonald T, Gann GD, Jonson J, Dixon KW (2016) International standards for the practice of ecological restoration—including principles and key concepts. Society for Ecological Restoration, Washington. https://seraustralasia.com/wheel/image/SER_International_Standards.pdf. Accessed 5 Oct 2020

Meyers N (1996) The biodiversity crisis and the future of evolution. Environmentalist 16:37–47

Muller G (1969) Index of geoaccumulation in sediments of the Rhine River. GeoJournal 2:108–118

Petchey OL, Gaston KL (2007) Dendrograms and measuring functional diversity. Oikos 116:1422–1426

Pouyat RV, Pataki KT, Belt PM, Groffman JH, Band LE (2007) Effects of urban land-use change on biogeochemical cycles. In: Canadell JG, Patak DE, Pitelka LF (eds) Terrestrial ecosystems in a changing world. Springer, Berlin, pp 45–58

Powell GVN, Powell AHX (1986) Reproduction by Great white herons Ardea herodias in Florida bay as an indicator of habitat quality. Biol Conserv 36:101–113

Qian Y, Feng FH, Gallagher FJ, Zhu Q, Wu M, Liu CJ, Jones K, Tappero R (2017) Risk assessment and interpretation of heavy metal contaminated soils on an abandoned urban brownfield site. Environ Sci Pollut Res 24:23549–23558

Robinson JG (2006) Conservation biology and real-world conservation. Conserv Biol 20:658–669

Ross SM (1994) Toxic metals in soil and plant systems. Wiley, New York

Salisbury A, Reinfelder JR, Gallagher FJ, Grabosky JC (2017) Long-term stability of trace element concentrations in a spontaneously vegetated urban brownfield with anthropogenic soils. Soil Sci 182–2:69–81

Salisbury A, Gallagher FJ, Caplan J, Grabosky JC (2018) Maintenance of photosynthetic capacity despite high load of trace metals in contaminated anthropogenic soils. Sci Total Environ 625:1615–1627

Salisbury A, Gallagher FJ, Holzapfel C (2020) Plant diversity continues to increase after 40 years at an urban brownfield. Urban Ecosyst. https://doi.org/10.1007/s11252-020-01018-x

Sax DF, Gaines SD (2003) Species diversity: from global decreases to local increases. Trends Ecol Evol 18:561–566

Schäfer KVR, Clark KL, Skowronski N, Hamerlynck EP (2010) Impact of insect defoliation on forest carbon balance as assessed with a canopy assimilation model. Glob Chang Bio 16(2):546–560

Texas Instruments (TI) (1976) LSP ecological study. Prepared for the Port Authority of New York and New Jersey

Tyler G, Pahlsson AMB, Bengtsson G, Baath E, Tranvik L (1989) Heavy-metal ecology of terrestrial plants, microorganisms and invertebrates: a review. Water Air Soil Pollut 47:189–215

United States Army Corps of Engineers (2004) Hudson-Raritan Estuary Environmental Restoration Study, LSP, Integrated Environmental Resource Inventory and Draft Feasibility Study. Appendix A, p 24

USDA. NRCS soils survey. https://soilseries.sc.egov.usda.gov/OSD_Docs/L/LADYLIBERTY.html

Van der Maarel E, Sykes MT (1993) Small-scale plant species turnover in a limestone grassland: the carousel model and some comments on the niche concept. J Veg Sci 4:179–188

Frank J. Gallagher For over 40 years, **Frank J. Gallagher** has explored the connection between people and landscape through both land management and academic research. His current appointment as Assistant Professor of Professional Practice and the Director of the Environmental Planning and Design Program within the Department of Landscape Architecture (Rutgers, The State University of New Jersey, USA) focuses on urban sustainability. His research attempts to establish new references associated with the urban environment that focus on ecological function and soil contamination mitigation. He has published extensively in both scientific journals and venues of general interest, on topics generally ranging from phytostabilization of contaminated soils to the ethics of ecosystem function monetization.

Strategies for Increasing Biodiversity Conservation in Cities Using Wastelands: Review and Case Study

Michael L. McKinney

Abstract Urban wastelands have great potential for promoting biodiversity. As many cities shrink in size, these unused areas will continue to increase while available revenues for land maintenance will decrease. Managing such wastelands in ways that promote biodiversity, such as encouraging spontaneous vegetation and ecological succession, reduces maintenance costs while also providing many environmental and social benefits to the urban population. I reviewed the literature since March 2014 and found 31 studies of factors that influence biodiversity in urban wastelands. There is a strong geographic bias toward studies in North America (15) and Europe (10), and a taxonomic bias toward arthropods (16), plants (11), and birds (6). These studies often show that wastelands contain as much, and often more, biodiversity than other urban green spaces. Consistent with a previous review, local factors (area, age, soil, microclimate, and previous and current land use) are dominant influences on biodiversity in wastelands, with landscape factors often playing a secondary role. Next, I discuss how these factors have played a role in the Knoxville Urban Wilderness (KUW) project which has been very successful in acquiring and managing a variety of wasteland parcels to achieve economic and ecological goals. Examples of their application include increasing land acquisition (more area and spatial habitat diversity including soils and microclimates), increasing connectivity (greenways and trails), and maintaining the parcels in different stages of ecological succession. These efforts have undoubtedly contributed significantly to the biodiversity of this urban region. For example, although the KUW occupies a tiny fraction (688 ha) of Knox County, roughly two-thirds of Knox County bird species utilize the KUW, and about 13% of known threatened species in Knox County have been recorded in the KUW. I also briefly discuss how the KUW has successfully coordinated the acquisition and management of many parcels of vacant land, worked with many stakeholder groups, greatly increased public exposure to biodiversity, and contributed over $8 million to the local economy each year.

Keywords Wasteland · Vacant land · Brownfield · Green infrastructure · Biodiversity · Urban ecology

M. L. McKinney (✉)
University of Tennessee, Knoxville, USA
e-mail: mmckinne@utk.edu

F. Di Pietro and A. Robert (eds.), *Urban Wastelands*, Cities and Nature,
https://doi.org/10.1007/978-3-030-74882-1_3

1 Introduction

Globally, at least one-sixth of the world's cities have decreasing human populations (Haase 2008), with roughly 25% of these cities occurring in the United States (Haase 2013). Driven by fundamental shifts in economic and other socio-political trends, these population declines have reduced demand for land in these cities leading to building demolition and increasingly vacant urban lands. For example, about 16.7% of large US cities' land area is vacant (Newman et al. 2016). At the same time, most of these shrinking cities suffer from decreasing tax revenues which exacerbates the economic burden of maintaining these vacant lands. Consequently, the reuse of vacant urban land has become a significant policy field in urban planning (Johnson et al. 2014). These vacant urban lands have been called several other terms including wastelands, brownfields, abandoned or derelict land (Bonthoux et al. 2014; Mathey et al. 2015). For simplicity, I will call them wastelands throughout.

Wastelands are often viewed by the public and urban governments as discarded and unattractive urban parcels that should be re-developed as soon as possible (Pagano and Bowman 2000). However, as abandoned areas with little or no landscape management, wastelands often become highly vegetated. For example, 62% of vacant lots in New York City have either shrubs or trees (Kremer et al. 2013) and vacant lots in Roanoke, Virginia average 30.6% tree cover (Kim et al. 2016). Consequently, there is a growing realization in many cities of the world that these wastelands can be incorporated into emerging efforts to increase urban green infrastructure to provide many useful ecosystem services (Mathey et al. 2015; McKinney and VerBerkmoes 2020). The ecosystem services provided by green infrastructure, including wastelands, are numerous and valuable: human health and recreation, microclimate regulation, air and water quality improvements, and stormwater regulation (Haase 2013; Palliwoda et al. 2020).

As an important element in urban green infrastructure, wastelands can also make major contributions to biodiversity conservation (Bonthoux et al. 2014). Indeed, increasing biodiversity often coincides with increasing urban ecosystem services (Adler and Tanner 2013; Haase 2013). Thus, promoting biodiversity in wastelands not only has conservation value but additionally enhances many ecosystem services. In this paper, I explore the importance of wastelands in biodiversity conservation in two ways. One, I review studies published since the review of factors affecting wasteland biodiversity by Bonthoux et al. (2014). I did this review because interest and research in wasteland biodiversity have increased in recent years, and I wanted to see if any different patterns were emerging. Two, I will apply these findings to a case study, the Knoxville Urban Wilderness (KUW) project, which has been assembled from vacated lands into a growing and successful entity providing many economic and ecological benefits to the general public. I integrate these findings to discuss strategies to design and manage wastelands in practical ways that can be applied in the many cities experiencing the rapid growth of wastelands.

Box 1: Methods

Following Bonthoux et al. (2014), my goal was to find papers documenting mechanisms (factors) that influence biodiversity in urban wastelands. Following their methods, I searched the Web of Knowledge (www.webofknowledge.com) by entering the following key words: (urb*) and (wasteland or waste land or brownfield or abandoned area or abandoned land or vacant area or vacant land or derelict area or derelict land or semi natural area or semi natural land or neglected area or neglected land or ruderal area or ruderal land or spontaneous urban vegetation or urban wild*) and (biodiv* or ecol* or species richness or species abundance).

The search was carried out during June 20–23, 2020 and included articles that were published from March 2014 until June 23, 2020. 431 records were located. Abstracts of each record were examined, and 311 were excluded as lacking relevance to this study. The full texts of the remaining articles were examined and 31 articles were ultimately selected as meeting the goals of this review. These articles (1) compared wasteland biodiversity to that of other urban habitats (12 articles), and (2) documented local and/or landscape factors influencing the biodiversity of wastelands (19 studies).

2 Wastelands and Biodiversity: A Literature Review

2.1 Wasteland Studies by Location and Taxa

Of the 31 studies identified (Tables 1 and 2), almost half (15) were from North America with 14 from the USA and 1 from Canada. The USA studies were entirely from eastern US, probably reflecting the chronology of settlement, with older cities occurring there. Also, just a few cities are represented, mostly large older industrial cities (Cleveland, Chicago, Toledo, and Baltimore). Ten of the 31 studies are from Europe, with Germany making up the largest proportion (5 studies). The remaining studies were located in Latin America (4), Australia (1), and India (1).

Interestingly, these results contrast with Bonthoux et al. (2014) who found that urban wasteland studies were almost entirely limited to European cities. It seems likely that the increase in North American (and Latin American) studies represents a growing realization of the importance of wastelands for conservation goals (e.g., Riley et al. 2018a; Zefferman et al. 2018). Another interesting pattern in the 31 studies here (Tables 1 and 2) is that this field continues to be dominated by just a few research groups working in just a few cities in Europe and North America.

The 31 studies (Tables 1 and 2) also show a strong taxonomic bias: arthropods (16 studies, plants (11 studies), and birds (6 studies). No mammals, reptiles, or other animal groups are included. Bonthoux et al. (2014) also found that these former groups were among the most commonly found in wasteland studies.

Table 1 Summary of the 12 papers comparing wasteland biodiversity to other urban green spaces

Source	Country	City	Taxa	Response variables	Comparative habitats
Burkman and Gardiner (2015)	USA	Cleveland	Spiders	Species abundance, richness, and traits	Vacant lots, community gardens, planted prairies
Da Rocha et al. (2020)	Brazil	Ribeirao	Bees, wasps	Species richness	Forest, wasteland, residential
Deak et al. (2016)	Hungary	Debrecen	Plants	Species richness, diversity, traits	Vacant lots, parks, peri-urban grasslands
Moorhead and Philpott (2013)	USA	Toledo	Spiders	Species richness, tax composition, activity	Vacant lots, gardens and forests
Philpott et al. (2014)	USA	Toledo	Arthropods	Abundance, species richness, trophic struct	Vacant lots, forest fragments, community gardens
Ramalingam and Rajan (2017)	India	Bangalore	Beetles	Species richness	Remnant forest, campuses, parks and vacant plots
Rega-Brodsky et al. (2018)	USA	Baltimore	Birds, plants	Vegetative structure, bird species richness	Emergent, planted and remnant vacant lots
Riley et al. (2018a, b)	USA	Cleveland	Trees	Abundance, diversity, size class tree species	Vacant, residential, suburban lots
Roche et al. (2016)	Australia	Philip Island	Lapwings	Nesting frequency and success	Vacant, residential and vacation lots
Sivakoff et al. (2018)	USA	Cleveland	Bees	Species richness, abundance, composition	Vacant lots and urban farms
Villasenor et al. (2020)	Chile	Santiago	Birds	Abundance and species richness	Vacant lands, urban parks and residential areas
Zuniga-Palacios et al. (2020)	Mexico	Pachuca	Plants, birds	Species richness, composition, structure	Small vacant lots, large natural areas

Table 2 Summary of the 19 papers analyzing the factors affecting wasteland biodiversity

Source	Country	City	Taxa	Response variables	Local factors	Landscape factors
Anderson and Minor (2019)	USA	Chicago	Plants	Species richness and evenness	Trash, socioeconomic factors	
Anderson and Minor (2020)	USA	Chicago	Plants	Taxic composition, species richness	Management	
Blouin et al. (2019)	Canada	Montreal	Plants	Species richness and beta diversity	Urbanization level	500 m buffer per lot
Buchholz et al. (2018)	Germany	Berlin	Arthropods	Species diversity and traits	Management, plant diversity	Isolation, urbanization level
Buchholz et al. (2020)	Germany	Berlin	Bees	Taxic and functional diversity	Management, richness, flowers	Isolation, urbanization level
Cabral et al. (2017)	Germany	Leipzig	Plants	Species richness	Management and garden type	
De la Flor et al. (2020)	USA	Cleveland	Spiders	Functional diversity, species richness	Mowing	Isolation, impervious area
Eckert et al. (2017)	Germany	Berlin	Grasshoppers	Functional diversity, richness, evenness	Age, herb cover, area, soil, habitat	Distance habitat, imperv area
Figueroa et al. (2020)	Chile	Santiago	Plants	Species richness	Age, area, pop density, seed bank	Distance urban ctr, boundary
Fischer et al. (2016)	Germany	Berlin	Bees	Species richness	Management, flower density	Isolation, urbanization level
Johnson et al. (2015)	USA	Baltimore	Plants	Taxic, functional, phylo diversity	Prior land use, soil, area	Distance to nearest lot
Johnson et al. (2018)	USA	Baltimore	Plants	Taxic, functional composition	Prior land use, age, soil	Distance to nearest lot, canopy

(continued)

Table 2 (continued)

Source	Country	City	Taxa	Response variables	Local factors	Landscape factors
Melliger et al. (2017)	Switzerland	Basel	Plants, moths	Species richness, traits	Habitat size	Landscape composition
Rega-Brodsky and Nilon (2017)	USA	Baltimore	Birds	Nesting success	Vegetation characteristics	Forest cover within 100 m
Rega-Brodsky and Nilon (2016)	USA	Baltimore	Birds	Abundance and species richness	Vegetation characteristics	Forest cover within 100 m
Twerd et al. (2017)	Poland	Bydgoszcz	Wasps	Abundance and species richness	Successional stage, soil	
Twerd and Banaszak-Cibicka (2019)	Poland	Bydgoszcz	Bees	Species richness, abundance, diversity	Area, successional stage, land use	Distance nearest patch, center
Vacht et al. (2019)	Estonia	Tallinn	Mites	Abundance and species richness	Successional stage	
Wadhwa (2018)	USA	Newark	Isopods	Abundance, diversity, health	Soil metal content	

2.2 *Wasteland Biodiversity: Comparison with Other Urban Habitats*

Urban wastelands often contain a significant proportion of species found in a city or region (Bonthoux et al. 2014). For example, a sample of just 24 urban wastelands in Berlin, Germany, contained 21 species of grasshoppers representing 45% of the entire grasshopper fauna of the city, including many species of conservation concern (Eckert et al. 2017). Similarly, a study of the urban grasslands (including wastelands) of Berlin found 62 wild bee species belonging to over 20% of the bee taxa known for that city (Fischer et al. 2016). Finally, a study of the wastelands in the city of Bydgoszcz, Poland, found they contained 42% of all wild bees known from the entire country of Poland (Twerd and Banaszak-Cibicka 2019).

Bonthoux et al. (2014) found 9 studies that compared the biodiversity of wastelands to that of other habitats, with 4 of these 9 studies indicating higher diversity in wasteland ecosystems. I found 12 studies that have since been published, or otherwise not included in their study (Table 1). All of these studies show that, where similarly sized areas are compared, wastelands have as much, and usually more, biodiversity than other urban green spaces such as residential lawns, city parks, community gardens, and urban forests.

Regarding plant diversity, Rega-Brodsky et al. (2018) found that remnant vacant lots had a greater vegetative structure compared to other lot origins. In Cleveland, Ohio, inner-city vacant lots support three times as many trees and a greater tree diversity than inner-city and suburban residential lots, with the plurality of trees being naturally regenerated saplings (Riley et al. 2018b). In Debrecen, Hungary, urban parks have lower plant diversity than vacant lots (Deak et al. 2016). On the other hand, in the city of Pachuca, Mexico, small vacant lots have less plant diversity than large urban forests (Zuniga-Palacios et al. 2020).

Animal diversity in wastelands is also typically comparable to, if not higher than, other urban green spaces. Most animal studies focus on invertebrates, especially arthropods. Spiders have more diversity in vacant lots in some studies (Philpott et al. 2014), but not in others (Moorhead and Philpott 2013; Burkman and Gardiner 2015). For ants, beetles, bees, and wasps, studies indicate a similar or greater diversity in wastelands compared to other urban green spaces (Da Rocha et al. 2020; Philpott et al. 2014; Ramalingam and Rajan 2017; Sivakoff et al. 2018). Birds are the only other animal group studied in this context and in all cases, bird diversity in wastelands and vacant lots is comparable to or higher than in other urban green spaces (Rega-Brodsky et al. 2018; Villasenor et al. 2020; Zuniga-Palacios et al. 2020).

In addition to higher diversity, wastelands also often contain a greater abundance of individuals than other green spaces. Spiders (Moorhead and Philpott 2013; Burkman and Gardiner 2015), bees (Sivakoff et al. 2018), and birds (Villasenor et al. 2020) are all more abundant in wasteland or vacant lot habitats. Philpott et al. (2014) found that spiders, true bugs, and ants are all more abundant in vacant lots in Toledo, Ohio, than in other green spaces.

It is, of course, not always the case that animal diversity or abundance is greater in urban wastelands compared to other urban green spaces. This is illustrated in an exemplary study by Philpott et al. (2014) which (unlike most studies) compared abundance and diversity in several arthropod taxa among urban gardens, forests, and vacant lots. Their results show that the diversity and abundance of these groups in these three urban habitats vary widely. For example, isopods were more than three times as abundant in gardens as in forests, and 15 times as abundant in forests than in lots. Spider species richness was highest in lots whereas ant richness was higher in forests than in gardens.

These findings indicate that while wastelands have great potential as habitat for urban biodiversity, other kinds of urban green spaces must be preserved to provide a diversity of habitats which, in turn, adds to the diversity of species. As there are many kinds of urban green spaces, Threlfall and Kendal (2018) provide a useful typology that classifies them on the basis of human involvement in community assembly and the novelty of the community composition. At one extreme are green spaces of undisturbed remnant ecosystems, while the other extreme consists of highly managed gardens and sports fields with mostly novel species composition. Intermediate to these two extremes are many kinds of urban green spaces, including wastelands, which are classified as relatively unmanaged and very novel in species composition (Threlfall and Kendal 2018).

A key role of urban wastelands is their value in supporting populations of species from surrounding habitats by providing temporary resources for nesting, food, water, and shelter (Bonthoux et al. 2014). There are relatively few studies of this type of transitory wasteland usage, but they are probably most common in more mobile groups such as birds. For example, in Philip Island, Australia, the ground-nesting bird, the masked lapwing (*Vanellus miles*) was found to prefer vacant urban lots rather than residential lots (Roche et al. 2016). Similarly, Villasenor et al. (2020) found that although vegetation in vacant land in Santiago, Chile, was dominated by exotic herbaceous plants, they provided important resources for native birds. Widespread usage of vegetation in vacant lots by birds in Baltimore, Maryland, for nesting and food has been well documented by Rega-Brodsky and Nilon (2016, 2017), Rega-Brodsky et al. (2018).

An important but understudied factor in comparing wastelands to other green spaces is that wastelands themselves will vary in their physical, chemical, and biological conditions depending on their previous land uses. For example, wild bees are most attracted to post-industrial wastelands created by extractive industries, such as sand and clay pits, whereas wastelands in areas directly influenced by the chemical industry are the least attractive to bees (Twerd and Banaszak-Cibicka 2019).

In addition to their value for increasing local biodiversity (and thus promoting local improvements in ecosystem services noted above), wastelands can have substantial value for conserving global biodiversity by providing habitats for species that are rare, declining, and at risk of extinction (Bonthoux et al. 2014). For example, a study of urban grasslands (including vacant lots and wastelands) in Berlin found 61 wild bee species of which 11 (18%) were Red-Listed (Buchholz et al. 2020). Similarly, a study of grasshopper diversity in urban wastelands in Berlin found 21 species of which 7 (33%) are classified as vulnerable or near-threatened (Eckert et al. 2017). Twerd et al. (2017) found that soda ash dumping grounds in a Polish city contained 64 species of digger wasps (28% of the domestic Polish fauna) of which 23.3% are rare or endangered.

For many rare and endangered species, which often occupy narrow ecological niches, wastelands provide urban "habitat analogues" to disappearing natural habitats. In the case of the vulnerable and near-threatened grasshoppers in Berlin, wastelands are especially important where they provide early successional grassland habitat (Eckert et al. 2017). For many of the rare and endangered digger wasps of Berlin, the soda ash dumping grounds provide needed open habitats with calcium-rich soils for these thermophilic species (Twerd et al. 2017). It is worth noting, however, for some rare and endangered species urban wastelands are more than just habitat analogues that substitute for disappearing natural habitats. In some cases, the urban wastelands are the original habitats that are in the process of being degraded by human activity. For example, the Berry Cave salamander (*Gyrinophilus gulolineatus*) is an endangered species whose cave habitats have been degraded by limestone quarries and urban growth in Knoxville, Tennessee (Zefferman et al. 2018). Under these conditions, conservation efforts need to focus on preserving the original habitat as opposed to creating a habitat analogue.

Considering their common land use legacy of extreme environmental disturbances, it is not surprising that urban wastelands often contain an abundant and diverse array of non-native species (Bonthoux et al. 2014). Regarding the floristic composition of wastelands, vacant lots often have a higher proportion of alien plants than urban parks (Deak et al. 2016) and inner-city vacant lots have more trees and more exotic trees than residential lots (Riley et al. 2018b). The seed bank in vacant lots in Santiago, Chile, contained 38 exotic species compared to just 3 native species (Figueroa et al. 2020). Such proportions of non-native plants in wastelands are relevant because they often provide poorer habitat for native animals, e.g., wild bees (Twerd and Banaszak-Cibicka 2019). In contrast, for birds the dominance of non-native species in wastelands is less clear. Vacant lots in Santiago, Chile, contain significantly fewer exotic bird species than urban parks or residential areas (Zuniga-Palacios et al. 2020). However, small vacant lots in Pachuca, Mexico, have more invasive birds, although this is relative to large natural areas (Zuniga-Palacios et al. 2020).

2.3 *Factors Affecting Biodiversity in Wastelands*

I will now examine local factors (e.g., area, age, soil, microclimate, disturbance, and vegetative structure) and landscape factors (e.g., connectivity).

Local factors

In their review of biodiversity in wastelands, Bonthoux et al. (2014) found 6 groups of local factors that influenced biodiversity: area size, age, soil, microclimate, human and domestic animal disturbances, and the vegetative structure and its effects on animals. I found 19 studies published since their review that examine factors of biodiversity in wastelands, and these studies found that many of the same factors have important influences on wasteland biodiversity (Table 2).

Relative to landscape variables, local factors tend to have greater influence in many biological communities including wastelands (Bonthoux et al. 2014). In their multi-group study of arthropods in urban habitats, Philpott et al. (2014) found that local factors (1–100 m scale) accounted for 80% of the variation of community composition, with landscape factors (>100 m scale) accounting for 20%. Furthermore, most of the significant correlations for local factors related to vegetation or ground cover predictors (24 of the 28 local predictors) rather than other characteristics specifically associated with urbanization (e.g., buildings or concrete). Moorhead and Philpott (2013) found that local habitat differences explained 50.3–90.9% of the spider community composition for that same city. Similarly, grasshopper diversity in Berlin was best explained by local factors (Eckert et al. 2017). However, a major caveat to this dominant role for contemporary local factors is noted by Johnson et al. (2015) who found that previous land use can have a stronger plant community structuring effect than contemporary environmental variation. Plant community compositional variation was primarily explained by differences in human legacies with no measurable effects of selected metrics of local environmental variation (abiotic soil characteristics) or environmental context (lot area, proximity to other vacant lots, and tree canopy) on compositional variation. Other studies, discussed below, reveal a similarly pronounced influence of land use legacies.

Area size is a factor noted in several wasteland studies (Table 2). Bee species diversity is positively correlated with wasteland area (Twerd and Banaszak-Cibicka 2019), and grasshopper diversity is correlated with the area of herb cover (Eckert et al. 2017). Melliger et al. (2017) found that species richness of plants is positively related to the size of the ruderal area, whereas species richness of grasshoppers and butterflies is correlated with meadow size. Wild bee diversity is reported to be positively related to the area of herbal flower coverage (Buchholz et al. 2020).

Age and ecological succession. Habitat age and succession are, not surprisingly, often studied as important local variables in accounting for diversity and composition (Table 2). In the case of age, the time allowed for a wasteland ecosystem to recover from industrial activity will alter and often create more amenable physical and chemical habitats for plants and animals. For example, freshly created soda ash dumps are very poor habitat for digger wasps because they are too moist and salty. However, as ecological succession occurs over time, these wastelands become drier and less salty and create high-quality habitats for digger wasps, with later successional states being the best habitat (Twerd et al. 2017).

The process of succession itself creates more diverse and complex vegetation which produces habitat for a greater diversity of animal species (Muratet et al. 2007).

Grasshopper diversity in wastelands, for instance, is highest at later successional stages (Eckert et al. 2017). Vacant lots in Chicago, Illinois, were found to have very different plant communities in the lot often-mowed interior compared to the plant communities of the unmowed fence line (Anderson and Minor 2020). Because they were not mowed, the fence line plant communities experienced ecological succession and had significantly more native and woody species than the lot interior. These examples illustrate that succession can be an important process in the creation of species diversity in wastelands. Indeed, allowing ecological succession to occur is an important management strategy of the growing trend to "rewild" vacant, wasteland ecosystems in urban areas (discussed below). It is important to note, however, that it is also important to maintain early successional communities in many urban areas, where they often contain species and assemblages with conservation or ecosystem service value. In such cases, mowing or other methods of arresting succession may be called for. For example, many spider species (especially smaller ones) tend to flourish in early successional habitats (Philpott et al. 2014) and tend to benefit from periodic mowing of vacant lots (de la Flor et al. 2020). For some taxa, the role of age and succession on wasteland sites may be much less important. In a study of mites as bioindicators in urban brownfields, Vacht et al. (2019) found some changes in community structure but no significant changes in species abundance or richness with successional stages.

Soil. Wasteland soils, like many urban soils, are highly varied with physical and chemical properties that are likely determined more by previous "legacy" human activities than by natural soil developmental processes (Bonthoux et al. 2014). My review found a study showing that soils in wastelands created by extractive industries may be rich in sand and clay, whereas wastelands from chemical industries will be more chemically altered with lower biodiversity value (Twerd and Banaszak-Cibicka 2019). Many industrial wasteland soils suffer from metal pollution. Wadhwa (2018) found that despite substantial metal pollution in a New Jersey brownfield, the overall diversity of soil arthropods was not negatively affected by soil metal contamination. Although some negative impacts were seen on some species, especially isopods, the results suggest that arthropods are generally developing an adaptive tolerance (Wadhwa 2018).

Seed banks in wasteland soils are a major determinant of the spatial and temporal abundance and diversity of the spontaneous vegetation. In vacant lots of Santiago, Chile, the species frequency in the upper seed bank was significantly positively correlated with the species frequency for above-ground vegetation (Figueroa et al. 2020).

Microclimate. Microclimate impacts on wasteland biodiversity can potentially be very important but, unfortunately, this factor is understudied. The previous review of this factor (Bonthoux et al. 2014) found only two studies. My review found none.

Human and domestic animal disturbances. The impacts of this factor on biodiversity include currently ongoing human disturbances such as trampling, trash, and mowing as well as domestic animals (e.g., cats and dogs). As with microclimate, few studies have been done on this topic, perhaps because wastelands tend to be less commonly visited or managed than other urban green spaces (Bonthoux et al. 2014).

However, I did find one notable study that does illustrate the potential importance of human disturbances in terms of wasteland biodiversity of vacant lots in Chicago, Illinois. Anderson and Minor (2019) found that plant species richness and vegetation height were better explained by social models and that the amount of trash in a lot was the most significant variable: lots with higher amounts of trash had higher species richness and evenness, and less vegetated area. In this case, trash (litter, debris) is not likely the major determinant of these biodiversity metrics but is a proxy indicator of human disturbance. Another study of vacant lots, in New Orleans, Louisiana, shows that increasing amounts of litter and debris correlate with rats, diseases, and other synanthropic species which also have a negative impact on native species biodiversity, as well as human health (Peterson et al. 2020). Deak et al. (2016) found the greatest proportion of weeds, disturbance tolerating and cosmopolitan plants in urban habitats with the highest intensity of trampling and soil disturbances.

Mowing is probably the most widespread disturbance in urban areas, especially in the USA where suburban lawns are very popular (Lerman et al. 2018). However, mowing is also common on vacant and abandoned lots to increase aesthetic appeal and reduce the appearance of neglect. Mowing is also often carried out on industrial landfills, brownfields, and wastelands for those same reasons, and also to arrest succession and maintain a grassland to prevent the growth of woody plants and the deep root growth of trees into the soil cap that covers many industrial waste sites (Kwit and Collins 2008). The impact of all this mowing on biodiversity is easily measured by observing changes that occur when mowing is reduced. Sehrt et al. (2020) found that when mowing in urban grasslands was reduced from every few weeks to 1–2 times per year, there was a 30% increase in the number of plant species after 6 years. Furthermore, plant community composition became much more spatially heterogeneous (increased beta diversity) and shifted from disturbance-tolerant species to more typical meadow species. Similarly, reduced mowing produced very substantial increases in the diversity and abundance of bees (Lerman et al. 2018) and many kinds of arthropods (Mody et al. 2020).

Vegetative structure effects on animals. Studies of vegetative structure and animal communities have focused mainly on birds and arthropods. For birds, Zuniga-Palacios et al. (2020) found that small vacant lots had significantly lower values of vegetation structural complexity than large vacant lots or urban forests but, interestingly, did not have reduced bird diversity. However, the smaller vacant lots did have more invasive exotic bird species. Studies of vacant lots in Baltimore showed that nesting success in three bird species was highest in lots with high shrub densities (Rega-Brodsky and Nilon 2016). Another study of the Baltimore area showed that, for five bird species, tree canopy cover and canopy height also have an effect on the bird community composition (Rega-Brodsky and Nilon 2017).

Effects of vegetative structure on several kinds of arthropods in urban green spaces (vacant lots, gardens, and forests) in Toledo, Ohio, were well described by Philpott et al. (2014). They found that millipede abundance was positively correlated with shrub height and forb abundance. Grasshoppers were more abundant in plots with more trees and more vegetation cover with taller herbaceous vegetation. Harvestmen were more abundant with higher woody plant and lower forb richness and with taller vegetation. Ant species richness was positively correlated with shrub abundance and herbaceous vegetation height. Beetle species richness was significantly lower with low levels of canopy cover, whereas spider species richness increased with fewer trees.

Landscape factors

In addition to the local habitat variables listed above, landscape characteristics beyond the local habitat often impact biodiversity, especially for highly mobile species which utilize multiple habitats on a regular basis. Thirteen of the 19 studies found in my review included an examination of at least one landscape variable (Table 2). In most cases, these studies found at least some role for landscape factors influencing local biodiversity, although, where they were quantified and compared, local factors were more important than landscape factors, e.g., Moorhead and Philpott (2013) for spiders and Philpott et al. (2014) for arthropods.

Over half (7) of the studies focused on arthropods, with several finding some landscape effects. Sivakoff et al. (2018) found that the proportion of impervious surface and number of green space patches in the surrounding landscape strongly influenced bee assemblages in Cleveland, Ohio. At a local scale (100 m radius), colonization of vacant lots and urban farms was limited by patch isolation via urbanization. However, at a larger landscape scale (1,000 m radius), increasing urbanization increased the concentration of bees in vacant lots and urban farms. This shows that maintaining green spaces provides important habitat, even within highly urbanized landscapes. Patch isolation is a very common theme in these arthropod studies. Patch isolation reduced bee diversity in urban brownfields in Bydgoszcz, Poland (Twerd and Banaszak-Cibicka (2019), and Berlin, Germany (Buchholz et al. 2020). For grasshoppers in Berlin, patch isolation not only reduced diversity but also reduced the conservation value of remaining grasshopper communities (Eckert et al. 2017). Similarly, patch isolation reduced the diversity of spider communities (Buchholz et al. 2018).

While increasing patch isolation via urbanization often reduces community diversity, it can promote spatial heterogeneity. In their study of spiders in vacant lots of Cleveland, Ohio, de la Flor et al. (2020) note that increasing the amount of impervious surface in cities produces more spatially distinct spider communities at a landscape scale. On the other hand, for plants, increasing urbanization had the opposite effect on vacant lots in Montreal and Quebec City where highly urbanized lots had more homogenized plant communities, reducing beta diversity (Blouin et al. 2019). This difference may be derived from the observation that seed dispersal is a more important driver than local environmental conditions of plant community assembly in vacant lots (Johnson et al. 2018).

In addition to patch separation, another recurring landscape influence on wasteland biodiversity is the distance to specific habitat features. Philpott et al. (2014) found that, as a group, arthropod abundance is negatively correlated with one landscape feature: open area within 2 km of the study site. They also found that some specific groups are affected by landscape features while others are affected only by local features. For example, spider abundance increased with more developed area within

200 m, while spider species richness increased with more developed area within 200 m and less open space within 200 m. Melliger et al. (2017) found that, for butterflies and moths living in ruderal habitats, the percentage cover of ruderal area in the closer surroundings was a better predictor of species richness than habitat size per se. In the case of bird abundance in Baltimore, Maryland, forest cover within 100 m of a vacant lot was the best predictor, rather than lot characteristics (Rega-Brodsky and Nilon 2017).

3 Strategies to Increase Biodiversity in Wastelands with a Case Study

I will now discuss how the biodiversity-enhancing factors above can be utilized in the design and the planning of wasteland habitats. In particular, I focus on examples from the KUW, which is a collection of land parcels owned by city, county, and state governments, non-profit foundations, and private landholders (Fig. 1). Historically, these lands have been utilized for farming, homesteading, logging, mining, and industrial activities, but these land uses have been discontinued in the last several decades. Consequently, these land parcels have undergone many decades of ecological recovery and represent the kinds of vacated wastelands that are characteristic of the urban wilderness movement (Kowarik 2018; Zefferman et al. 2018).

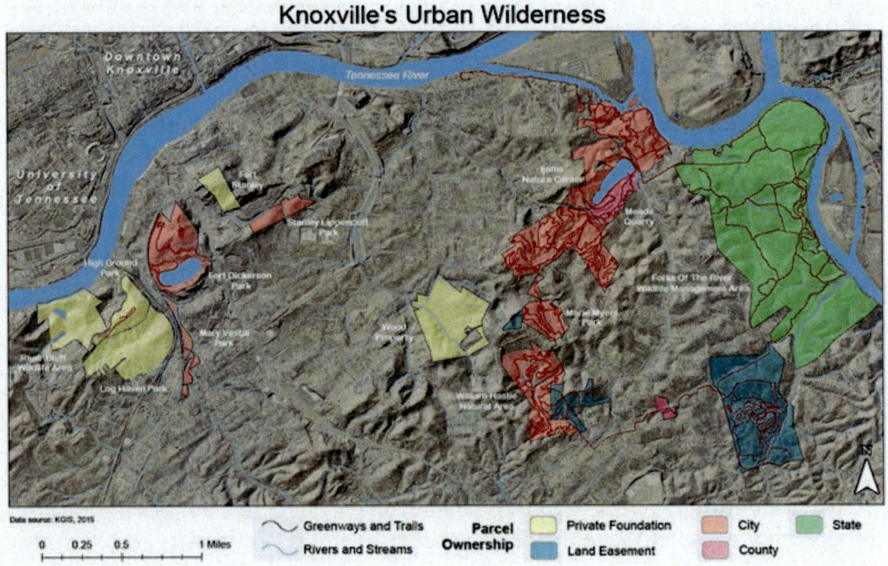

Fig. 1 Extent of Knoxville's urban wilderness. Properties are owned by city, county, and state government, non-profit foundations, and private landowners

Design strategies

Because local factors are the dominant influence on biodiversity in so many cases, it seems clear that local factors should be the starting point for design considerations. Area size, age and successional stage, soil, microclimate, anthropogenic disturbance, and vegetative structure are among the most critical features to consider for the biodiversity potential of any specific wasteland habitat. Regarding area, the KUW parcels comprise approximately 688 ha (over 1700 acres) just south of the Tennessee River (Fig. 1). This amount of area makes the KUW one of the largest "urban wilderness" areas in the USA (Zefferman et al. 2018).

In addition to area, all of the other local factors affecting biodiversity are exemplified by the KUW. As shown in Table 3, the individual parcels comprising the KUW represent a diversity of previous land uses and anthropogenic disturbances. Thus, these parcels also differ in their age (successional stage), anthropogenic disturbance, soil, microclimate, and vegetative structure (Table 3). For example, the Meade quarry is an abandoned limestone mine that contains a man-made lake surrounded by calcium-rich soils that are populated by mid-successional calciphilic vegetation (Fig. 2). In contrast, the Wood property (now called Baker Creek Preserve) is a recently vacated area that was cleared for logging and farming with a very early successional mixed hardwood forest (Fig. 3). In addition, several wastelands in areas bordering the KUW have soils that are heavily altered by smelting and other industrial activity and are populated by pollution-tolerant grasses and herbs. This diversity of parcel age, soil, disturbance, and microclimate, and other local factors increases spatio-temporal habitat heterogeneity and thus has the cumulative effect of promoting biodiversity. A total of 193 bird species have been reported in the KUW, which is *two-thirds* of the total 288 bird species reported in Knox County since 1951 (Zefferman et al. 2018). About 13% (7/54) of recorded listed threatened species in Knox County occur in the KUW (Zefferman et al. 2018).

As local factors and habitat needs can differ greatly among taxa, such as among spiders, ants, and other arthropod groups (Philpott et al. 2014), a key question for the biodiversity design of any urban green space is therefore what species or group(s) of organisms is the habitat intended to promote? Where wastelands are used to promote the preservation of a specific species or group, as in the case of threatened species, this is a straightforward matter of designing (and managing) them to meet the habitat needs of that species or group (e.g., Buchholz et al. 2020). For example, as noted above, the Berry Cave salamander (*Gyrinophilus gulolineatus*) is an endangered species living in the KUW where its habitat is being protected.

Despite the large role of local habitat features in promoting urban biodiversity, there is also a role for landscape considerations, especially for species that are highly mobile (e.g., birds, butterflies, and bees) or have wide dispersal mechanisms (e.g., some plants). In such cases, connectivity between wastelands is often critical to species maintenance, e.g., the "enormous importance of corridors, stepping stones, and flightpaths (especially in high-rise districts) for wild bees in urban environments" (Buchholtz et al. 2020). Again, the KUW provides a good example. Since the early 1990s, there has been a continuing effort to physically link the parcels of the KUW

Table 3 Major units of the Knoxville Urban Wilderness (KUW). Previous land uses and time of the last major disturbances are based on historical research from known records (newspapers, city records). See Fig. 1 (map) for locations (some smaller parcels of the KUW, that lack adequate documentation, are not shown in this table)

Parcel	Previous Land Uses	Time of last major disturbance	Current ownership	Current uses
Forks of the River Wildlife Refuge	Logging	>25 years	Tennessee Wildlife Resources Agency	Hiking, mountain biking, hunting
Fort Dickerson	Civil War Fort, rock mining	At least several decades	City of Knoxville	Hiking, swimming (quarry lake), history and culture (historic Civil War earthen fort)
Fort Stanley	Civil War trenches	>100 years	Private Foundation	Not open to public
High Ground Park	Civil War trenches	>100 years	Private Foundation	Hiking, history and culture (historic Civil War fort)
Ijams Nature Center	Farming, homesteads, mining	>60 years	Jointly owned by non-profit organization and City of Knoxville	Hiking, mountain biking, nature education, climbing/ziplining
Log Haven Property	Logging, homesteads	>70 years	Private foundation	Not open to public
Meade Quarry	Mining	>30 years	Knox County	Hiking, mountain biking, swimming, paddling (quarry lake)
River Bluff Park	Civil war trenches, logging, farming, homesteads	50 years	Private Foundation	Hiking
William Hastie Natural Area	Logging	>40 years	City of Knoxville	Hiking, mountain biking
Wood property (now Baker Creek Preserve)	Logging, farming	5 years	Private Foundation	Hiking, mountain biking

by a local non-profit group, Legacy Parks Foundation, and more recently the Urban Wilderness Alliance. This linkage has taken the form of greenways and trails, which currently connect most of the eastern parcels (Fig. 1), with major plans to increase connectivity within and outside the KUW. This includes the expansion of the KUW to encompass two large state parks, House Mountain and Seven Islands Wildlife Refuge, by connecting the current parcels via the construction of more greenways.

Fig. 2 Meade Quarry. This abandoned limestone mine is now a 25-acre lake that is a popular recreational area for canoeing and kayaking. It is surrounded by rugged terrain with an early successional forest used by hikers and bikers, and includes an interpretative nature area

Planning strategies

By many metrics, the KUW has been a major success at achieving many of the multifunctional goals it has set (Zefferman et al. 2018). Biodiversity conservation has been promoted by enhancing many of the biodiversity factors noted above: habitat area has been increasing as more unused land is added, different habitat parcels are being maintained with different microclimates and soils and at different stages of ecological succession ranging from grasslands to long-term afforestation. Human disturbances from walking and biking are regulated by a trail system that seeks to minimize impacts, and the construction of an extensive greenway system that connects all of the land parcels in the KUW and provides ecological corridors. In addition to direct conservation value, this biodiversity-friendly approach has great value for educating the general public, with many thousands of visitors each week. Also, several "citizen science" (i.e., participatory science) events are held here each year to assist in the growing ecological inventory of the KUW and surveys indicate that a large majority, over 98%, of visitors wanted to see the KUW expanded in size (Zefferman et al. 2018).

Importantly, there have been substantial economic benefits directly derived from the biodiversity-friendly approach to these areas. Although exact estimates are not available, land maintenance costs are relatively low as much of the KUW

Fig. 3 Baker Creek Preserve. This recently vacated area was cleared for logging and farming with a very early successional mixed hardwood forest

land is undergoing "rewilding" via natural succession with spontaneous vegetation (personal communication, Knox County Parks and Recreation Department). Conversely, economic benefits are increasing as the KUW is becoming a regional destination for recreational activities, especially mountain biking because of the rugged terrain. Public access to all parcels of the KUW is free of charge but the revenue generated by visitors to the KUW the local economy (e.g., restaurants and lodging) is over $8 million per year and this is expected to grow substantially in coming years (Sims et al. 2015). In addition, each year the KUW provides an estimated $3.4 million in ecosystem services such as carbon sequestration, waste assimilation, flood mitigation, and other services (Table 4). Waste assimilation and wildlife habitat are the two ecosystem services that provide the highest dollar value, and the forest is the land cover providing the most services, by far (Table 4). Note that these values in Table 4 also include largely vacated lands adjacent to the current KUW parcels that are under consideration for addition to the KUW project. These will bring the total to 8861 acres.

The KUW example is just one model of long-term wasteland management, but it is worth noting two reasons for its success. One is the creation of the Urban Wilderness Alliance, which has allowed the many stakeholders (e.g., City of Knoxville, Knox County, Legacy Parks Foundation, Tennessee Wildlife Resource Agency, Appalachian Mountain Bike Club, and Ijams Nature Center) to communicate, plan, and coordinate the management of their separate land parcels. Another reason for

Table 4 Ecosystem service values in the Knoxville Urban Wilderness (and surrounding area—see text) by land cover. Note that these are *annual* dollar values. Missing values (n.a.) are where a sufficient number of valuation studies have not yet been performed. Sources: Costanza (2006), Nowak et al. (2006). Calculations by the Baker Center, UTK

Ecosystem service	Acres	Forests	Wetlands	Pasture/crops	Riparian	Total
Water supply	1424	$228,689	$3,088	n.a	$33,425	$265,202
Waste assimilation	1473	$1,175,714	$1,676	$2,992	n.a	$1,180,382
Flood mitigation	94	n.a	$6,643	n.a	$1,531	$8,174
Wildlife habitat	1499	$1,294,969	$226	$78,114	n.a	$1,373,309
Pollination	1565	$227,286	n.a	$1,636	n.a	$228,922
Pollution removal	1403	$290,379	n.a	n.a	n.a	$290,379
Carbon sequestration	1403	$69,924	n.a	n.a	n.a	$69,924
Total	8861	$3,286,960	$11,633	$82,724	$34,956	$3,416,291

success is geographic and geological: the KUW occupies a single large area of abandoned and disused land in South Knoxville because this area has a very rugged topography that makes the land difficult to develop and urbanize. Specifically, this entire area is karstic terrain, i.e., underlain by limestone widely permeated by caves and sinkholes that make it very difficult to develop for building (Zefferman et al. 2018). Instead, the land has historically been used for mining, logging, and small-scale farming that is no longer economically viable. Thus, most of the current and planned KUW parcels are lands that have little potential for economic development. As a result, many of these disused sites are located in the same general vicinity and are therefore able to be organized into a collective management unit, the KUW (Zefferman et al. 2018).

As the KUW is widely used by the public and has been a growing economic asset, it seems likely that its planned expansion into surrounding wasteland areas in South Knoxville will continue. For this continued growth to be successful, it is important to maintain public acceptance of these formerly abandoned urban green spaces (Ives et al. 2017). Fischer et al. (2020) found that establishing near-natural urban green spaces was strongly supported by people who know about biodiversity and the benefits of "wild-looking" green spaces. Similarly, in the specific case of public perceptions of wasteland vegetation, perceptions tend to be negative when viewed as abandoned, but positive when the habitat is viewed as natural (Brun et al. 2018; Mathey et al. 2018). Although such "wild" unmanaged wastelands can be perceived as neglected areas (Mathey et al. 2018), incorporating minimal "cues of care" of wasteland habitats is one solution (Hostetler 2020). Examples would be intermittent litter clean-ups, pruning, and well-maintained buffer strips that can make natural "wild" urban habitats more appealing and aesthetically acceptable (Riley et al. 2018a; Mathey et al. 2018).

A significant obstacle to public acceptance of wastelands for promoting biodiversity is the tendency of many people to view public green spaces only or mainly in terms of recreational usage, especially walking. For example, the large majority of visitors to the KUW use this area for recreation, either walking (74.4%) or biking (19.9%) (Zefferman et al. 2018). On the one hand, such high recreational usage is helpful for public acceptance of conservation goals because studies show that people who frequently visit urban green spaces are more supportive of efforts to promote urban biodiversity (Fischer et al. 2020). On the other hand, recreational activities are a major source of disturbance in natural areas, often leading to many negative physical, chemical, and ecological impacts (Ballantyne and Pickering 2015). A compromise solution to allowing access while also minimizing harm is to have the most ecologically valuable parts of the green space set aside and isolated from trail use. Thus, even though the KUW is visited by hundreds of walkers and mountain bikers daily, the vast majority of these visitors utilize only a few heavily traveled trails that cover a very small portion (less than 2%) of the overall wilderness area (Zefferman et al. 2018).

4 Conclusions and Research Recommendations

Urban wastelands are a rapidly growing part of the landscape in many regions of the world. This "counter-urbanization" trend (Peterson et al. 2020), driven by a combination of shrinking cities and decreasing funds to maintain these lands, has great potential for biodiversity conservation for economic, educational, and ecological reasons. Economically, using wastelands for biodiversity requires less management and is thus typically much cheaper than other land uses. Educationally, the high density of people in urban areas provides the most effective way to educate (and inspire) people to appreciate and conserve biodiversity. This is especially important for people who live in areas without easy access to wilderness and undisturbed natural systems (Kowarik 2018), and wastelands are often most common in areas where such populations live (Peterson et al. 2020). Ecologically, the last several decades have seen increasing recognition that cities often have very high levels of biodiversity, largely due to a combination of diverse habitat heterogeneity and anthropogenic importation of species (Adler and Tanner 2013). In many cases, this urban biodiversity contains high levels of native and threatened species. Most of this rich biodiversity of modern cities has been an unintended consequence of human activities with little or no forethought to conservation goals.

The growing availability of "unwanted" lands provides an opportunity, for the first time in urban history, to incorporate basic ecological principles on a large scale for the purpose of conserving regional and global biodiversity. This review has expanded upon an earlier review (Bonthoux et al. 2014) to document that urban wastelands often have high levels of biodiversity, including threatened species, owing to their diverse land use histories and low levels of management that allow succession and other natural processes to occur. I further document the key roles of area, age, soil, microclimate, current anthropogenic disturbances, previous land use, connectivity, and other landscape variables in promoting wasteland biodiversity. Perhaps more importantly, I discuss how these factors are applicable to the KUW project, which has been very successful in "repurposing" and linking a variety of diverse wasteland parcels to achieve economic and ecological goals. Specifically, I have noted how increasing land acquisition (more area and spatial habitat diversity including soils and microclimates), increasing connectivity (greenways and trails), and maintaining the parcels in different stages of ecological succession have significantly improved urban biodiversity. While the precise ecological success of this project is difficult to measure because an ecological inventory of the KUW is very incomplete (Zefferman et al. 2018), information from well-studied taxa indicates that the KUW has much conservation value: roughly two-thirds of Knox County bird species utilize the KUW and about 13% of known threatened species in Knox County are recorded from the KUW.

As with any urban green space, wasteland conservation designs must gain public acceptance before they can be implemented. Fortunately, increasing urban biodiversity often coincides with increasing many other ecosystem services (Adler and Tanner 2013; Haase 2013). There are, however, at least three barriers to public acceptance of using wastelands for biodiversity conservation. One is the negative view of wastelands as being unattractive. This can be mitigated by "cues of care" and other aesthetic indicators of management (tidiness). Public education on the benefits of "wild" biodiversity has also been effective. The second barrier is also perceptual: the perception that wastelands are harmful to human health, e.g., wildlife impacts, spreading diseases, and promoting crime. This can often be mitigated by public education and modest improvements in maintenance and sanitation (Riley et al. 2018a, b). Finally, the third barrier is the common public view that urban green space should mainly be valued for recreation. While wastelands are indeed mostly used for walking and other forms of recreation, it is not difficult to design green spaces so that human activities are compartmentalized around hot spots of trails and visitation areas with other areas set aside for conservation goals (Zefferman et al. 2018). This also helps with aesthetic goals if the highly visited areas include landscaped buffer zones and other "cues of care" whereas the natural areas are allowed to undergo ecological succession. The KUW is obviously just one model for using urban wastelands for biodiversity goals, but it has been very successful in its multi-functional approach. In addition to achieving conservation goals, generating an increasingly large revenue for local businesses, increasing ecosystem services including health and recreation, and contributing enormously to public education of biodiversity, the KUW also illustrates a management model whereby a diverse group of stakeholders can coordinate (via the Urban Wilderness Alliance) to work together toward these goals.

Future research. The few dozen studies published on wasteland biodiversity factors are the first steps in what is likely to be a rapidly emerging field of research given that the global trends driving "counter-urbanization" are unlikely to change in the foreseeable future. As such, there are many glaring gaps in this field of study that remain to be filled. Perhaps most obvious are the geographic biases of previous studies, which have focused largely on European and, more recently, North American cities. The results herein do indicate a growing rate of interest in a few US cities. But these represent only a few cities by a few research groups in eastern US. Clearly, much more work is needed in many other geographic areas, especially Asia and Latin America. Another major bias here is the taxonomic bias entirely toward arthropods, plants, and birds. The absence of mammals and reptiles is especially notable.

Another major deficit is that most of the existing studies tend to focus on a single group of organisms. Studies that compare the habitat factors of several groups in the same wasteland habitats (e.g., Philpott et al. 2014) will be extremely valuable. Given the importance of public acceptance of urban biodiversity conservation, we need much more research to expand our understanding of what drives the green space preferences of urban dwellers. An important review by Botzat et al. (2016) outlines the many major gaps in our understanding of gender, age, culture, and other socio-demographic factors that influence how people perceive wastelands and other green spaces.

Finally, and in some ways most importantly, we need a much better understanding of how to sustainably manage urban wastelands for long periods of time. While cities are very dynamic, biodiversity conservation requires long-term planning at least on the scale of decades and perhaps centuries. The KUW is but one example of an attempt to set aside, expand, and manage a "wild" ecosystem on such a long time scale. Similar interest in emerging urban forests is growing in other cities (e.g., Kowarik et al. 2019), and it remains to be seen how these ecosystems will be managed (and funded) in evolving urban settings with multiple stakeholders.

References

Adler FR, Tanner CJ (2013) Urban ecosystems. Cambridge Univ Press. Cambridge, UK, 345 pp

Anderson EC, Minor ES (2019) Assessing social and biophysical drivers of spontaneous plant diversity and structure in urban vacant lots. Sci Total Env 653:1272–1281. https://doi.org/10.1016/j.scitotenv.2018.11.006

Anderson EC, Minor ES (2020) Management effects on plant community and functional assemblages in Chicago's vacant lots. Applied Veg Sci 23:266–276. https://doi.org/10.1111/avsc.12480

Ballantyne M, Pickering CM (2015) The impacts of trail infrastructure on vegetation and soils: current literature and future directions. J Envir Manag 164:53–64. https://doi.org/10.1016/j.jenvman.2015.08.032

Blouin D, Pellerin S, Poulin M (2019) Increase in non-native species richness leads to biotic homogenization in vacant lots of a highly urbanized landscape. Urban Ecosyst 22:879–892. https://doi.org/10.1007/s11252-019-00863-9

Bonthoux S, Brun M, Di Pietro F, Greulich S, Bouche-Pillon S (2014) How can wastelands promote biodiversity in cities? A review. Landsc Urb Plan 132:79–88. https://doi.org/10.1016/j.landurbplan.2014.08.010

Botzat A, Fischer LK, Kowarik I (2016) Unexploited opportunities in understanding liveable and biodiverse cities. A review on urban biodiversity perception and valuation. Global Env Change 39:220–233. https://doi.org/10.1016/j.gloenvcha.2016.04.008

Brun M, Di Pietro F, Bonthoux S (2018) Residents' perceptions and valuations of urban wastelands are influenced by vegetation structure. Urban for Urban Green 29:393–403. https://doi.org/10.1016/j.ufug.2017.01.005

Buchholz A, Hannig K, Moeller M (2018) Reducing management intensity and isolation as promising tools to enhance ground dwelling arthropod diversity in urban grasslands. Urban Ecosyst 12:1139–1149. https://doi.org/10.1007/s11252-018-0786-2

Buchholz S, Gathof AK, Grossmann AJ, Kowarik I, Fischer LK (2020) Wild bees in urban grasslands: urbanisation, functional diversity and species traits. Landsc Urb Plan 196. Article Number: 103731. https://doi.org/10.1016/j.landurbplan.2019.103731

Burkman CE, Gardiner MM (2015) Spider assemblages within greenspaces of a deindustrialized urban landscape. Urban Ecosyst 18:793–818. https://doi.org/10.1007/s11252-014-0430-8

Cabral I, Keim J, Engelmann R, Kraemer R, Siebert J, Bonn A (2017) Ecosystem services of allotment and community gardens: a Leipzig, Germany case study. Urban for Urban Green 23:44–53. https://doi.org/10.1016/j.ufug.2017.02.008

Costanza R (2006) The value of New Jersey's ecosystem services and natural capital. New Jersey Department of Environmental Protection, Division of Science, Research, and Technology, Trenton, NJ

Da Rocha LC, Montagnana PC, Boscolo D, Garofalo CA (2020) Green patches among a grey patchwork: the importance of preserving natural habitats to harbour cavity-nesting bees and wasps (Hymenoptera) and their natural enemies in urban areas. Biodiv Conserv 29:2487–2514. https://doi.org/10.1007/s10531-020-01985-9

De la Flor YDA, Perry KI, Turo KJ, Parker DM, Thompson JL, Gardiner MM (2020) Local and landscape-scale environmental filters drive the functional diversity and taxonomic composition of spiders across urban greenspaces. J Appl Eco. https://doi.org/10.1111/1365-2664.13636EarlyAccess

Deak B, Huse B, Tothmeresz B (2016) Grassland vegetation in urban habitats—testing ecological theories. Tuexenia 36:379–393. https://doi.org/10.14471/2016.36.017

Eckert S, Moller M, Buchholz S (2017) Grasshopper diversity of urban wastelands is primarily boosted by habitat factors. Insect Conserv Biodiv 10:248–257. https://doi.org/10.1111/icad.12221

Figueroa JA, Saldias G, Teillier S, Carrera E, Castro SA (2020) Seed banks in urban vacant lots of a Latin American megacity are easily germinable and strongly dominated by exotic flora. Urban Ecosyst. https://doi.org/10.1007/s11252-020-00986-4

Fischer LK, Eichfeld J, Kowarik I, Buchholz S (2016) Disentangling urban habitat and matrix effects on wild bee species. PEERJ 4. Article Number: e2729. https://doi.org/10.7717/peerj.2729

Fischer LK, Neuenkamp L, Lampinen J, Tuomi M, Alday JG, Bucharova A et al (2020) Public attitudes toward biodiversity-friendly greenspace management in Europe. Conserv Lett Early. Article Number: e12718. https://doi.org/10.1111/conl.12718

Haase D (2008) Urban ecology of shrinking cities: an unrecognized opportunity? NAT + CUL 3:1–8. https://doi.org/10.3167/nc.2008.030101

Haase D (2013) Chapter 12: shrinking cities, biodiversity and ecosystem services. Urbanization, biodiversity and ecosystem services: challenges and opportunities: a global assessment. Springer: Dordrecht, The Netherlands

Hostetler M (2020) Cues to care: future directions for ecological landscapes. Urban Ecosyst. https://doi.org/10.1007/s11252-020-00990-8

Ives CD, Oke C, Hehir A, Gordon A, Wang Y, Bekessy SA (2017) Capturing residents' values for urban green space: mapping, analysis and guidance for practice. Landsc Urb Plan 161:32–43. https://doi.org/10.1016/j.landurbplan.2016.12.010

Johnson AL, Borowy D, Swan C (2018) Land use history and seed dispersal drive divergent plant community assembly patterns in urban vacant lots. J Appl Ecol 55:451–460. https://doi.org/10.1111/1365-2664.12958

Johnson MP, Hollander J, Halluli A (2014) Maintain, demolish, re-purpose: policydesign for vacant land management using decision models. Cities 40:151–162. https://doi.org/10.1016/j.cities.2013.05.005

Johnson AL, Tauzer EC, Swan CM (2015) Human legacies differentially organize functional and phylogenetic diversity of urban herbaceous plant communities at multiple spatial scales. Appl Veg Sci 18:513–527. https://doi.org/10.1111/avsc.12155

Kim G, Miller P, Nowak D (2016) The value of green infrastructure on vacant and residential land in Roanoke, Virginia. Sustainability 8. Article Number 296. https://doi.org/10.3390/su8040296

Kowarik I (2018) Urban wilderness: supply, demand and access. Urban for Urban Green 29:336–347. https://doi.org/10.1016/j.ufug.2017.05.017

Kowarik I, Hiller A, Planchuelo G, Seitz B, von der Lippe M, Buchholz S (2019) Emerging urban forests: opportunities for promoting the wild side of the urban green infrastructure. Sustainability 11. Article Number 6318. https://doi.org/10.3390/su11226318

Kremer P, Hamstead ZA, McPhearson T (2013) A social-ecological assessment of vacant lots in New York City. Landsc Urb Plan 120:218–233. https://doi.org/10.1016/j.landurbplan.2013.05.003

Kwit C, Collins B (2008) Native grasses as a management alternative on vegetated closure caps. Envir Manag 41:929–936. https://doi.org/10.1007/s00267-008-9090-x

Lerman SB, Contosta AR, Milam J, Bang C (2018) To mow or to mow less: lawn mowing frequency affects bee abundance and diversity in suburban yards. Biol Conserv 221:160–174. https://doi.org/10.1016/j.biocon.2018.01.025

Mathey J, Arndt T, Banse J, Rink D (2018) Public perception of spontaneous vegetation on brownfields in urban areas-results from surveys in Dresden and Leipzig (Germany). Urban for Urban Green 29:384–392. https://doi.org/10.1016/j.ufug.2016.10.007

Mathey J, Rossler S, Banse J, Lehmann I, Brauer A (2015) Brownfields as an element of green infrastructure for implementing ecosystem services into urban areas. J Urb Plan Dev 14:A4015001-1-A4015001–13. https://doi.org/10.1061/(ASCE)UP.1943-5444.0000275

McKinney M, VerBerkmoes A (2020) Beneficial health outcomes of natural green infrastructure in Cities. Curr Landsc Ecol Rep. https://doi.org/10.1007/s40823-020-00051-y

Melliger RL, Rusterholz HP, Baur B (2017) Habitat- and matrix-related differences in species diversity and trait richness of vascular plants, Orthoptera and Lepidoptera in an urban landscape. Urban Ecosyst 20:1095–1107. https://doi.org/10.1007/s11252-017-0662-5

Mody E, Lerch D, Muller AK, Simons NK, Bluethgen N, Harnisch M (2020) Flower power in the city: replacing roadside shrubs by wildflower meadows increases insect numbers and reduces maintenance costs. PLOS ONE 15. Article Number: e0234327. https://doi.org/10.1371/journal.pone.0234327

Moorhead LC, Philpott SM (2013) Richness and composition of spiders in urban green spaces in Toledo, Ohio. J Arachnol 41:356–363. https://doi.org/10.1636/P12-44.1

Muratet A, Machon N, Jiguet F, Moret J, Porcher E (2007) The role of urban structures in the distribution of wasteland flora in the greater Paris area, France. Ecosystems 10:661–671. https://doi.org/10.1007/s10021-007-9047-6

Newman GD, Bowman AO, Lee RJ, Kim B (2016) A current inventory of vacant land in America. J Urban Des 21:302–319. https://doi.org/10.1080/13574809.2016.1167589

Nowak DJ, Crane DE, Stevens JC (2006) Air pollution removal by urban trees and shrubs in the United States. Urban for Urban Green 4:115–123

Pagano MA, Bowman AO (2000) Vacant land in cities: an urban resource. Brookings Institution, Center on Urban and Metropolitan Policy, Washington, DC, USA

Palliwoda J, Banzhaf E, Priess JA (2020) How do the green components of urban green infrastructure influence the use of ecosystem services? Examples from Leipzig, Germany. Landsc Ecol. https://doi.org/10.1007/s10980-020-01004-w

Peterson AC, Ghers, BM, Campanella R, Riegel C, Lewis JA, Blum MJ (2020) Rodent assemblage structure reflects socioecological mosaics of counter-urbanization across post-Hurricane Katrina New Orleans. Landsc Urb Plan 195. Article Number: 103710. https://doi.org/10.1016/j.landurbplan.2019.103710

Philpott SM, Cotton J, Bichier P, Friedrich RL, Moorhead LC, Uno S, Valdez M (2014) Local and landscape drivers of arthropod abundance, richness, and trophic composition in urban habitats. Urban Ecosyst 17:513–532. https://doi.org/10.1007/s11252-013-0333-0

Ramalingam R, Rajan PD (2017) Effects of various urban land-uses on the epigeic beetle communities in Bangalore city, India. Tropical Ecol 58:295–305

Rega-Brodsky CC, Nilon CH (2017) Forest cover is important across multiple scales for bird communities in vacant lots. Urban Ecosyst 20:561–571. https://doi.org/10.1007/s11252-016-0614-5

Rega-Brodsky CC, Nilon CH (2016) Vacant lots as a habitat resource: nesting success and body condition of songbirds. Ecosphere 7. Article Number: e01578. https://doi.org/10.1002/ecs2.1578

Rega-Brodsky CC, Nilon CH, Warren PS (2018) Balancing urban biodiversity needs and resident preferences for vacant lot management. Sustainability 10. Article Number: 1679. https://doi.org/10.3390/su10051679

Riley CB, Herms DA, Gardiner MM (2018b) Exotic trees contribute to urban forest diversity and ecosystem services in inner-city cleveland, OH. Urban for Urban Green 29:367–376. https://doi.org/10.1016/j.ufug.2017.01.004

Riley CB, Perry KI, Ard K, Gardiner MM (2018a) Asset or liability? Ecological and sociological tradeoffs of urban spontaneous vegetation on vacant land in shrinking cities. Sustainability 10. Article Number: 2139. https://doi.org/10.3390/su10072139

Roche DV, Cardilini APA, Lees D, Maguire GS, Dann P, Sherman CDH, Weston MA (2016) Human residential status and habitat quality affect the likelihood but not the success of lapwing breeding in an urban matrix. Sci Total Env 556:189–195. https://doi.org/10.1016/j.scitotenv.2016.03.008

Sehrt M, Bossdorf O, Freitag M, Bucharova A (2020) Less is more! Rapid increase in plant species richness after reduced mowing in urban grasslands. Basic Appl Ecology 42:47–53. https://doi.org/10.1016/j.baae.2019.10.008

Sims C, Davis B, Kim B (2015) Economic potential of South Knoxville's urban wilderness. Baker Center Policy Paper 2:15. University of Tennessee, Knoxville, TN

Sivakoff FS, Prajzner SP, Gardiner MM (2018) Unique bee communities within vacant lots and urban farms result from variation in surrounding urbanization intensity. Sustainability 10. Article Number: 1926. https://doi.org/10.3390/su10061926

Threlfall CG, Kendal D (2018) The distinct ecological and social roles that wild spaces play in urban ecosystems. Urban for Urban Green 2:348–356. https://doi.org/10.1016/j.ufug.2017.05.012

Twerd L, Krzyzynski M, Waldon-Rudzionek B, Olszewski P (2017) Can soda ash dumping grounds provide replacement habitats for digger wasps (Hymenoptera, Apoidea, Spheciformes)? PLOS ONE 12. Article Number: e0175664. https://doi.org/10.1371/journal.pone.0175664

Twerd L, Banaszak-Cibicka W ((2019) Wastelands: their attractiveness and importance for preserving the diversity of wild bees in urban areas. J Insect Conserv 23:573–588. https://doi.org/10.1007/s10841-019-00148-8

Vacht P, Niglas H, Kuu A, Koff T, Kutti S, Raamets J (2019) Oribatid mite (Acari: Oribatida) communities of urban brownfields in Tallinn, Estonia, and their potential as bioindicators of wasteland successional stage. Acarologia 59:26–32. https://doi.org/10.24349/acarologia/20194310

Villasenor NR, Chiang LA, Hernandez HJ, Escobar MAH (2020) Vacant lands as refuges for native birds: an opportunity for biodiversity conservation in cities. Urban For Urban Green 49. Article Number: 126632. https://doi.org/10.1016/j.ufug.2020.126632

Wadhwa S (2018) Community and population characteristics of soil arthropods along a heavy metal gradient in urban brownfields. Rutgers University Community Repository. https://dx.doi.org.proxy.lib.utk.edu:90/10.7282/T3X63QW1

Zefferman EP, McKinney ML, Cianciolo T, Fritz BI (2018) Knoxville's urban wilderness: moving toward sustainable multifunctional management. Urban for Urban Green 29:357–366. https://doi.org/10.1016/j.ufug.2017.09.002

Zuniga-Palacios J, Zuria I, Moreno CE, Almazan-Nunez RC, Gonzalez-Ledesma M (2020) Can small vacant lots become important reservoirs for birds in urban areas? A case study for a Latin American city. Urban For Urban Green 47. Article Number: 126551. https://doi.org/10.1016/j.ufug.2019.126551

Michael L. McKinney is Professor of Earth & Planetary Sciences at the University of Tennessee, Knoxville, Tennessee (USA). He is also the Director of the Environmental Studies Program. His research and teaching focus on urban ecology with a specific interest in utilizing vacant and abandoned urban areas as "urban wildernesses" and other viable public spaces. He is currently exploring the use of drones to map and analyze vegetation, soil, and other geospatial data to create ecological inventories of urban wastelands and urban wilderness areas. Much of his work has centered on the Knoxville Urban Wilderness (KUW), which has received considerable public attention and support.

Urban Wastelands' Contribution to Ecological Connectivity

Marion Brun and Francesca Di Pietro

Abstract Urban wastelands can play a significant role in habitat continuity in fragmented landscapes such as cities, as they can act as stepping-stones for urban green networks that should be considered in ecological planning. This study aims to characterize urban wasteland contributions to the green network in two urban areas in France by studying their structural connectivity, to test the link between urbanization and wasteland plant communities, and to assess the extent that potential connectivity is considered in ecological planning. We suggest that spatial proximity, and especially urban similarity, influence wasteland plant communities: increasingly different urban characteristics result in increasingly dissimilar plant communities. We also show that plant community diversity is influenced by urban characteristics: plant communities are more similar in low urbanization contexts and more diversified in highly urbanized areas. The comparison between potential wasteland connectivity mapped at the city scale and the current ecological plan shows that urban wasteland potentialities for ecological connectivity are considered in ecological planning. In addition to integrating specific urban areas in ecological mapping, green networks should consider the importance of temporary spaces of nature, such as wastelands, since their spatiotemporal dynamics may contribute greatly to biodiversity conservation within cities.

Keywords Connectivity · Dissimilarity · Urbanization · Ecological planning · Urban wastelands · Plant communities

M. Brun (✉) · F. Di Pietro
UMR CITERES-CNRS, Université de Tours, Tours, France
e-mail: marion.brun@univ-tours.fr

F. Di Pietro
e-mail: francesca.dipietro@univ-tours.fr

F. Di Pietro and A. Robert (eds.), *Urban Wastelands*, Cities and Nature,
https://doi.org/10.1007/978-3-030-74882-1_4

1 Introduction

Urbanization leads to fragmented landscapes in cities: habitat patches are isolated from each other and are surrounded by complex urban structures, generally hostile to biodiversity dispersal (McKinney 2002; Kattwinkel et al. 2009). Considering this growing phenomenon, it is essential to study connectivity, i.e., "the degree to which urban matrix promotes or prevents species dispersal between two patches" (Taylor et al. 1993), in its multiple arrangements. Urban green networks consist of habitats favorable for biodiversity (Pickett and White 1985) which take the form of corridors (i.e., continuous semi-natural habitat) or of stepping-stones (i.e., discontinuous patches which are close enough to allow species dispersal). Furthermore, the contribution of different types of urban habitat patches in green networks should be facilitated. Indeed, if connectivity conservation mainly focuses on corridors, large and contiguous areas are rare in urban areas. Therefore, small habitat patches should be included in the urban mosaic as valuable spaces for biodiversity conservation (Kattwinkel et al. 2011; Muratet et al. 2013). In addition, urbanization constantly transforms the urban matrix. On the one hand, urbanization causes rapid land use modifications and/or semi-natural habitat destruction. On the other hand, extensive urbanization and urban renewal can create new types of urban habitats, which are defined as ephemeral and unstable in the city (Kattwinkel et al. 2011). Yet, these a priori unstable habitats offer a great variety of situations within urban areas that can foster rich plant diversity by creating potentially complementary communities within cities (Godefroid and Koedam 2007).

Among these urban habitats, urban wastelands are defined as temporary and abandoned spaces within urban areas that permit spontaneous vegetation to colonize (Bonthoux et al. 2014). These can appear or disappear according to urban planning decisions (Muratet et al. 2007), but can "offer opportunities for temporal biodiversity conservation" (Kowarik 2018). As wastelands can be scattered at multiple places in the city, they can be useful as stepping-stone habitats in fast-changing cities (Kattwinkel et al. 2011; Penone et al. 2012; Johnson et al. 2018). Thus, while habitat fragmentation can impede connectivity between wastelands, due to their temporary nature, wastelands are ignored in nature conservation measures (Harrison and Davies 2002; Lynch 2019). In France, an urban planning policy, the Green and Blue Plan (*Trame Verte et Bleue*, hereafter GBP) is intended to mitigate biodiversity loss from habitat fragmentation by preserving corridors or stepping-stones at the local level (regional and municipal scales): habitats are viewed as a network with connectivity that must be enhanced, including in urban areas.

This study aims to characterize the contribution of urban wastelands to the green network in two urban areas in France by analyzing their structural connectivity, "equated with habitat contiguity" (Tischendorf and Fahrig 2000), or connectedness, which "can be described from mappable elements" (Baudry and Merriam 1988). We assessed whether wasteland plant communities are linked to spatial proximity and urban environmental similarity among wastelands. Moreover, we tested the link between urbanization and wasteland plant communities. Finally, we compared potential wasteland connectivity, mapped at the city scale, with the current Green and Blue Plan.

2 Methodology: Urbanization and Flora Analysis

Our study sites were two urban areas with similar location but different sizes: the urban areas of Tours (167 km^2 with 250,000 inhabitants) and Blois (114 km^2 with 65,000 inhabitants), whose Green and Blue plan are among the most advanced in the Centre-Val de Loire region (39,151 km^2) in the center-west of France. The study sites present a degraded oceanic climate with moderate annual rainfall (around 660 mm) and mild temperatures (around 11.5 °C mean annual temperature). Both areas are composed of flood plains and hills which reach 150 m a.s.l. at their highest point; both cities originally developed along the Loire River, and now urban development is expanding progressively inland on the hills. The landscape contexts of Tours and Blois slightly differ: the urban area of Tours is wider and is more densely built-up than Blois, whereas remaining forests and agricultural lands are larger in the Blois urban area than in the Tours urban area.

2.1 Wasteland Landscape and Plant Data

Urban wastelands were identified following a specific methodology (Box 1 and Fig. 1). Within wastelands, vegetation sampling was carried out between May and July 2013; we recorded the presence of all vascular plant species, as detailed in Box 1.

Box 1: Wasteland identification and plant data collection.

We selected wastelands inside the urban matrix, where buildings are situated less than 200 m from each other (INSEE 2010; Loriot and Di Salvo 2008). We identified wastelands by photointerpretation of aerial photographs from the BDOrtho© database (French National Geographical Institute, IGN, 2010), and field observations. From a total of 240 identified wastelands, we selected only those with open access, larger than 400 m^2, and colonized by herbaceous or shrubby vegetation (i.e., no woodlands), to avoid confounding wastelands with urban forests. The final sample included 179 wastelands, 103 in Tours, and 76 in Blois. Their size ranged from 0.45 to 6.20 ha (mean = 0.76 ha, SD = 0.72). These wastelands present great variability in structure and habitat heterogeneity, as well as in environmental conditions, presented hereafter.

As wastelands harbor heterogeneous vegetation structures and differ in form and length, we sampled the vegetation along transects to capture community heterogeneity. For small and medium-sized wastelands, transects crossed the area from one side to the other. For larger wastelands, transect length was limited to 150 m (based on magnitude studies to explain the influence of the urban matrix on flora:

Kattwinkel et al. 2011; Muratet et al. 2007). Thus, transect lengths ranged from 24 to 150 m (mean: 102 m, SD: 37 m). In each wasteland, we surveyed vegetation in ten 1 × 2 m quadrats equidistantly located along each transect, resulting in a total of 1,790 quadrats sampled. In each quadrat, we recorded the presence of all vascular plant species. These results were summed for the ten quadrats to obtain the number of occurrences of each species per wasteland. Nomenclature followed the taxonomic repository TAXREF (Gargominy et al. 2013).

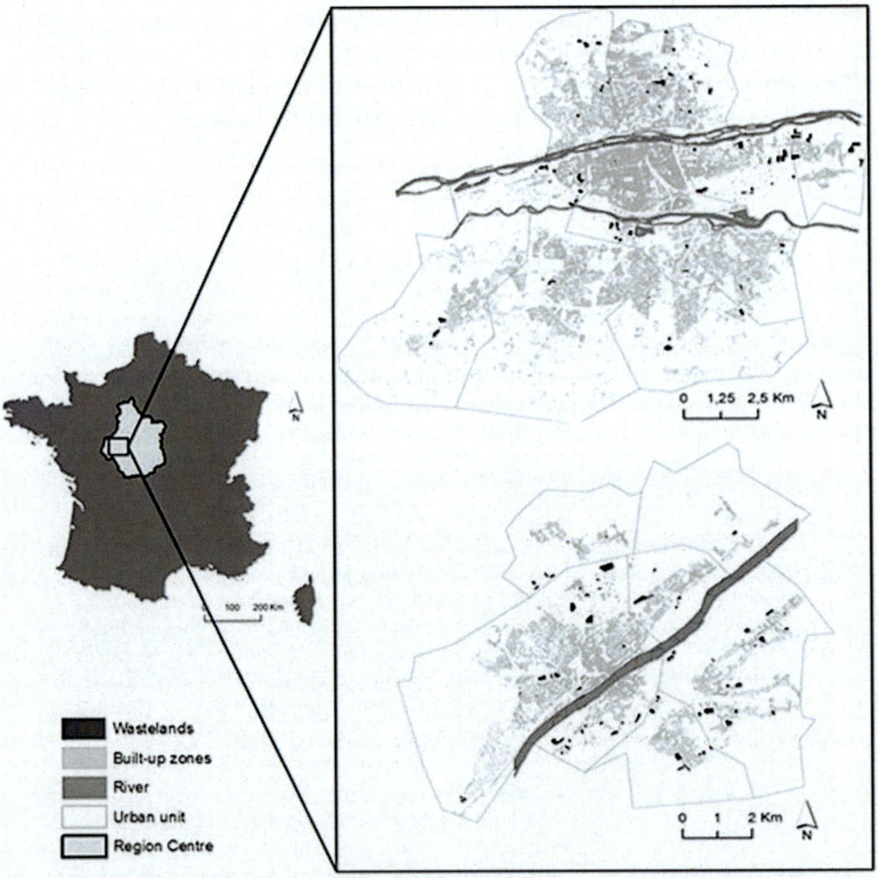

Fig. 1 Map of the wastelands within Tours, above (n = 103) and Blois, below (n = 76). M. Brun, ArcMap10, projection: RGF Lambert 93, *Sources* IGN

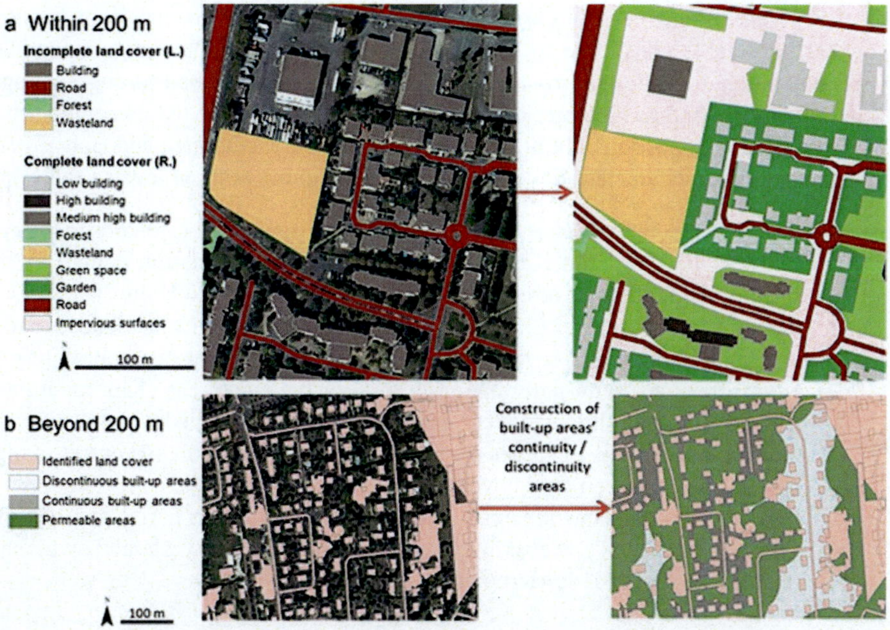

Fig. 2 Land cover data collection. **a** Within 200 m buffer; **b** beyond 200 m buffer

Urban landscape study was based on in-depth land cover analysis, described in Box 2 and Fig. 2.

Box 2: Land cover data collection.

To characterize urban land cover, we used the BD Topo® database (IGN, 2008 for Tours and 2010 for Blois) which partially informs land covers. However, because photointerpretation is very time-consuming, we identified buffer zones around the wastelands within which land cover photointerpretation was precise, and was less detailed beyond, as detailed below.

a. Within a 200 m radius buffer around each wasteland, we specified land cover by aerial photographs interpretation (BD Ortho®, IGN, 2008) and field visits. The choice of a 200 m radius was based on several studies assessing the influence of land covers on urban plant communities (Kattwinkel et al. 2009; Serret et al. 2014).
b. Beyond the 200 m radius buffers, we relied on the proximity of buildings and their height to identify continuous and discontinuous areas, by the erosion/dilatation technique (Loriot and Di Salvo 2008): when buildings were closer than 20 m, the built-up area was considered as dense (continuous), beyond and up to 50 m, the area was considered as less dense (discontinuous). Areas which presented no built-up area continuity (beyond 50 m between buildings) were not included in these buffers and were considered as favorable for species dispersal.

Even if the urban matrix has long been considered as homogeneous, it can be more or less permeable (favorable to species dispersal) and can influence connectivity between patches. Although urbanization has often been described through a single factor to understand its influence on plant communities (distance to city center, urban vs. rural areas, proportion of built-up areas around sites, McDonnell and Hahs 2008), we defined urban characteristics by a panel of descriptors at three levels (Table 1):

(i) the city level: urbanization gradient proxies: distance of wastelands to the city center (Godefroid and Koedam 2007) and human population density in the sub-municipal district of each wasteland (McDonnell and Hahs 2013);
(ii) the neighborhood level (200 m radius buffer centered on wastelands): buildings mean height and several land covers resulting from our photointerpretation: forested, semi-natural, agricultural or built-up areas (Kindlmann and Burel 2008); not all land covers were used to describe this level due either to too small variations in values of occurrence among wastelands or to too high correlations between land covers;
(iii) local level: wasteland area and past cover, identified on aerial photographs (Muratet et al. 2007; Strauss and Biedermann 2008), and wasteland ownership (Bonthoux et al. 2014; Marco et al. 2008).

Table 1 Urban descriptors of wastelands

Scale	Descriptor	Type	Limits/Modalities
City	Distance to city center	Quantitative	0.2–9.8 km. mean: 3.8 ± 1.7
	Population density	Quantitative	65–10,345 persons per km^2. mean: 1149 ± 1553
Neighborhood (200 m buffer)	% of built-up areas	Quantitative	0.2–44.6%. mean: 11 ± 7.3
	% of semi-natural areas	Quantitative	5.9–84.3%. mean: 42.4 ± 16.6
	% of forest areas	Quantitative	0–76.5%. mean: 12.4 ± 12.8
	% of agricultural areas	Quantitative	0–60.2%. mean: 9.6 ± 14.4
	Buildings mean height	Quantitative	2.5–13.3 m. mean: 5.6 ± 1.8
Local	Wasteland area	Quantitative	0.04–6.2 ha. mean: 0.76 ± 0.82
	Wasteland ownership	Qualitative	Public, private, company
	Wasteland past cover	Qualitative	Built-up, open-semi-natural, agricultural

2.2 Connectivity Metrics

At the city level, wasteland connectivity was assessed by comparing flora composition with dissimilarity metrics, and by dispersal cost mapping at the city scale.

At the habitat level, connectivity was assessed with dissimilarity-based metrics. To find out if wastelands are ecologically connected to each other and, if so, to what extent this connection is influenced by urban descriptors, we calculated the three metrics detailed in Box 3:

(i) Plant dissimilarity, based on vegetation sampling, to account for the taxonomic differences of plant communities between wastelands;
(ii) Euclidean distance among wastelands;
(iii) Urban dissimilarity, based on the dissimilarity of urban quantitative descriptors between wastelands.

To test the hypothesis that wasteland plant communities are linked to spatial proximity, we compared plant dissimilarity to Euclidean distance by using the Mantel test (Legendre and Legendre 2012). To test the hypothesis that wasteland plant communities are linked to urban characteristics, we compared plant dissimilarity to urban dissimilarity, by using the same test. Moreover, we tested the influence of urbanization (described using multivariate analyses on urban descriptors) on wasteland plant communities (represented by the mean plant dissimilarity between wastelands) with a linear regression model.

Table 2 Dispersal costs allocated to each land cover, ranging from 1 (favorable to species dispersal) to 100 (not favorable). Built-up area data were compiled according to building presence, density, and height (CLB: continuous low building; DLB: discontinuous low building; CMB: continuous medium height building; DMB: discontinuous medium height building; CHB: continuous high building; DHB: discontinuous high building)

Land cover	Score	Land cover	Score	Land cover	Score
Low building (<3 m)	80	Forest	25	Parking	60
High building (>8 m)	100	Graveyard	40	Meadow	5
Medium height building (3–8 m)	90	Watercourse	50	Road	70
CLB	80	Farming	25	Bare soil	40
DLB	70	Water	50	Constructed sport facility	90
CHB	100	Vacant lot	1	Open sport facility	50
DHB	90	Green space	10	Impervious area	100
DMB	80	Garden	10		

2.3 *At the City Scale: Mapping the Potential Connectivity of Wastelands*

To explore to what extent wasteland potential for plant dispersal has been taken into account in local policies, and particularly by the Green and Blue Plan, we assessed potential connectivity at the city scale, based on a land cover permeability index and consequent accumulated dispersal costs (Verbeylen et al. 2003), as shown in Table 2 (see Box 3 for methodological details).

Box 3: Connectivity metrics.

At the habitat scale: dissimilarity-based metrics

(i) Plant dissimilarity. To measure plant dissimilarity, we computed the occurrence of each species in the wastelands depending on the number of quadrats where the species is observed. We then calculated plant dissimilarity based on a Bray-Curtis dissimilarity index that reports the difference of species assemblages and frequency between two habitats (Bray and Curtis 1957).

(ii) Euclidean distance. To measure distance among wastelands, we calculated the ordinary straight-line distance between two wastelands (polygon centroids).

(iii) Urban dissimilarity. To measure the urban dissimilarity index, we computed the dissimilarity of urban quantitative characteristics between wastelands (at city and neighborhood scales, as previously shown in Table 1), using Gower's method (Gower and Legendre 1986), which accounts for the way the descriptors are arranged between each wasteland: two wastelands will have a greater urban dissimilarity if they don't share similar proportions of land cover (urban dissimilarity at the neighborhood level) or if they show different distances to the city center or different human population densities (urban dissimilarity at the city level). Even if this method does not measure the contribution of each descriptor to plant dissimilarity, it summarizes efficiently the influence of urban descriptors on plant communities (Legendre and Fortin 2010).

At the city scale: potential connectivity of wastelands. A dispersal cost index, i.e., an index defining the ability to let overall plant species pass through, has been intuitively assessed and mapped by assigning a score to land covers identified in the BD Topo® database (IGN, 2008 for Tours and 2010 for Blois) throughout the two cities, ranging from 1 (favorable to dispersal of a large panel of plant species, mostly land covers known to host plant communities) to 100 (not favorable, e.g., land covers recognized as having a barrier effect on plant communities), as shown in Table 2. For built-up areas, such scores were based on building presence, height, and density (when buildings were closer than 20 m, the built-up area was considered as continuous, with a higher score, beyond and up to 50 m, the area was considered as discontinuous, as previously shown in Box 2, with a lower score). Graph-based connectivity metrics were calculated on the basis of the land cover accumulated dispersal costs, based on least-cost distances (Foltête et al. 2012), in order to define habitats, i.e., buffer zones around wastelands, supposedly favorable for plant species persistence or suitable for plant species dispersal, whose radius depends on the dispersal cost index. By analyzing the overlap of habitats around wastelands in the study sites, it is possible to identify which wastelands are potentially connected to each other.

Methods **to compare the potential connectivity of wastelands to the** Green and Blue Plan (GBP). We then identified the intersections between urban wastelands and the GBP, following several steps: 1. Mapping of the GBP's biodiversity reservoirs and networks, and creation of a hexagonal grid summarizing the information about the various reservoirs and networks (Fig. 3a, b); 2. Identification of intersections between wastelands and the GBP (Fig. 3c); 3. Identification of probable favorable habitats for plant species persistence linked to the GBP (i.e., areas identifying contiguous sets of wastelands, Fig. 3d); 4. Identification of suitable habitats for plant species dispersal linked to the GBP (larger than favorable habitats, representing potentially interconnected habitats, Fig. 3d).

To date, the cities of Tours and Blois do not include specifically urban green networks in their GBP. Therefore, we aimed to cross the identified connections between wastelands, necessarily participating in the green network, with the official GBP. After mapping the official GBP biodiversity reservoirs and networks (wooded, wetland or grassland networks), we identified its intersections with habitats around wastelands, following several steps (Box 3 and Fig. 3).

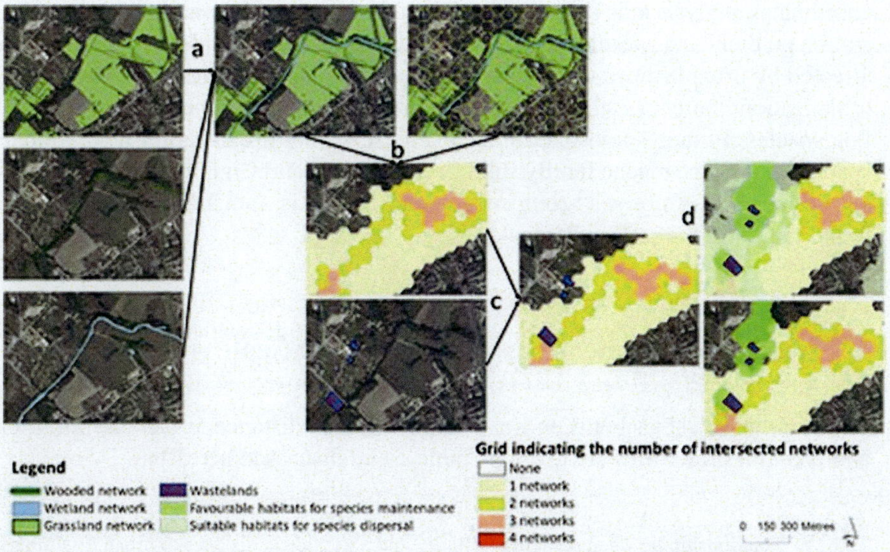

Fig. 3 Method of identification of wastelands intersecting the Green and Blue Plan in the city of Blois (here the example for a grassland network); **a** Networks identified in the GBP; **b** Construction of a hexagonal grid to identify the number of intersected networks; **c** Identification of wastelands intersecting the hexagonal grid; **d** Identification of favorable habitats for species persistence and of suitable habitats for species dispersal, intersecting the hexagonal grid

3 Wasteland Urbanization Context and Flora

The urbanization gradient shown by a Principal Component Analysis of urban descriptors (Table 1), opposes high urbanization, characterized by high percentages of built-up areas and buildings mean height, and low urbanization presents open and agricultural land covers and past uses. Wastelands belonging to companies are mainly situated in highly urbanized areas, whereas those owned by private people are linked to a low urbanization.

We identified 544 plant species, representing 26% of the regional species pool (Cordier et al. 2010), with 473 species found in Tours and 384 in Blois. Among them, 311 were shared by both cities. The average species richness per wasteland was 58.91 (standard deviation, SD = 15.35). The most frequently found species (i.e. found on more than 80% of wastelands) were the same for both cities: *Picris hieracioides, Daucus carota,* and *Vicia sativa* (found in 167, 164, and 159 wastelands out of 179, respectively). 160 species were only observed once. Multivariate analysis (projection of a correspondence analysis of hosted plant communities on a principal coordinates analysis axis representing urban descriptors) allowing to link the urbanization gradient and wastelands plant communities, shows that plant communities are affected by urbanization: while many Poaceae and shrub species are favored in low urbanization context (such as Dactylis glomerata, Elytrigia repens, Holcus lanatus, Poa species, Prunus spinosa, Sambucus nigra), strong urbanization favors pioneer species of the Asteraceae family (like Crepis, Picris, and Cirsium species). Invasive species such as Ambrosia artemisiifolia and Stenactis annua are also more likely found in high urbanization contexts.

4 Wasteland Connectivity at the Habitat Scale

The correlation values between wasteland Euclidean distance, urban dissimilarities, and plant dissimilarities are given in Table 3 and discussed hereafter.

Table 3 Relationship between plant dissimilarity and a Euclidean distance, b urban dissimilarity

	Correlation of plant dissimilarity with:	r (Mantel randtest)	P-Value
a	Euclidean distance in Tours	0.107	0.41
	Euclidean distance in Blois	0.148	0.001
b	Urban dissimilarity at neighborhood level (land cover proportions)	0.132	0.003
	Urban dissimilarity at city level (human population density and distance to city center)	0.141	0.004

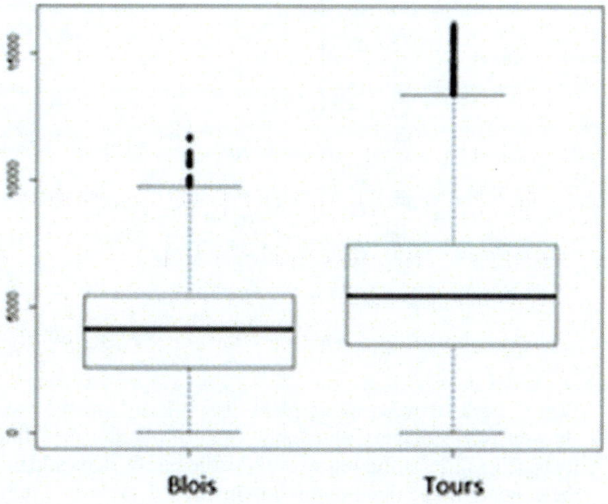

Fig. 4 Mean distances between wastelands in the two urban areas

4.1 *Relationship Between Euclidean Distance and Plant Dissimilarity*

Plant dissimilarity increases with Euclidean distances in Blois, wastelands hosting more similar communities when they are close to each other, but not in Tours (Table 3a, in Blois: $r^2 = 0.15$, $P = 0.41$; in Tours: $r^2 = 0.11$, $P < 0.001$). We can conclude that while connectivity between wastelands is the case in Blois, it is not in Tours. Either wastelands close to each other do not share the same communities, or wastelands are too distant for connectivity to occur. Indeed, Student's test reveals that the mean distances between wastelands are higher in Tours than in Blois ($P < 0.01$, Fig. 4).

4.2 *Relationship Between Urban and Plant Dissimilarities*

The dissimilarity-based analysis comparing plant and urban dissimilarities shows that plant dissimilarity increases with urban dissimilarity (Table 3b). The more different urban characteristics are (in terms of land cover proportions, human population density, and distance to city center), the more dissimilar plant communities are. Plant communities are therefore influenced by urban descriptors.

Figure 5 shows that the plant dissimilarity between wastelands correlates with the urbanization gradient, revealed by a Principal Component Analysis of urban descriptors (Table 1). Mean plant dissimilarity decreases when urbanization is lower (linear regression, $r^2 = 0.01$, $P < 0.0001$). Plant communities are therefore more similar in low urbanization contexts and vice versa, they are more diversified in dense urban contexts.

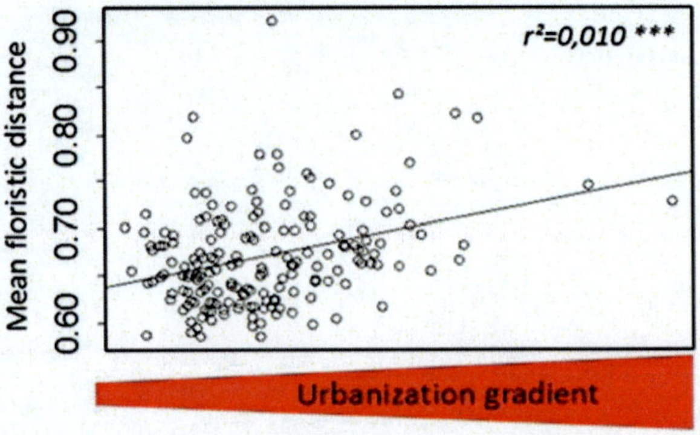

Fig. 5 Link between mean plant dissimilarity among wastelands and the urbanization gradient showed by a Principal Component Analysis of urban descriptors (Table 1). Circles represent habitats (wastelands), r^2 is the regression coefficient of determination (a statistical measure of how close the data are to the fitted regression line)

5 Wasteland Potential Connectivity at the City Scale

The cartographic analysis of connectivity identifies wastelands that are potentially related to each other by detecting habitats favorable to species persistence and ones suitable for plant species dispersal. Figure 6 presents the number of wastelands that can be considered as connected, i.e., whose habitats favorable for species persistence overlap. Figures 7 and 8 highlight that these wasteland overlaps are consistent and form an urban network of wastelands.

In Blois, we can observe that many wastelands are potentially connected to each other on the outskirts of the city, but also within the city center, despite the building density. For example, we can see in the north of the city, in an industrial area, that six wastelands constitute a small network (rectangle 1 in Fig. 7). Many wastelands

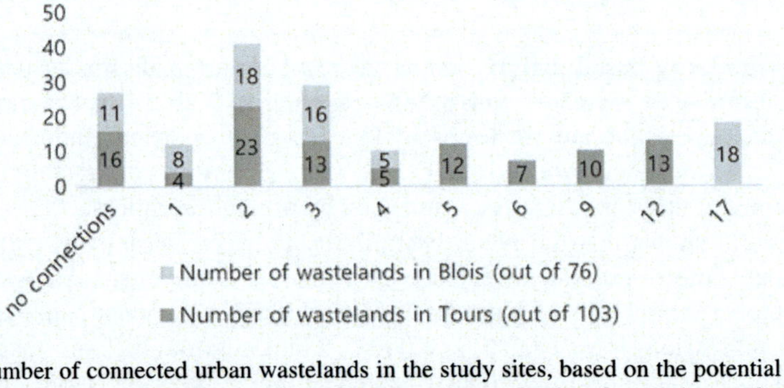

Fig. 6 Number of connected urban wastelands in the study sites, based on the potential dispersal of species in the urban area. The number of connected wastelands refers to the number of other wastelands with which a wasteland makes up a network. Example: 13 wastelands in Tours have connections with 3 other wastelands around

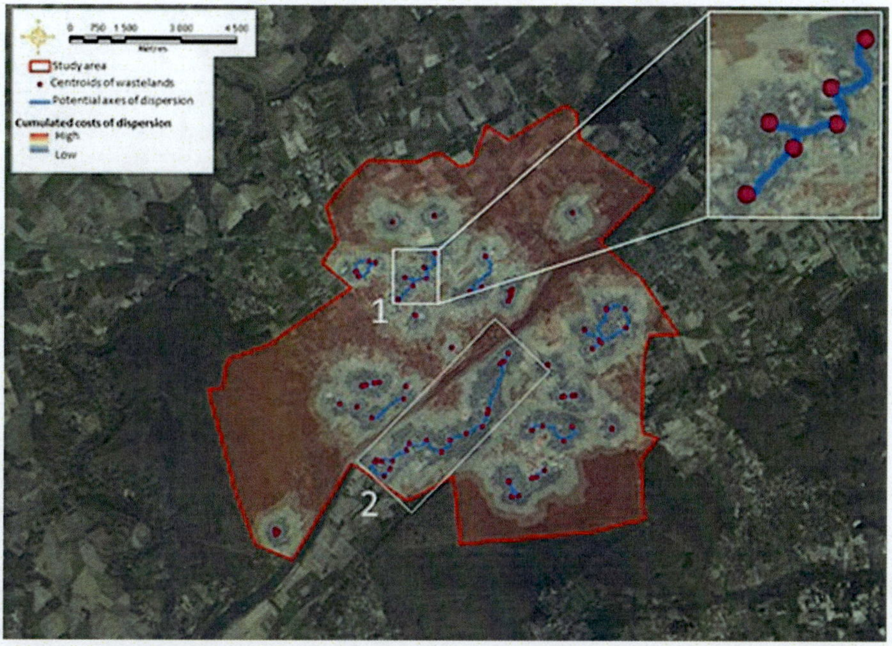

Fig. 7 Cumulated dispersal costs and potential axes of dispersal for wasteland plant communities in Blois (see Box 3 and Fig. 3 for methodological details). Blue links between wastelands indicate that wastelands are potentially connected to each other, and the habitats around wastelands indicate potential difficulty for species to disperse within the urban matrix. Biotope, ArcMap10, projection: RGF Lambert 93, *Sources* IGN

(n = 18) potentially connected to each other, form an urban green network in the southwest part of the city, along a country road (rectangle 2 in Fig. 7). As some wastelands could not be included in this study (i.e. wastelands whose access was inaccessible), this potential green network could be underestimated.

In Fig. 8, we see that wastelands are more scattered in Tours than in Blois. This spatial distribution of wastelands can explain why the relationship between Euclidian distance and plant dissimilarity is not significant (as previously shown in Table 3a). Wastelands may be too far apart for the similarity of plant communities to be explained by spatial proximity. Rather, it is urban differences that determine plant dissimilarity between wasteland communities. Despite this, we can see groups of wastelands connected to each other: at both the east and west ends of the city, between the two rivers (rectangles 1 and 2 in Fig. 8). Unlike in Blois, we do not observe an obvious wasteland network in Tours, particularly in densely built-up areas. Connectivity between wastelands is less complete, due to the greater distances between wastelands. We can also observe that in the north of the city, potential dispersal habitats around wastelands nearly overlap, but we cannot infer a proper axis of dispersal.

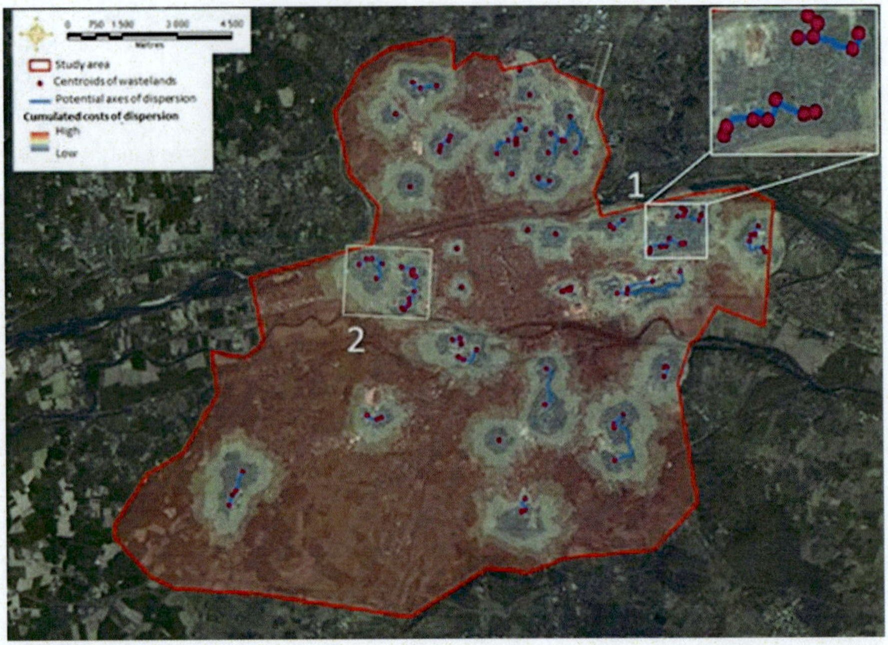

Fig. 8 Accumulated dispersal costs and potential axes of dispersal for wasteland plant communities in Tours (see Box 3 and Fig. 3 for methodological details). Blue links between wastelands indicate that wastelands are potentially connected to each other, and the habitats around wastelands indicate the potential difficulty for species to disperse within the urban matrix. Biotope, ArcMap10, projection: RGF Lambert 93, *Sources* IGN

Some wastelands do not show any connections with other ones (11 out of 76 in Blois, 14.5%; 16 out of 103 in Tours, 15.5%). As can be seen in Fig. 1, the city of Tours shows higher built-up density than Blois (55% vs. 35%) and wastelands are probably isolated because of this density. In Blois, it seems that wastelands remain isolated because of large agricultural (in the north and in the southeast of the city) or forest (in the southwest of the city) areas which separate them. Even though agricultural and forest habitats are more permeable than built-up ones, they are isolating urban wastelands due to their large areas.

6 Relationship with the Green and Blue Plan

As many wastelands are connected to each other and plant dissimilarity is influenced by Euclidean distance in Blois but not in Tours, we only identified interactions between the Green and Blue Plan (GBP) and wastelands, habitats favorable for species persistence or habitats suitable for species dispersal, in the city of Blois.

Sixteen wastelands (21%) intersect a biodiversity reservoir or the GBP networks (Fig. 9). Moreover, 33 areas (43%) recognized as favorable habitats for species persistence intersect the GBP (not shown), and habitats suitable for species dispersal of 61 wastelands (80%) also interact with the GBP (Fig. 10).

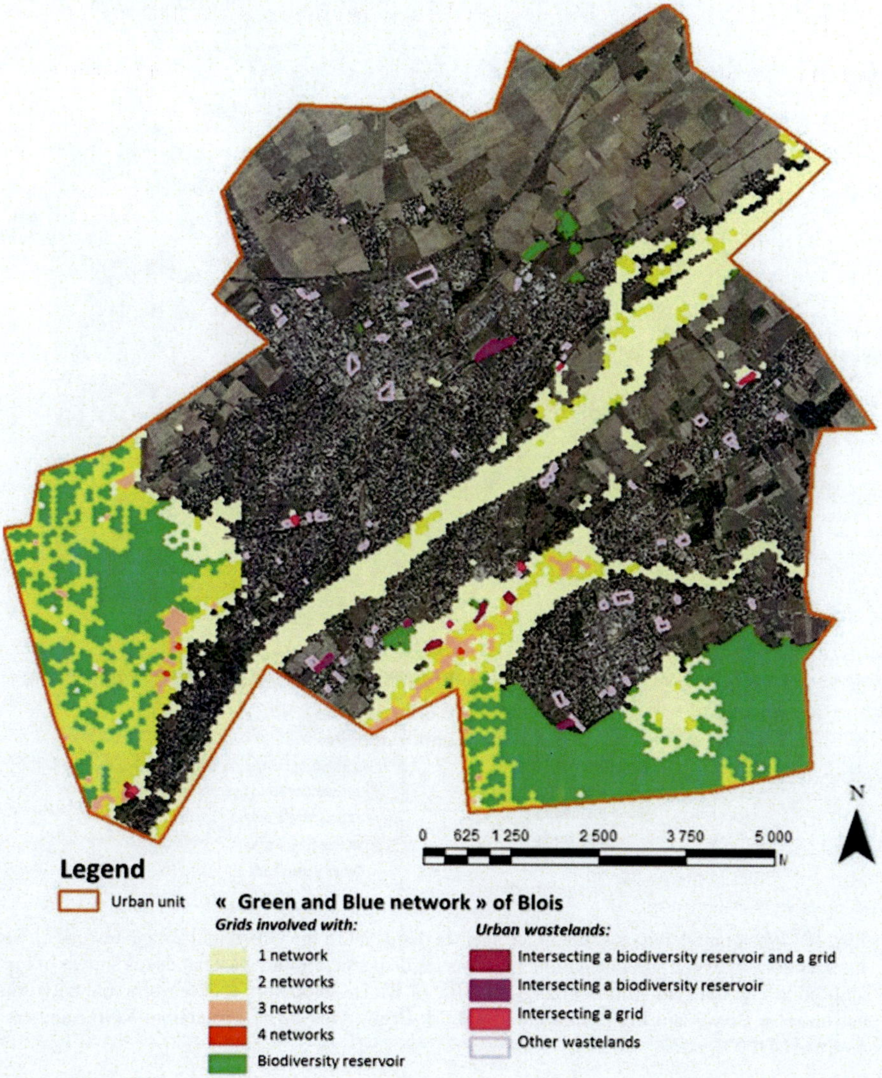

Fig. 9 Wasteland positions in the Green and Blue Plan of Blois. Intersections between wastelands and the hexagonal grid (summarizing the GBP of Blois—biodiversity reservoirs and networks—as shown in Box 3 and Fig. 3) are identified. M. Brun, ArcMap10, projection: RGF Lambert 93, *Sources* CDPNE, IGN

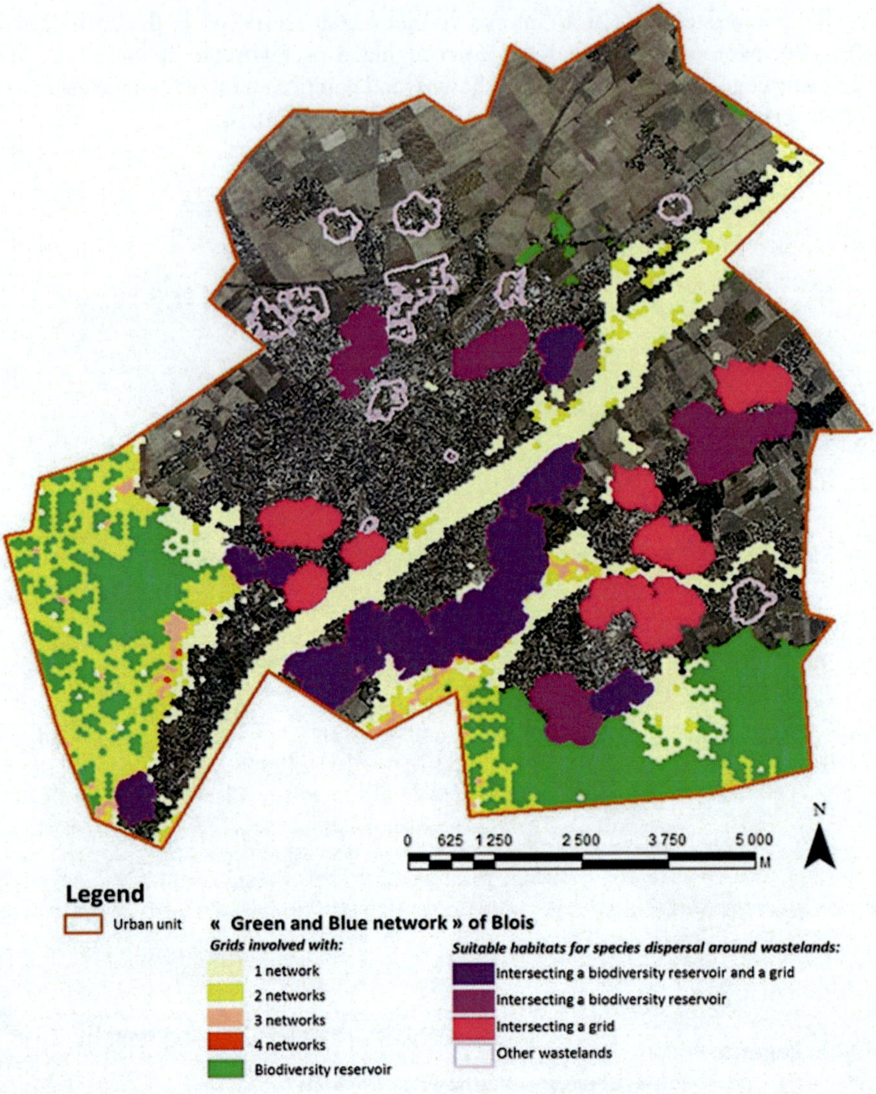

Fig. 10 Wasteland habitats suitable for species dispersal in the Green and Blue Network of Blois. Intersections between suitable habitats for species dispersal around wastelands (as shown in Fig. 7) and the hexagonal grid (summarizing the GBP of Blois—biodiversity reservoirs and networks—as shown in Box 3 and Fig. 3) are identified. M. Brun, ArcMap10, projection: RGF Lambert 93, *Sources* CDPNE, IGN

7 Discussion and Conclusion

Our study was based on the use of several land cover-based metrics to characterize not only wasteland urban contexts and their relationships with plant species (dissimilarity metrics based on precise photointerpretation within 200 m around wastelands), but also to identify potential connectivity on a larger scale (citywide), with regards to habitats suitable for species dispersal and potential links to the official Green Network Plan. We analyzed wasteland plant communities and urban dispersal and dealt with multiple spatial scales. This study is based only on plant communities: in further studies, it would be interesting to combine these results with information about birds and/or flying insects, as species both crucial for services as pollination and sensitive to connectivity (Brückmann et al. 2010).

Study of an urbanization gradient reveals that plant community diversity follows variations in urban characteristics: plant communities are more similar in low urbanization contexts and more diversified in highly urbanized areas. This is likely because of the relatively greater variability of environmental conditions and ages of wastelands. Thus, the commonly noted process of biotic homogenization (i.e., the tendency for communities to be more similar) with increased urbanization is not observed in urban wastelands (Kühn and Klotz 2006). Considering that "increasing vegetation diversity and complexity enhances species diversity in small green spaces [and] maximizing the plant diversity of stepping-stones can make them more effective at providing refuge services and consequently at facilitating animal dispersal" (Lynch 2019), the importance of the presence of wastelands within highly urbanized areas is crucial.

In both cities, we find that wastelands not ecologically connected to each other, but geographically close, are the ones separated by isolating land covers such as built-up areas, confirming that urban fragmentation reduces connectivity (Figs. 7 and 8; Fahrig 2003). Indeed, as highlighted by Tian et al. (2017), the presence of green spaces does not necessarily improve connectivity if their layout is not appropriately arranged to facilitate connectivity. The differences between plant communities in wastelands are influenced not only by urban characteristics such as land covers, but also by the distance between wastelands: plant community dissimilarity is higher when wastelands are far from each other. This result is only valid in Blois, suggesting that wastelands are connected to each other only in that city, as shown in Fig. 7. On the other hand, in Tours, isolating land covers are more numerous and prevent connectivity between wastelands. Indeed, as shown in Fig. 8, the higher share of isolating elements (such as the two rivers and a higher proportion of built-up areas) and higher distances between wastelands impede potential connectivity.

These results suggest that green networks are more evident in smaller cities, where even small-scale actions to promote species dispersal could significantly help to strengthen urban green networks. The example of habitats suitable for species dispersal (Fig. 10) showing that potential connectivity between wastelands covers a large proportion of the urban area is compelling: wastelands may play a major role in plant dispersal within urban areas and constitute a significant element for green network policies. Some groups of wastelands are close to each other and could create a non-linear network and reinforce connectivity if isolating elements were reduced, as observed by Lynch (2019): "(...) *stepping-stones should be closer in environments where the matrix is difficult to traverse, such as an urban center*." Tracking these under-exploited potential connections between wastelands could enhance their contribution to the global green networks. Yet, as we can see in Fig. 9, not a single green space (public one or wasteland) is included in the Green and Blue Plan, as this policy, conceived at a regional scale, aims to identify ecological networks at a large scale even in its local implementation, which induces a lack of precision on a finer scale, especially in urban areas. Moreover, the Green and Blue Plan does not focus specifically on urban areas, which are considered as globally hostile to biodiversity. In general, only formal and public spaces are considered for nature conservation and informal or private habitats hosting ordinary biodiversity are often excluded from conservation measures (Lemoine 2013). This is a frequently observed situation "*Large vacant lots and brownfields have characteristics that could make them effective stepping-stones, but this role of informal green spaces is understudied outside of transportation verges and local residents may find unmanaged areas undesirable or even unsafe*" (Lynch 2019). However, these habitats can shorten the distances between formal urban green spaces and thus increase the global connectivity of urban areas, and further research on ecological connectivity should therefore be conducted on a case-by-case basis, according to each urban area's specificity and depending on the diversity of potentially connected habitats (Tian et al. 2017).

In conclusion, environmental policies should not only focus on the reduction of fragmenting elements, but also on the protection and extension of connecting ones such as wastelands, even within urban areas. Even though wastelands do not show continuous connectivity (e.g., corridors), they can act as stepping-stones for urban green networks that should be considered in ecological planning. Indeed, an increasing number of studies are examining the role of small habitats within urban areas in facilitating species dispersal, as spatially disconnected spaces can be functionally connected. Therefore, as highlighted by Lynch (2019) in a recent review, "cities are moving away from corridor strategies such as greenways and toward stepping-stones" as corridors are not always possible to establish, stepping-stones offer a more realistic implementation and planning (Clergeau and Blanc 2013). In addition to integrating specific urban areas into ecological mapping, green networks should also consider the importance of temporary spaces of nature (Kattwinkel et al. 2011), such as wastelands, since their spatiotemporal dynamics may contribute greatly to biodiversity conservation within cities.

References

Baudry J, Merriam G (1988) Connectivity and connectedness: functional versus structural patterns in landscapes. In: Schreiber K-F (ed) Connectivity in landscape ecology (Proceedings of the 2nd international seminar of the "International association for landscape ecology"). Münstersche Geographische Arbeiten, pp 23–28

Bonthoux S, Brun M, Di Pietro F, Greulich S, Bouché-Pillon S (2014) How can wastelands promote biodiversity in cities? a review. Landsc Urban Plan 132:79–88. https://doi.org/10.1016/j.landurbplan.2014.08.010

Bray JR, Curtis JT (1957) An ordination of the upland forest communities of Southern Wisconsin. Ecol Monogr 27(4):325–349. https://doi.org/10.2307/1942268

Brückmann SV, Krauss J, Steffan-Dewenter I (2010) Butterfly and plant specialists suffer from reduced connectivity in fragmented landscapes. J Appl Ecol 47:799–809. https://doi.org/10.1111/j.1365-2664.2010.01828.x

Clergeau P, Blanc N (2013) Trames vertes urbaines : de la recherche scientifique au projet urbain [Urban green networks: from scientific research to the urban project]. Le Moniteur, Paris

Cordier J, Dupre R, Vahrameev P (2010) Catalogue de la flore sauvage de la région Centre [Catalog of the wild flora of the Central region]. Symbioses 26:36–84

Fahrig L (2003) Effects of habitat fragmentation on biodiversity. Annu Rev Ecol Evol Syst 34(1):487–515. https://doi.org/10.1146/annurev.ecolsys.34.011802.132419

Foltête JC, Clauzel C, Vuidel G, Tournant P (2012) Integrating graph-based connectivity metrics into species distribution models. Landsc Ecol 27(4):557–569

Gargominy O, Tercerie S, Daszkiewicz P, Régnier C, Ramage T, Dupont P, Poncet L (2013) TAXREF v6. 0, référentiel taxonomique pour la France. Méthodologie, mise en œuvre et diffusion [Taxonomic repository for France. Methodology, implementation and dissemination]. MNHN, Service du Patrimoine Naturel, rapport SPN, 7: 92

Godefroid S, Koedam N (2007) Urban plant species patterns are highly driven by density and function of built-up areas. Landsc Ecol 22(8):1227–1239. https://doi.org/10.1007/s10980-007-9102-x

Gower JC, Legendre P (1986) Metric and Euclidean properties of dissimilarity coefficients. J Classif 3(1):5–48. https://doi.org/10.1007/BF01896809

Harrison C, Davies G (2002) Conserving biodiversity that matters: practitioners' perspectives on brownfield development and urban nature conservation in London. J Environ Manag 65(1):95–108

Johnson AL, Borowy D, Swan CM (2018) Land use history and seed dispersal drive divergent plant community assembly patterns in urban vacant lots. J Appl Ecol 55(1):451–460

Kattwinkel M, Strauss B, Biedermann R, Kleyer M (2009) Modelling multi-species response to landscape dynamics: mosaic cycles support urban biodiversity. Landsc Ecol 24(7):929–941. https://doi.org/10.1007/s10980-009-9371-7

Kattwinkel M, Biedermann R, Kleyer M (2011) Temporary conservation for urban biodiversity. Biol Conserv 144(9):2335–2343. https://doi.org/10.1016/j.biocon.2011.06.012

Kindlmann P, Burel F (2008) Connectivity measures: a review. Landsc Ecol 23(8):879–890. https://doi.org/10.1007/s10980-008-9245-4

Kowarik I (2018) Urban wilderness: supply, demand, and access. Urban For Urban Green 29:336–347

Kühn I, Klotz S (2006) Urbanization and homogenization—comparing the floras of urban and rural areas in Germany. Biol Conserv 127(3):292–300. https://doi.org/10.1016/j.biocon.2005.06.033

Legendre P, Fortin M-J (2010) Comparison of the Mantel test and alternative approaches for detecting complex multivariate relationships in the spatial analysis of genetic data. Mol Ecol Resour 10(5):831–844. https://doi.org/10.1111/j.1755-0998.2010.02866.x

Legendre P, Legendre LF (2012) J Numer Ecol [s.l.]. Elsevier

Lemoine G (2013) La biodiversité aménage-t-elle les territoires ? De la présence d'espèces protégées et/ou patrimoniales…, ou comment passer d'une logique de contraintes à celle de projets de territoire [Does biodiversity shape territories? From the presence of protected and/or heritage species..., or how to move from a logic of constraints to that of territorial projects?]. Développement Durable Territ 4:1–6

Loriot P, Di Salvo M (2008) Détermination d'un MOS et calcul d'une tache urbaine à partir de la BD TOPO® de l'IGN : étude expérimentale [Determination of land use patterns and calculation of an urban stain from the IGN's TOPO® database: experimental study]. CERTU

Lynch, AJ (2019) Creating effective urban greenways and stepping-stones: four critical gaps in habitat connectivity planning research. J Plan Lit 34(2):131–155. https://doi.org/10.1177/0885412218798334

Marco A, Dutoit T, Deschamps-Cottin M, Mauffrey J-F, Vennetier M, Bertaudiere-Montes V (2008) Gardens in urbanizing rural areas reveal an unexpected floral diversity related to housing density. C R Biol 331(6):452–465. https://doi.org/10.1016/j.crvi.2008.03.007

McDonnell MJ, Hahs AK (2008) The use of gradient analysis studies in advancing our understanding of the ecology of urbanizing landscapes: current status and future directions. Landsc Ecol 23(10):1143–1155. https://doi.org/10.1007/s10980-008-9253-4

McDonnell MJ, Hahs AK (2013) The future of urban biodiversity research: moving beyond the 'low-hanging fruit.' Urban Ecosyst 16(3):397–409

McKinney ML (2002) Urbanization, biodiversity, and conservation. Bioscience 52(10):883–890. https://doi.org/10.1641/0006-3568

Muratet A, Machon N, Jiguet F, Moret J, Porcher E (2007) The role of urban structures in the distribution of Wasteland Flora in the Greater Paris Area, France. Ecosystems 10(4):661–671. https://doi.org/10.1007/s10021-007-9047-6

Muratet A, Lorrillière R, Clergeau P, Fontaine C (2013) Evaluation of landscape connectivity at community level using satellite-derived NDVI. Landsc Ecol 28(1):95–105. https://doi.org/10.1007/s10980-012-9817-1

Penone C, Machon N, Julliard R, Le Viol I (2012) Do railway edges provide functional connectivity for plant communities in an urban context? Biol Conserv 148(1):126–133. https://doi.org/10.1016/j.biocon.2012.01.041

Pickett ST, White PS (1985) The ecology of natural disturbance and patch dynamics. Academic Press, Orlando, Fla

Serret H, Raymond R, Foltête J-C, Clergeau P, Simon L, Machon N (2014) Potential contributions of green spaces at business sites to the ecological network in an urban agglomeration: the case of the Ile-de-France region, France. Landsc Urban Plan 131:27–35. https://doi.org/10.1016/j.landurbplan.2014.07.003

Strauss B, Biedermann R (2008) Fit for succession–community structure and life strategies of leafhoppers in urban brownfields. Ecol Entomol 33(1):107–118

Taylor PD, Fahrig L, Merriam G (1993) Connectivity is a vital element of landscape structure. Oikos 68:571–573

Tian Y, Liu Y, Jim CY, Song H (2017) Assessing structural connectivity or urban green spaces in metropolitan Hong Kong. Sustainability 9(9):1653. https://doi.org/10.3390/su9091653

Tischendorf L, Fahrig L (2000) On the usage and measurement of landscape connectivity. Oikos 90(7):19

Verbeylen G, De Bruyn L, Adriaensen F, Matthysen E (2003) Does matrix resistance influence Red squirrel (Sciurus vulgaris L. 1758) distribution in an urban landscape? Landsc Ecol 18(8):791–805. https://doi.org/10.1023/B:LAND.0000014492.50765.05

Marion Brun After a PhD on the benefits of urban wastelands for ecological continuities and plant diversity in cities, linking urban ecology and urban planning, **Marion Brun** is currently an Associate Researcher at the University of Tours (France), UMR CITERES. Her research themes are focused on the multidisciplinary study of urban biodiversity, from a socio-ecological point of view, on issues of urban ecology and representations and perceptions of nature.

Francesca Di Pietro As Associate Professor of Environmental Geography at the University of Tours (France), in the Planning & Environment Department, **Francesca Di Pietro** works on plant biodiversity in anthropogenic environments, at first in agricultural landscapes, later in urban ones. Her research focuses on plant communities of urban habitats (public green spaces, urban wastelands, roadsides, and community and individual gardens). Since 2011 she has worked on urban wastelands, coordinating work on their plant communities, informal uses, and how they are perceived.

Wasteland, a Refuge for Biodiversity, for Humanity

Audrey Muratet, Myr Muratet, Marie Pellaton, Marion Brun, Mathilde Baude, Assaf Shwartz, and Colin Fontaine

Abstract This chapter presents the results of our transdisciplinary work since 2001 on 100 wastelands in the greater Paris area, using ecological, geographical, and artistic approaches. This research subject, which we addressed with a diversity of methodologies and sensibilities, had hitherto been little studied: wastelands were unmapped and often disregarded. Our aim has been to understand the ecological, artistic, and political interrelationships of these spaces and their role as refuges for human and non-human beings.

Keywords Urban ecology · Traveling species · Ecological interaction · Human activity

A. Muratet (✉)
UMR LIVE-CNRS, Université de Strasbourg, Strasbourg, France
e-mail: amuratet@unistra.fr

M. Muratet · M. Pellaton
Paris, France
e-mail: myrmuratet@free.fr

M. Pellaton
e-mail: mariepellaton@free.fr

M. Brun
UMR CITERES-CNRS, Université de Tours, Tours, France
e-mail: marion.brun@univ-tours.fr

M. Baude
LBLGC-INRAE, Université d'Orléans, Orléans, France
e-mail: mathilde.baude@univ-orleans.fr

A. Shwartz
Faculty of Architecture and Town Planning, Technion - Israel Institute of Technology, Haïfa, Israël
e-mail: shwartzA@tx.technion.ac.il

C. Fontaine
CESCO (Centre d'Ecologie et des Sciences de la Conservation), Paris, France
e-mail: colin.fontaine@mnhn.fr

F. Di Pietro and A. Robert (eds.), *Urban Wastelands*, Cities and Nature,
https://doi.org/10.1007/978-3-030-74882-1_5

1 Study Area—Wastelands in the Greater Paris Area

This chapter presents the results of studies conducted separately or collectively by the authors on wastelands located in the Greater Paris area. Two groups of wastelands were specifically analyzed: one group of 98 wastelands located in the Hauts-de-Seine department studied between 2001 and 2005 and one group of 17 wastelands located in the Seine-Saint-Denis department studied in 2010 (Fig. 1). The artistic part of the study expanded into other wastelands and is still running.

1.1 Characterization

An urban space becomes a vacant lot or wasteland once humans stop using it formally. Industrial zones, overgrown gardens, ruined dwellings, abandoned areas alongside channels of transportation, roads, rivers, and canals, all these sites, these lands to which we attach the label "waste" because they no longer have an assigned purpose, with time become covered with a mass of plants which soon conceal all vestiges of their anthropic past.

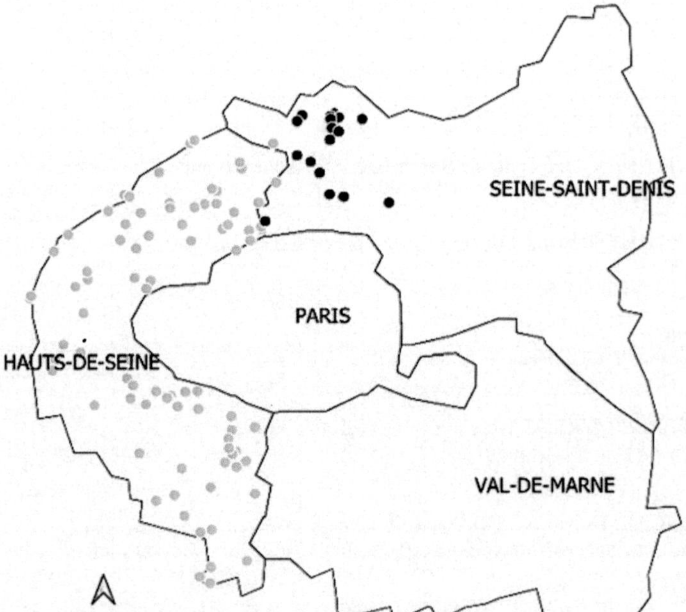

Fig. 1 Wastelands studied. In gray, the group of 98 wastelands located in the Hauts-de-Seine department; in black, the group of 17 wastelands located in the Seine-Saint-Denis department

Wastelands inexorably metamorphose through an ecological process of spontaneous colonization by species-rich animal and plant communities. Lawns give seeds, brambles, shrubs, and leafy undergrowth cover the grassland and provide shelter for trees, which grow, spread, and in time become a forest.

Vacant lots and wastelands do not constitute a specific environment, but rather an assemblage of multiple habitats. Their hosted plant life is determined by the nature of the soil, which may be polluted by heavy metals or hydrocarbons, by the proximity of water, by the age of the wasteland, and by the latest disruption, such as mechanical turning of the soil.

Nature finds free expression on wastelands. The diversity of plants and animals and the rich interactions between them create refuges for exuberant life, which is neither managed nor domesticated. Wastelands are refuges not only for flora and fauna but also for humans, alone or in groups, who settle there so as to lead their lives out of sight, while other people use them for illegal dumping.

Such human usage inevitably leads to the intervention of public or private bodies, which evacuate settlements, erect more fencing, and implement various measures to dissuade resettlement, for example, by bulldozing the site, chipping and shredding, digging deep trenches, and blocking access points with rockfill. These successive human activities regularly disrupt the landscape and the plant community dynamics of urban wastelands, thereby highlighting their pioneering and paradoxical nature: wastelands are lush and soiled, wild, inhabited, and disrupted, troubling, and tranquil.

1.2 History and Structure

Wastelands reflect urban dynamics. They result from the closure of industries, the ending of professions and careers, from economic recession. Wholesale abandonment of sites, of neighborhoods, sometimes leads to the extinction of whole cities. This phenomenon of de-urbanization has affected North American cities like Detroit, Philadelphia, Buffalo, and Cleveland, which since the 1950s have experienced a decline in the steel and automobile industries that has led to the depopulation of whole neighborhoods, which have been spontaneously recolonized by flora and fauna typical of the wastelands of these places (Ramalho and Hobbs 2012). In contrast to North American cities, wastelands in the Greater Paris area are continuously shrinking as the population, economic activity, and land use needs increase. Wastelands are therefore much more sparsely spread within the city. In the Hauts-de-Seine department, we qualified 98 wasteland sites in view of more managed urban spaces as parks, cemeteries, embankments, or more natural spaces like reforested areas, reed beds, and riverbanks (Muratet et al. 2007). Wastelands are small: 2,000 m^2 on average versus 6,000 m^2 for the other spaces identified. The oldest and smallest (over 21 years old and 808 m^2, on average) accounted for 50% of the wastelands inventoried and were located on the rights of way of built-up areas, residential areas, amenities, etc. The largest and youngest wastelands (4,500 m^2 and under 13 years, on average) were located on vacant lots and building sites (i.e., 30% of the wastelands sampled), and the wastelands intermediate in size and age (2,900 m^2 and 13–21 years, on average) were found in open urban spaces, parks, gardens, and sports fields (20% of wastelands). The link between age and surface area evidenced in this study is explained by land development pressures that rise as the site increases in size, meaning that the largest wastelands disappear first.

2 Wastelands and Their Inhabitants

Domestic and wild flora and fauna of diverse provenance flourish on wastelands. We found there human settlements and abandoned objects. More or less permanent constructions or caravans are sometimes installed, in isolation or in structured camps. Women and men, families, live there by developing an informal economy involving, among other things, the recovery and sale of scrap metal. Occasionally, wastelands also play host to non-mainstream artistic and music activities (graffiti, tagging, raves, etc.).

Wastelands are ecosystems clearly bounded by urban structures, buildings, roads, or other impervious linear structures. Although they have neither use nor assigned purpose, these places harbor thriving life. Ecological studies of urban wastelands have shown that they are important reservoirs of urban biodiversity (Angold et al. 2006; Herbst and Herbst 2006; Muratet et al. 2008; Bonthoux et al. 2014).

In the 98 wastelands located in the Hauts-de-Seine department, wild vascular plants were inventoried once between 2001 and 2005, over the whole site area, in spring, summer, or autumn. They harbored 365 plants, i.e., close to 60% of the department's flora (Muratet et al. 2007). In the 17 wastelands studied in the Seine-Saint-Denis department, we conducted standardized inventories in 2010. In each of the 40 quadrats (1 m × 1 m) systematically placed on the whole wasteland surface, the presence of all vascular plants was recorded once between May and July. Every bird species heard or seen along a transect of 200 m going through the wasteland was recorded during a 10-min period. This bird count was repeated 8 times for each wasteland between April and May. Butterflies were recorded along the same transect during periods of 15 min 7 times between July and August. These wastelands harbored 338 plant species, 42 birds, and 17 butterflies, i.e., nearly 1/3 of the whole biodiversity observed in this department (Muratet et al. 2011).

2.1 A Dynamic Assemblage

Flora, fauna, and human beings share wastelands in a haphazard and unique way. This urban ecosystem comprises an assemblage of varied habitats whose dynamics depend on natural processes in which the anthropic contribution is represented more or less spatially and temporally (Muratet et al. 2017). It virtually always harbors pioneer plants, those that grow in newly dug bare soil. Included in these pioneer species are annuals whose life cycle from germination to seed production extends over less than 1 year, such as poppies (*Papaver* sp.), mustard plants (*Sinapis* sp.), goosefoots (*Chenopodium* sp.), and pigweed (*Amaranthus* sp.). These pioneer plants invest great energy to produce their progeny rapidly. Trees and shrubs, butterfly bush (*Buddleja davidii*), willows (*Salix* sp.), poplars (*Populus* sp.), and trees of heaven (*Ailanthus altissima*) join these pioneer plant communities, facilitating the growth of biennial and perennial species, which progressively replace them. Biennial plants like mullein (*Verbascum* sp.), viper's bugloss (*Echium vulgare*), evening primrose (*Oenothera* sp.), burdock (*Arctium* sp.), and others develop their vegetative parts, notably leaves and roots, during the first year, before spending the winter in the form of a rosette of leaves with buds at ground level. The following year, a large part of their energy is allocated to the development of their reproductive parts, the stems grow, spectacular inflorescences form, fruiting follows, and then they die. Perennials, unlike annuals, are more social and become established over time. They can form a dense assemblage that pretty much obscures the soil. This is the case of most *Poaceae*, or grasses. Perennial plants allocate as much energy to the production of roots, stems, and leaves as they do to seeds. They live for several years, part of their vegetative apparatus surviving the winter. They form meadows, thickets, or even wooded areas in the absence of human intervention. These different successional stages form patches of leafy vegetation of varying height side-by-side and are mixed within the limited space that is a wasteland (Kowarik and von der Lippe 2018). This assemblage is not fixed, but evolves over time, the species being constantly on the move. These successions follow no plotted path toward an equilibrium or a climax, but rather trajectories that are constantly readjusted.

2.2 A Refuge for Species Intolerant of Urban Conditions

The nature of wastelands differs from that of gardens, parks, and squares (Strauss and Biedermann 2006; Small et al. 2002). Infrequent human presence on wastelands predisposes to a great diversity of birds, plants, and butterflies (Muratet et al. 2011). Wastelands tend to favor species intolerant of urban areas, such as the northern wheatear (*Oenanthe oenanthe*), which in Berlin nests on the ground and which, being very sensitive to human presence, finds refuge in the open habitats of wastelands (Meffert et al. 2012). In the Paris region, the same is true of the lesser whitethroat (*Sylvia curruca*), which finds refuge in tall weeds that hide their nests (Muratet et al. 2011). Squares harbor species that in urban settings are adapted to human presence, trampling, and mowing. Plant species like cultivated vetch (*Vicia segetalis*), red bryony (*Bryonia dioica*), cultivated parsnip (*Pastinaca sativa*), and false oat-grass (*Arrhenatherum elatius*) are often seen in wastelands (over 50% of sites) and have never been observed in Parisian squares. Conversely, creeping wood sorrel (*Oxalis corniculata*), scarlet pimpernel (*Lysimachia arvensis*), shaggy soldier (*Galinsoga quadriradiata*), lady's thumb (*Persicaria maculosa*), and smooth hawksbeard (*Crepis capillaris*) have been noted in over 50% of squares and never in wastelands. It is important to conserve and promote this complementarity between these two types of spaces as it enables urban biodiversity comprising species more or less well adapted to different anthropic influences to become established and to move around in the city. Depending on their degree of tolerance to the management applied to urban spaces, flora and fauna find a refuge suited to their ecological requirements.

2.3 A Welcome for Newcomers

Cities are a haven for traveling plants that originate in other parts of the world. The spontaneous proportion of these alien plants is substantial in wastelands compared with other urban spaces (20% vs. 17% over the whole of the Hauts-de-Seine department, Muratet et al. 2007; 24% vs. 19% in Brussels, Godefroid and Koedam 2007). Traveling plants that successfully form spontaneous populations in Europe originate from zones of the globe with equivalent climatic conditions and a high volume of trade exchange. For the most part, these plants have been introduced intentionally for the purposes of ornamentation, horticulture, and agriculture, which account for 70% of the reasons for the introduction of these plants (Pyšek et al. 2009).

In northern France, there are species from Asia, such as Japanese knotweed (*Reynoutria japonica*), tree of heaven (*Ailanthus altissima*), summer lilac (*Buddleja davidii*), and birdeye speedwell (*Veronica persica*), from North America, like Canada goldenrod (*Solidago canadensis*), least pepperwort (*Lepidium virginicum*), black locust (*Robinia pseudoacacia*), and evening primrose (*Oenothera* sp.), and from South Africa, as narrow-leaved ragwort (*Senecio inaequidens*).

Some ecologists do not welcome these newcomers, which they see as disrupting natural ecosystems. Knowledge of these species is marred by ideological values that bias scientific reasoning and experimentation. These alien species are treated with mistrust and scorn, even disgust (Brown and Sax 2004), and large sums are disbursed to eradicate them (Wittmann and Flores-Ferrer 2015). The discourse is emotional and punctuated by metaphors devoid of scientific argument. Medical and surgical approximations abound: "cancer," "plague," "invader," "killer," "scourge." These terms imply an attack on the integrity of a body that was never constituted in this way in the first place. The ecosystem is considered as a "pure" body free of any contribution from outside (Rémy and Beck 2008).

On reading and hearing these widespread assertions, it is hard not to see a parallel between policies of protection against exotic species and immigration policies.

Some scientists see evolutionary potential in these traveling plants. Others call for a measured response (Tassin 2014) and are accused of denialism (Russell and Blackburn 2017), because they are outnumbered in the face of myriad conferences, colloquia, and articles warning the general public of the "dangers" involved. The debate rages on within the scientific community.

3 Interactions

Biodiversity varies within wastelands, depending on local and landscape parameters, on human communities, and on interacting communities of plants, birds, butterflies, and pollinators.

3.1 Human Traces and Plants

Stucco columns strewn down the slope of a mound of earth, a pistol hidden behind a broken TV set, a moonlike ball on a carpet of ivy, a ticking clock still telling the right time in a deserted clearing, a shopping trolley trapped in a thornbush, an orange office chair propped against a tree of heaven, a pair of once-red ballet shoes bleached to pink in the summer sun, a bucket of solidified oil-based paint, a blue, obese soft toy rat slumped on a heap of refuse, a tin can eroded by the rain… Just a few of the many objects found on wastelands. There are the most conspicuous, the recently dumped, balanced as if they've just been tossed aside. Others, nestling in the ground, are being eaten away by a relentless process of decomposition and revegetation, or of mineralization. Most objects are thrown away or abandoned, but some are used or carefully positioned and bear witness to past or present human activity in or alongside wastelands. Objects abound on the fringes of wastelands, tipped into pits or piled on the ground. Some wastelands harbor very specific objects: computer equipment, children's balls, tires, and the metal carcasses of cars.

When we were starting out, no research had been done on fly-tipping and what people dump on wastelands. Everything remained to be done in terms of classification and analyses. Confronted by the chaotic heaps of garbage and scattered objects, we used a quadrat sampling method in our study. Using systematic sampling, we analyzed 40 quadrats of 1 m^2 at each of 17 wastelands in the Seine-Saint-Denis department (Brun 2012).

We studied two types of influences associated with human activities: (i) human presence, which is greater on the approaches to wastelands and reflected by the quadrats studied on the wasteland margins, and (ii) proven uses, reflected by the numbers and materials of objects present in a quadrat. Within each quadrat, we also qualified the type of habitat (bare soils, meadows, thickets, and wooded areas) and identified the plants growing there. The aim was to look for the relation between these human traces and the ecological characteristics of the associated flora.

Edges of wastelands

Anthropic influences are greater on the fringes of wastelands than at their center. These disruptions are physical in nature, like trampling and turning or exposure of the soil, and lead to the appearance of flora comprising a higher proportion of pioneer species, like wall barley (*Hordeum murinum*) and annual meadow grass (*Poa annua*). Disruption is also chemical, with enrichment or pollution of the soil by waste, urine, feces, and so forth. The most nitrophilous species, like the common nettle (*Urtica dioica*), are best able to resist these disruptions.

Objects: numbers and materials

The presence of objects has no significant effect on the specific richness or average rarity of the species in a quadrat but is correlated with the traits of the species. An increase in the number of objects, reflecting greater use, is associated with a rise in the proportion of annual species that are specialists of urban habitats. In quadrats containing metal objects, there are more specialist and traveling species. In quadrats with stone objects, the proportion of annual species is increased.

These results show that human activity acts as a filter to select the pioneer plants of primary successions that are resistant to stress and adapted to soils rich in nutrients (McKinney 2008).

3.2 *Insects and Plants*

Plant-pollinator networks shed light on the interactions between plant and pollinator species and help to identify which plants depend on which pollinators for their reproduction and which insects depend on which plants for their food. In a sample of 6 of the 17 urban wastelands in the Seine-Saint-Denis department, we inventoried four times in the month of July insects foraging on flowers in three 15 × 2 m transects positioned in open habitats of each wasteland. Nearly 1,400 insects collected belonged mostly to two groups: *Hymenoptera* (45%: wild and domesticated bees, bumblebees…) and *Diptera* (32%: flies, hoverflies…), and to a lesser degree *Coleoptera* (11%: beetles…), *Hemiptera* (5%: true bugs…), and *Lepidoptera* (4%: butterflies). The relative proportions of these different groups of insects observed on wastelands are comparable to those seen in natural habitats in Europe. Like residential and community gardens (Baldock et al. 2019), wastelands are habitats favorable to the maintenance of populations of pollinators in the urban setting. Most pollinating insects were found on hawkweed oxtongue (*Picris hieracioides*), wild carrot (*Daucus carota*), and creeping thistle (*Cirsium arvense*). The abundance of these flowers in wastelands and their non-negligible nectar production (Baude et al. 2016) mean that these species are crucial for the nutrition and maintenance of pollinating insects in urban areas.

4 Permeability

At 98 wastelands in the Hauts-de-Seine department, we measured geographical distances, defined as the Euclidian distances between two wastelands, floristic distances by measuring their dissimilarities in species compositions, and environmental distances corresponding to the number of divergent environmental parameters between these wastelands: light exposition, slope, type of parent rock, and site geomorphology (Muratet et al. 2007). Looking at the correlations between the floristic, geographical, and environmental distances for different areas, we found no evidence of a relation between the floristic and environmental distances. Local environmental factors therefore seem not to influence the floristic composition of a site. This may also be due to measurements more relevant in large natural landscapes than in fragmented urban areas. In wastelands, the ground is usually leveled and covered with man-made materials or excavated deep soil horizon (Vergnes et al. 2017) so that the effects of parent rock, slope and exposure are no longer visible. In contrast, the floristic distance was positively correlated with the geographical distance, but only at sites greater in area than 2,500 m^2. This result suggests that the plants migrate between these large habitat patches, which thus form a functional network, whereas smaller sites seem to be isolated in the urban matrix. At any time during its life cycle, a living organism needs to move to feed itself, reproduce, and spread. The daily and seasonal movements and dispersion of individuals from one habitat to another are greatly limited in urban areas. The connectivity of the landscape is a major factor in reducing obstacles and facilitating movement within the mineral matrix. The distance between wastelands, their size, and the degree of vegetation cover of the urban matrix are important factors in the conservation of a wasteland network. Their plant communities could be connected at a maximum distance of 2 km if the urban matrix is sufficiently permeable to their movements (Muratet et al. 2013).

5 A Refuge for All

In large cities and their suburbs, wastelands are never totally abandoned to wild flora and fauna, and harbor varying degrees of anthropic activity. Apart from their function as clandestine dumping grounds, wastelands are often used for housing and informal activities by communities or by isolated individuals who find refuge there. Hidden from the outside world, these people build discreet shelters concealed among the luxuriant vegetation of wastelands. Very quickly, and when the geographical, political, and economic circumstances are propitious, shantytowns arise, generally built by people from the same country, region, or even city, thus forming a community. It also happens, if the space allows, that these people are joined by other individuals or communities of different origins. The shantytowns comprise dozens of shacks or are more erected around one or more alleyways. Depending on their size, shantytowns may also have common rooms, water points, dry toilets, churches, even shops. Made of salvaged materials, rafters, paneling, carpets, and more, the shacks tend to be somewhat similar, with about 12 m^2 of floor space and a roof apex at 2.5 m. They are equipped with wood-burning stoves for heating in winter and for cooking. These stoves are made of old large metal drums cut in two onto which is crimped sheet metal to serve as a cooktop pierced by a pot-sized hole and a connector for the smokestack. More advanced stoves are fitted with a hinged door in the front for the loading of wood. Activities are numerous in urban wastelands and one of the most important is the recovery of scrap, in which various metal materials are salvaged from the streets, worksites, and basements, and sorted upon return to the wasteland according to the type of metal (iron, aluminum, brass, and copper) for sale by weight to specialized companies. The fringes of wastelands may host activities such as auto-mechanics, fast-food sales, barbecues, and so on. Wastelands are also repeatedly used for self-expression through graffiti and wall paintings, works that are all the more remarkable given that their only viewers are the street artists themselves and the people living on the sites. Some urban wastelands, notably north of Paris, serve as gathering points for large communities of drug users and have shelters used for the purposes of prostitution. More anecdotally, there are also garden spaces. Occupation of wastelands is usually a result of hardship, and the occupants likely do not consider nature and its preservation to be their prime concern. The floral density of wastelands decreases sharply once the number of human occupants exceeds a few individuals, and flora even disappears entirely with the erection of large shantytowns. Conversely, plant growth seems more vigorous once humans abandon the site. Because of their wildness and filth and their occupation by outsiders, wastelands worry local residents and are frowned upon by society, which senses unspeakable dangers lurking in their anarchic luxuriance.

6 Place of Free Expression

Wastelands are not places for the conservation of rare or threatened species, but their biodiversity is not without interest. Studies of these spaces over several years have revealed their functional and heritage role. Wastelands are the main reservoirs of biodiversity in built-up areas and the nature that flourishes in them is little affected by human turmoil, and the dynamics of nature there run free. Their ephemeral character prompts reflection on how to conserve them without being overprotective. Wastelands are complex, dynamic, constantly evolving systems that are the stage for interactions that are ceaselessly made and unmade. This uncontrollable complex can only be protected by dynamic city-level oversight. The aim is to preserve the freedom left to these spaces so that the development, spread, adaptation, and resilience of their biodiversity are given full rein.

This teaches or reminds us that we should absolutely not intervene in the biodiversity of wastelands. Non-intervention is in itself a form of management that allows the originalities of each environment to be expressed through the communities of plants and animals that thrive there. It enables the preservation of ecological singularities and therefore supports the fight against the standardization of nature that is ongoing in all the world's cities.

However, the floristic diversity of wastelands has been declining in France in recent times, and the area they occupy in large built-up areas like Greater Paris has been halved in the last 30 years (Zucca et al. 2019). Our work has been done in close collaboration with the local authorities of the Seine-Saint-Denis department and has enabled greater inclusion of these spaces in urban planning and, notably, in the layout of ecological corridors. It is essential to inform and to raise awareness among urban populations of the richness, originality, vital necessity, and beauty of these environments so that they are no longer just considered as places for fly-tipping by some and land reserves by others, but are accorded their true ecological status.

References

Angold PG, Sadler JP, Hill MO et al (2006) Biodiversity in urban habitat patches. Sci Total Environ 360:196–204

Baldock KCR, Goddard MA, Hicks DM et al (2019) A systems approach reveals urban pollinator hotspots and conservation opportunities. Nat Ecol Evol 3:363–373

Baude M, Kunin WE, Boatman ND et al (2016) Historical nectar assessment reveals the fall and rise of floral resources in Britain. Nature 530:85–88

Bonthoux S, Brun M, Di Pietro F (2014) How can wastelands promote biodiversity in cities? a review. Landsc Urban Plan 132:79–88

Brown JH, Sax DF (2004) An essay on some topics concerning invasive species. Austral Ecol 29(5):530–536

Brun M (2012) Des plantes et des objets: analyses des friches urbaines comme zones de refuges et d'interactions entre les hommes et la nature [Plants and objects: analyses of urban wastelands as zones of refuge and interaction between humans and nature]. Internship report of research master 2. Pierre et Marie Curie University

Godefroid S, Koedam N (2007) Urban plant species patterns are highly driven by density and function of built-up areas. Landsc Ecol 22(8):1227–1239

Herbst H, Herbst V (2006) The development of an evaluation method using a geographic information system to determine the importance of wasteland sites as urban wildlife areas. Landsc Urban Plan 77:178–195

Kowarik I, von der Lippe M (2018) Plant population success across urban ecosystems: a framework to inform biodiversity conservation in cities. J Appl Ecol 55(5):2354–2361

McKinney ML (2008) Effects of urbanization on species richness: a review of plants and animals. Urban Ecosyst 11(2):161–176

Meffert PJ, Marzluff JM, Dziock F (2012) Unintentional habitats: value of a city for the wheatear (Oenanthe oenanthe). Landsc Urban Plan 108(1):49–56

Muratet A, Fontaine C, Shwartz A et al (2011) Terrains vagues en Seine-Saint-Denis [Vacant lots in Seine-Saint denis]. Journal Published by Plaine Commune & Natureparif

Muratet A, Lorrillière R, Clergeau P et al (2013) Evaluation of landscape connectivity at community level using satellite-derived NDVI. Landsc Ecol 28(1):95–105

Muratet A, Machon N, Jiguet F et al (2007) The role of urban structures in the distribution of wasteland flora in the greater Paris area, France. Ecosystems 10:661–671

Muratet A, Muratet M, Pellaton M (2017) Flore des friches urbaines [Flora of urban wastelands]. Editions Xavier Barral

Muratet A, Porcher E, Devictor V et al (2008) Evaluation of floristic diversity in urban areas as a basis for habitat management. Appl Veg Sci 11:451–460

Pyšek P, Lambdon PW, Arianoutsou M (2009) Alien vascular plants of Europe. In: DAISIE (eds) Handbook of alien species in Europe. Springer, Dordrecht, pp 43–61

Ramalho CE, Hobbs RJ (2012) Time for a change: dynamic urban ecology. Trends Ecol Evol 27(3):179–188

Rémy E, Beck C (2008) Allochtone, autochtone, invasif: catégorisations animales et perception d'autrui [Allochthonous, indigenous, invasive: animal categorizations and perception of others]. Politix 2:193–209

Russell JC, Blackburn TM (2017) The rise of invasive species denialism. Trends Ecol Evol 32(1):3–6

Small EC, Sadler JP, Telfer MG (2002) Carabid beetle assemblages on urban derelict sites in Birmingham, UK. J Insect Conserv 6:233–246

Strauss B, Biedermann R (2006) Urban brownfields as temporary habitats: driving forces for the diversity of phytophagous insects. Ecography 29(6):928–940

Tassin J (2014) La Grande invasion: Qui a peur des espèces invasives ? [The Great Invasion: Who is afraid of invasive species?] Odile Jacob. Paris

Vergnes A, Blouin M, Muratet A, Lerch TZ, Mendez-Millan M, Rouelle-Castrec M, Dubs F (2017) Initial conditions during Technosol implementation shape earthworms and ants diversity. Landsc Urban Plan 159:32–41

Wittmann A, Flores-Ferrer A (2015) Analyse économique des espèces exotiques envahissantes en France [Economic analysis of invasive alien species in France]. Ministère de l'écologie, du développement durable et de l'énergie

Zucca M, Loïs G, Muratet A, Ricci O (2019) Panorama de la biodiversité francilienne [Panorama of the Paris Region's biodiversity]. ARB îdF/IAU îdF, Paris

Abandoned railway, Porte de la Chapelle, Paris

African refugee camp in an industrial wasteland, Aubervilliers

Datura stramonium, industrial wasteland, Gennevilliers

Young man in the remains of a shantytowm evacuated from the lower side of the north ring road,Porte d'Aubervilliers, Paris

Car carcass, industrial wasteland, Aubervilliers

Knives, shantytowm recently evacuated, Champ-sur-Marne

Shantytown in an industrial wasteland, Porte d'Aubervilliers, Paris

Abandoned camp in a wasteland, Stains

Interior, shantytown in an abandoned railway, Gennevilliers

Stove, shantytown in the lower side of the north ring road, Porte d'Aubervilliers, Paris

Caravan, wasteland along the canal Saint-Denis, Aubervilliers

Party, shantytown in an industrial wasteland, Porte d'Aubervilliers, Paris

Rosa canina, industrial wasteland Gennevilliers

Audrey Muratet is a Lecturer at the University of.Strasbourg (France). As an ecologist, she studies the interactions among human and non-human parts of biodiversity in cities. She especially explores local and landscape human influences on the composition and dynamics of plant communities in wastelands.

Myr Muratet's work involves cities, both those he lives in and those he visits, and he carries it out from within their very sinews. The comings and goings in the places that he observes multiply according to his encounters with the people he photographs. This work has been under way for several years and has no determinate duration; in this way, he has created *Paris-Nord, Wasteland, CityWalk*, series of photographs about the people who lived or survived in discrete interstices of the city and about the mechanisms put in place to coerce and contain them.

Marie Pellaton, a graduate of the Basel School of Design (Switzerland), is a freelance graphic designer. She mainly works for cultural institutions. Her personal research, often very methodical, focuses on everyday observations. As part of a wasteland study, she photographed the traces left by human beings and identified abandoned objects, as clues to life and past stories.

Marion Brun After a PhD on the benefits of urban wastelands for ecological continuities and plant diversity in cities, linking urban ecology and urban planning, **Marion Brun** is currently an Associate Researcher at the University of Tours (France), UMR CITERES. Her research themes are focused on the multidisciplinary study of urban biodiversity, from a socio-ecological point of view, on issues of urban ecology and representations and perceptions of nature.

Mathilde Baude is a Lecturer in Ecology at the University of Orléans (France). Her research topics focus on the interactions between pollinating insects and flowering plant communities. In the urban context, she is investigating the impact of urbanization on the diversity of pollinators, the role of spontaneous versus ornamental plants in feeding pollinators, and the importance of some urban habitats such as wastelands in harboring pollinator diversity in cities.

Assaf Shwartz is a Professor of Ecology and Conservation at Technion—Israel Institute of Technology. He specializes in theoretically grounded applied research into how human activities influence biodiversity, but also how people can benefit from biodiversity conservation. His interdisciplinary research aims to find sustainable solutions for planning and managing human-dominated landscapes for the mutual benefit of people and nature.

Colin Fontaine is a community ecologist mostly interested in the way species interact to form complex interaction networks and in the effects of such interdependence on the response of species and ecosystems to perturbations. He has studied interactions between plants and pollinators in an urban wasteland, a perturbed habitat within an even more perturbed urban matrix.

Dwelling in an Urban Wasteland: Struggles for Resources

Cécile Mattoug

Abstract Wastelands in a dense urban context form temporary residual spaces including several resources. In Northern Paris, urban wastelands are used by inhabitants for different purposes: gardens, farms and settlements without a legal status, as well as business transactions in the gray economy. These inhabitants' temporary uses transform urban wastelands into self-organizing niches for subsistence and recreational needs. With the contemporary renewal of professional practices, merging ecological and democratic expectations increase the integration of inhabitants' uses into public green areas. Local authorities develop innovative management plans for these areas and are willing to delegate the management of urban open spaces to associations or to choose inhabitants. Stakeholders' diverging expectations for wastelands cause conflicts of interest due to oppositions, where urban wastelands appear as an ambivalent resource and figure. The starting point is a double ethnographic survey focused on social uses and the professional practices of wasteland transformation in Northern Paris. Thanks to a spatial survey based on the materiality of these uses and on several interviews, this paper examines dwelling forms of urban wasteland and the power relations that occur in urban development.

Keywords Participatory processes · Urban development · Urban open spaces · Marginalization · Inhabitants' uses · Gardens · Camps

1 Introduction

With the affirmation of democratization and the planning challenges of the greening of land, enthusiasm for nature in the city is now leading to a renewed consideration of uses and participation by inhabitants (Franchomme et al. 2013). The objectives of public policies in terms of nature conservation are becoming central to territorial and urban development and call for a reconfiguration of professional practices. Since the early 2000s and the evolution of sustainable development policies (Desjardins 2018),

---

C. Mattoug (✉)
UMR Géographie-Cités (EHGO), Paris I Panthéon-Sorbonne/ADEME, Paris, France
e-mail: cecile@mattoug.net

F. Di Pietro and A. Robert (eds.), *Urban Wastelands*, Cities and Nature,
https://doi.org/10.1007/978-3-030-74882-1_6

these issues have defined a triangulation between democracy, environment and planning (Soubeyran 2014) which has become essential. Indeed, development projects regularly generate power struggles and conflicts between inhabitants, or between civil society, developers and authorities, because of the impact of these transformations on living environments. In the context of metropolitan areas and urban densification, these environmental protection issues have been directly questioned by research since 2010. They testify the invalidity of a schematic representation in which the planned city and the inhabited city are opposed (Younes in Soubeyran 2014), but they call for these power relations to be considered as part of the city's production (Dorso 2008).

In this context of the renewal of representations, urban wastelands appear as privileged supports for the development of uses of nature for the inhabitants (Blanc 2016). For a long time considered as spaces of extension of the street, wastelands have been the subject of social science research for the presence of informal uses (Dorso 2016), working-class practices (Sauvadet and Bacqué 2011) and occupations by marginalized populations (Pichon 2007). In addition, since 2010, research conducted on wastelands, gaps and neglected areas has been analyzing the presence of nature (fauna and flora) and the uses it allows (Brun et al. 2019). Recent research tends to deconstruct the dualistic representations of the city, in which urban production over a long period of time is opposed to a minor urban factory (Hatzfeld 1998) that is also part of people's daily lives (Tonnelat 2003). The deconstruction of this dualistic perspective is accompanied by a desire to undo the marginalizing representations of the small players vis-à-vis the big players, turning wastelands into spaces for the development of economies and uses beyond norms. With the emergence of new professional practices, the evolution of professional planning practices and the development of non-militant participatory practices are helping to deconstruct this opposition between stakeholders (Gatta 2018), and forms of integration of inhabitants' uses in planning are asserting themselves, as for example in shared gardens (Demailly 2015) or in temporary occupation projects taking place on wasteland (Dubeaux 2017). Wastelands now appear more as resource spaces for a city production that integrates the uses of the inhabitants, where the uses of nature play a specific role (Scapino 2016; Toublanc and Bonin 2019). In addition to these temporary occupation projects, long-term landscaping projects are being organized according to the resources of the wasteland. They propose appropriate democratic and inclusive mechanisms. However, the integration of inhabitants' uses and the development of democratic approaches in the development of open spaces have many limitations. Indeed, does this mean that the processes of marginalization of the people who live on and occupy urban wasteland have come to an end? What forms of living allow the resources of urban wasteland? What is the evolution of professional landscaping practices on urban wasteland and what does this tell us about the evolution of the processes of marginalization of the inhabitants of the wasteland?

Our first hypothesis concerns the resources of urban wastelands. We will first demonstrate that inhabitants' uses are organized in a relatively autonomous way in living areas where the inhabitants mobilize to defend themselves. Our second hypothesis relates to landscaping projects which take urban wastelands as their site.

We will show that the integration of inhabitants' uses also proceeds to forms of standardization. Our third hypothesis relates to the process of transposing the resources of wastelands into project figures. We will demonstrate that natural forms such as gardens contribute to a dynamic of metropolitan revaluation of marginal spaces.

First of all, we will present the specificities of our field of study, the Tartres zone in the northern banlieue[1] of Paris, and the details of our research approach, the frameworks and tools deployed to examine our hypotheses. Next, we will present the uses of urban wastelands by the inhabitants, detailing their resources. After that, we will analyze the integration of the inhabitants' uses in the landscaping project. Finally, we will come back to the devices for translating the resources of the wastelands into project figures.

2 A Double Survey of the Tartres Zone North of Paris

2.1 The Tartres Zone and Its Fringes in the Northern Banlieue of Paris

Located in the heart of the northern banlieue of Paris and the Greater Paris Metropolis, the Tartres zone[2] is an interface between several municipalities and several contrasting districts. This 33-acre area is thus located on the edge of the communes of Saint-Denis, Pierrefitte-sur-Seine and Stains, in the heart of the public intercommunity of Plaine Commune (Territory of Plaine Commune). This former agricultural area in a state of wasteland at the time of the survey is commonly known as the Tartres zone. It comprises a core of vegetable growing land left fallow surrounded by a border of building site areas, temporary deposits and temporary occupations that are inhabited or are overseen by associations. Some operations carried out as part of the development and urban renewal project for the Clos Saint-Lazare—an adjacent district—have also taken place in the Tartres zone (Fig. 1).

The deferred development zone in which the Tartres zone is located was begun in the early 1970s, but operations really began in 2011 with the construction of public facilities. The Tartres zone is a concerted development zone (ZAC)[3] officially created in 2011 and whose project management is supported by the Territory of Plaine Commune. In 2019, the program for this ZAC included the construction of 2,300 housing units, several school facilities (secondary school and two school complexes) and a 16-hectare "green heart" designed to accommodate a mixed program of open

[1] "Banlieue" is kept in the French language because it has no equivalent in English. It means the first ring of metropolitan areas and is used to represent the post-industrial working-class neighborhoods.

[2] This is a literal translation of "la zone des Tartres." Tartres has been chosen by urban managers in the local toponymy to define the global perimeter of the urban development project. It may be related to the topographical and geological presence of a low knoll.

[3] "Zone d'aménagement concerté" (ZAC): a concerted development zone is a public development tool that defines a precise perimeter.

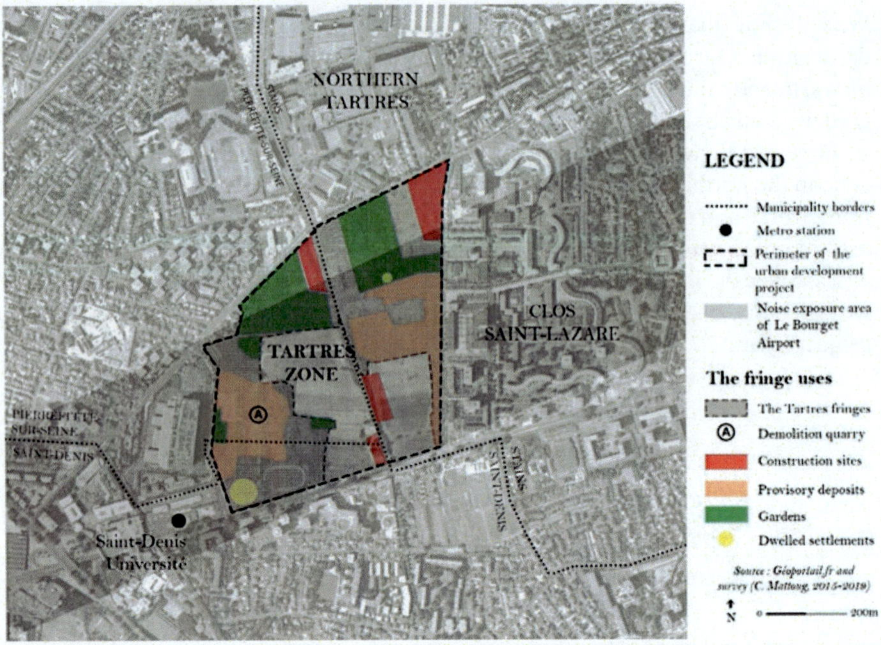

Fig. 1 Ground plan of uses in the Tartres zone in 2019. Map on aerial view (Géoportail) (C. Mattoug 2020)

spaces (vegetable growing, allotment gardens and temporary occupations by associations, to be completed in stages until the end of the development planned for 2027).

This field of study is characterized by its social geography, its context of urban densification and the metropolitan challenge of territorial change. The three cities of Pierrefitte-sur-Seine, Stains and Saint-Denis show a high rate of poverty, a high rate of migration and a very young demographic compared not only to France as a whole but also to the Paris region. In addition, the territory has a lack of green spaces and its urban areas are sporadically marked by the presence of migrant or Roma camps. The northern suburbs of Paris are part of a metropolitan densification territory. In the coming decades, this territory will experience numerous development projects linked to the creation of future facilities for not only the 2024 Paris Olympic Games but also numerous urban renewal projects in social housing districts, as well as mixed housing development zones (ZAC) and the creation of economic activity and office zones. This development dynamic meets public policy objectives in terms of a social mix for greater diversification of the social geography and the objectives of economic change toward the tertiary sectors.

2.2 Survey of Social and Planning Practices

The results we will present here are based exclusively on a double qualitative research survey conducted between 2015 and 2019 in the Tartres zone.[4] The anthropological approach to the transformation of the city (De Biase 2014) has been favored here, by combining an analysis of social practices with an analysis of professional urban planning practices. These two analyses have in common that they focus on the reality of the land and its materialities, the uses that the wastelands make possible and the representations that their transformations generate. By wasteland, we mean spaces in transition between two states, which are the subject of development projects and which are marked by a state of neglect by the public authorities or owners. We therefore consider that there are uses of wasteland, in the same way as there are inhabitants of wasteland. These inhabitants make their homes there or establish living areas, such as convivial spaces or gardens. These people are dwelling, in the sociological sense of the term: they establish systems of existential, individual, community or collective relationships with a place, in an aspiration which expresses needs, memories and imaginations (Ricoeur 1998). Thus, wasteland enables people to express a vital desire to build and appropriate the space (Ricoeur 1998). The investigation consisted in understanding the materialities and political relations which are exercised in the forms of inhabitation and in urban production.

The first part of the survey was carried out among the people who occupy the wasteland and focused in particular on the living spaces and uses of the inhabitants. We consider these forms of living as both social and spatial organizations that are part of the materiality of the wasteland. This anthropological and landscaping perspective has made it possible to study the power relations between the inhabitants. Anthropological immersion and multiple explorations of urban space (Box 1) shed light on these power relations. In the same way, interviews and informal exchanges provided an opportunity to understand the interests and representations of the people surveyed. A field notebook was drawn up, including in situ surveys of the occupations observed, landscape sketches and numerous notes detailing the material features of the places. The annotations and drawings were accompanied by a photographic survey. Exchanges with the inhabitants of the Tartres zone complete the observation work by an introduction to its uses. Thus, the meeting with many inhabitants has enabled a more precise study of the role that this set of sub-areas assumes for the people who live there (Box 2).

[4] The survey was conducted in the framework of a geography Ph.D. at the University Paris 1 Panthéon-Sorbonne, financed by the National Agency for the Environment and Energy (working on the implementation of public policy in the areas of the environment, energy and sustainable development) between 2017 and 2021.

Box 1: Limits of the anthropological approach to wasteland

The anthropological approach to fallow spaces is subject to many difficulties. First of all, contact with the inhabitants is extremely awkward, given the lack not only of legal status of the occupations studied, but also of the survey itself. In the same way that some people use the wasteland, the survey is carried out without legitimacy, or even in limited security conditions. A certain contingency must therefore be taken into account in this survey of the use of wasteland. The core of the results lies in the capture of the variety of situations and in the social and spatial experiences of wasteland.

The second part of the survey was carried out among those involved in a development project in the Tartres zone, which means in the same area as that occupied by the inhabitants. This part of the survey focuses more specifically on the pre-operational issues of the urban transformation. The anthropological survey has made it possible here to define the projected time (Boutinet 2012) that characterizes these forms of anticipation of the planning and development project. The monitoring of the development project of the ZAC des Tartres has made it possible to follow the changes in its programming between October 2015 and July 2019, by comparing the perspectives of the various development stakeholders. We distinguish four bodies of stakeholders in the development: the political leadership (elected representatives of the three municipalities of Pierrefitte-sur-Seine, Stains and Saint-Denis); the project management (technicians and managers of the Territory of Plaine Commune); the urban and landscape project management (agencies, support staff and study leaders) and the inhabitants and civil society. The semi-directive interviews and the exchanges obtained from these different stakeholders focused on two themes: the uses made by the inhabitants of the wastelands and the roles of the different bodies of stakeholders in drawing up a landscape guide plan, which was validated in 2019 (Box 2). We will therefore come back to the materialities that appear not only in the speeches, but also in the plans and other documents produced by the development stakeholders.

Box 2: People met and surveyed during the research

Among the inhabitants surveyed:

- The observed inhabitants are about 80 families or single persons. The inhabitants met were two single men between 30 and 50 years old and three families with children. Among the latter, the two single men were migrants from Somalia and the families were Roma people from Romania.
- The gardeners observed occupied about 140 garden plots, divided between three associations with legal status in Pierrefitte-sur-Seine (50, 30 and 40 units), and the others divided between conglomerate gardens in Stains (about 15 units) and Saint-Denis (about 6 units). The gardeners met were eight men and one woman (between 40 and 80 years old). The survey revealed that the gardens without a legal framework are exclusively occupied by men over 40 years of age and that there is also a higher proportion of men cultivating allotment gardens, even if the plots are always allocated to families.

Among the 18 development stakeholders surveyed:

- The elected representatives of the three municipalities and a town councilor;
- The technicians and managers of the Territory of Plaine Commune (from the Planning, Urban Renewal, Urban Ecology, Culture and Creation Workshop and Economic Development Department);
- The developer in charge of the ZAC des Tartres;
- Landscaping project management and an association of landscape mediators;
- The project leaders of associations mandated by the local authorities.

Overall, we conducted 16 semi-directive interviews with 13 different interlocutors and participated in 14 meetings and 5 events open to the public.

This double survey focused our attention on the power relations that are established around the resources of urban wastelands. As both material resources and figures in the design of the landscaping project, the resources of wastelands are at the root of inhabitants' organizations, the intentions of developers and power relations between them. These resources are above all domesticated or used by people: they are vehicles and supports in the balance of power between inhabitants or between inhabitants and planners. The survey focuses on the many representations which may seem opposed, even contradictory, and which we have chosen to reconstruct using extracts from the field notebook. The choice of an anthropological narrative led us to prioritize the description, analysis and finally interpretation. Moreover, the narrative allows us to restore, without betraying them too much, the richness of the representations used to elaborate the research.

3 Inhabited Resources: The Uses of Wasteland by Their Inhabitants

3.1 A Diversity of Uses for Wasteland Resources

The survey revealed a variety of uses relating to gray economies not only like the sale of goods or services, street mechanics, but also camps, gardens and the daytime presence of inhabitants. Some uses could not be determined with certainty. The economic uses identified are similar to those found in the urban space, such as street mechanics. Some day- or nighttime occupations are in fact simple extensions of the street space or public spaces in the vicinity of the wasteland. This presence is encouraged by the fact that these spaces are away from the urban centers. However, wastelands occupied by spontaneous vegetation accommodate more specific uses such as gardens and camp dwellings. These uses have required a long period of vacancy of the wastelands: they are part of living spaces, thanks to the interactions between the resources of the urban wastelands and the personal resources of inhabitants and gardeners.

Box 3: Story from the Field Diary: "Mehdi's Garden"

Over a period of 4 years, I was able to follow the developments and interactions that took place between a gardener who occupied a plot of land without a contract in the Tartres zone, a concrete demolition company and the inhabitants of two camps on Les Tartres. I met Mehdi[5] in October 2015 as I was walking through the vegetation on the southwestern fringe of the Tartres. He invited me into his garden, which he had been tending for 35 years, to show me his flowers and vegetable garden. His garden is in the clearing of profuse shrubby vegetation and is accessed through a concrete quarry from the north or by sneaking through a scrap iron gate on the Rue Toussaint Louverture. At that time, Mehdi gained direct access to his garden via the Rue Toussaint Louverture, so that the camp on the Avenue de Stalingrad, some 100 m away, formed a remote neighborhood for him. Mehdi explained to me that his relations with the inhabitants of the camp, whom he calls "Romanians," are distant, but rarely difficult.

In a subsequent interview, in April 2018, Mehdi told me how the camp's inhabitants were evicted the previous winter. He added that he was asked to move his garden as well, as the space it occupied was to be used for the demolition dump. He explained that he uprooted his trees, put away his tools and waited. The company then took possession of the space to turn it into a landfill and his old garden was turned over by the machines. Mehdi told me that he obtained permission from the head of the company to set up in a corner of the quarry, about ten meters away. He negotiated with the "boss" and, in exchange for the key to the large quarry gate, he watched the comings and goings and warned him when "Romanians" entered the quarry.

In June 2019, the slow installation of a Roma community on a new camp was taking place on the same plot of land as the one occupied 2 years earlier, about 20 meters from Mehdi's new garden. First of all, a small shack made of canvas and a prison appeared on the premises at the end of spring; then the camp was enlarged over the course of the summer. As the weeks went by, the discreet hammering of nails and the whirring of screwdrivers set the rhythm of the installations. One August morning when I set off to meet Mehdi in the quarry, the camp inhabitants watched me pass by. As I came back empty-handed because Mehdi's garden was deserted, two men were waiting for me, leaning on the concrete debris that mark out the path. They smiled at me and asked me in approximate French what I was looking for. I answered that I was looking for Mehdi, the gardener of the quarry, and I asked them if they had seen him pass by lately. One of them replied that he hadn't seen anyone. As the months passed, I noticed that the guard who questioned me holds the entrance to the camp and blocks the pedestrian access to the quarry. I entered again at the very end of the summer and went to Mehdi's garden. It was still deserted and his garden was beginning to disappear under the spontaneous vegetation.

Above all, wasteland is an available space, not only for living but also for gardening. Apart from allotment gardens (formerly known as worker's gardens), many gardens that are established on wasteland do not benefit from any institutional framework: they occupy land that has been left unattended in order to meet a need for subsistence or pleasure. In Mehdi's garden (mentioned above), spontaneous vegetation contributes to the development of a cultivable and productive environment (Fig. 2). A wooden fence, a hand-built fence or a wire mesh and a stratum of spontaneous trees border his plot, which also includes a hut built with salvaged materials, a

[5] For the sake of privacy, all names have been changed.

seating area with a table and a plot of cultivated vegetable garden. The walnut and fig trees that Mehdi received as a gift from his father, brought from Algeria, are planted in the ground and structure the space. In this garden, we can find different techniques for preserving and transforming resources: containers to keep rainwater and water the vegetable garden, doors and windows that are recovered and kept under cover of sheet metal and that can be used as walls or supports, a space to make a fire... Domesticated vegetation participates in the organization of the space by structuring the different uses of the garden.

The presence of resources on wasteland does not make it a space suitable for all uses. As far as housing is concerned, the vital need to sleep is barely satisfied by the wasteland's own resources. For these inhabitants, the wasteland is first and foremost an available space that allows them to build a daily life. People find there a sought-after place to live, and they enjoy the peace and quiet they are looking for. However, the presence of spontaneous vegetation produces other side effects. Indeed, the wasteland leaves people vulnerable to bad weather, as the sky is open and the soil can turn into mud with the rain, producing difficult living conditions depending on the seasons and over time. The inhabitants of community camps want built contexts to spread out, with flat, artificial ground enabling them to set up work sites. Some camp habitats have more material resources, where materials recovered here and there allow the construction of barracks. Solitary habitats are in a much more dilapidated state (Fig. 3). These habitats in wasteland areas are built in the

Fig. 2 Mehdi's second garden, Pierrefitte-sur-Seine. Photograph © C. Mattoug 2018

Fig. 3 Residual traces of a dwelling, Stains. Photograph © C. Mattoug, 2015

nooks and crannies, and their inhabitants take advantage of the materials found in the wasteland to build their habitat: ruins, walls, waste… materials present on the site contribute to the establishment of a living space.

These two uses have in common that they constitute niches inserted in the vegetation of the wasteland and are long term. The inhabitants take advantage of the resources of these spaces to create a living space in the spontaneous vegetation and to protect the privacy acquired in a neglected area. Indeed, the presence of vegetation constitutes a barrier that can be used by users. One of the first actions of the inhabitants is to create an envelope to protect their living space. This need for protection is described by the inhabitants as essential to avoid theft and aggression and to guarantee their peace and quiet. It is therefore also, thanks to the feeling of invisibility provided by the vegetation, that inhabitants feel safe (Fig. 4).

These living spaces thus flourish out of sight and take advantage of the invisibility produced by the plant cover to create economies and social organizations whose rules often escape the control of the public authorities. These organizations also take advantage of another essential resource of wasteland: the long time associated with urban development. This time saved is linked to indecision regarding land ownership (Serre 2017) or to the complexity of the political support for urban projects, which turn wasteland spaces into places of culture and the city's inhabited factories.

Fig. 4 Plants fence in southwestern fringe of the Tartres. Photograph © C. Mattoug, 2019

3.2 *From the Cohabitation of Uses to the Struggle for Living Space*

The different uses of brownfield sites do not simply coexist because relationships are established between users. The many power relationships that can be observed there testify to the role of these spaces as places to live. In many cases of cohabitation of uses, the need to share resources is a determining factor in the balance of power. Access not only to water but also to cultivable or habitable land is one of the first reasons for agreements or disagreements. The more specific situations of camp habitats present in the gardens appear to be symptomatic of this cohabitation of uses. In the Tartres zone, this cohabitation is prone to conflicts because the gardeners are afraid of being taken for camp inhabitants. In fact, they consider that the uses of the camps are related to the poverty of their inhabitants. Gardeners reside in adjacent communes, and they only come to the Tartres zone during evenings, weekends, holidays or when they are retired. They see this living space as a place of care, culture and conviviality. For them, the wasteland is not a place of necessity but a place to spend time with relative autonomy (Lion 2015).

This concern expressed by the gardeners appears to be a source of conflict when their living space is reduced. In the case presented above of Mehdi's garden, it is the operations of the concrete deconstruction company in the quarry which produce tensions between the camp inhabitants and the gardener. The object of the tension is quite simply the maintenance of the available land which is necessary for their subsistence practices. In turn, the users of the wasteland deploy strategies that enable them to settle and remain on the site. In 2018, Mehdi was negotiating to remain on the site in exchange for an informal role as a caretaker. In 2019, the inhabitants of the camp organized a guard at the entrance and controlled the passage of people. Lack of legal status makes several surveillance and control actions possible without any contribution from the public authorities. The inhabitants then deploy strategies similar to those of the gardeners: closing their living space and controlling access. These defense strategies demonstrate that the autonomy acquired on wasteland is always temporary.

The 4 years of the survey made it possible not only to isolate several situations of power relations between users in the Tartres zone, but also according to the progress of the urban project within the framework of the concerted development area (Fig. 5).

The balances of power that are established in these spaces are thus struggles to inhabit these resources that temporarily are freely accessible. They are silent struggles that have no banner, no party, no public mobilization… Their only form of expression is the materialization of their presence (Sassen 2006). The inhabitants' uses are nevertheless visible and are part of what is at stake for developers. The maintenance of certain uses to the detriment of others over this period of time shows that gardens without legal status have, on the whole, been able to remain on the site, whether they are communal or solitary. On the other hand, there have been

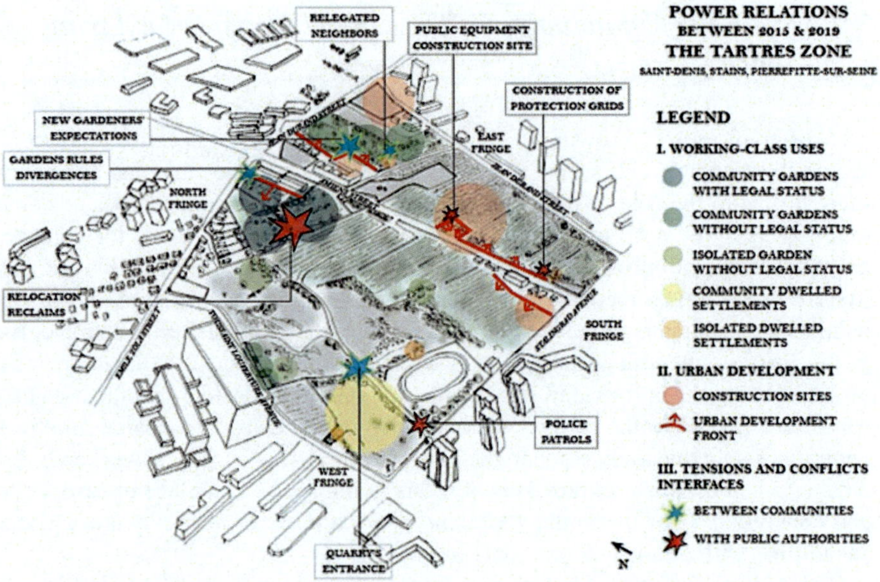

Fig. 5 Balance of power between users in the Tartres zone (2015–2019). Composed bird's-eye view map (C. Mattoug 2020)

almost systematic evictions from the settlements, as well as attempts to put an end to gray economy activities and to many social uses (such as gathering places for young adults). The other part of our survey of those involved in development highlights the reasons for maintaining certain uses compared to others.

4 Resources Made Public: Integration of Inhabitants' Uses in the Development of Open Spaces

4.1 Ambitious and Impeded Programming

When we take a detailed look at the programming of the Tartres zone development project, we can see that it has undergone many changes. In 2019, the Tartres zone was expecting the construction of 2230 housing units on the edge of a 16-acre "green heart." The program was part of a public policy objective of densifying the inner suburbs of Paris. The "green heart" in particular had been imposed by the state at the beginning of the 2010s, because it is located in the wake of the "Le Bourget noise cone:" under the air corridor for planes landing at Le Bourget airport, 4 km away. The third regulatory framework of the ZAC is linked to the Territorial Development Contract of the Territory of Plaine Commune, which points to the Tartres zone as a future intercommunal centrality, thanks to the creation of a cultural program. An educational and heritage farm is therefore to be set up on the former market gardening land.

Given the Tartres zone's position as an inter-municipal interface, the political responsibility for the ZAC is shared between the three mayors of the three communes of Saint-Denis, Pierrefitte-sur-Seine and Stains, while the technical responsibility belongs to the technical services of the Territory of Plaine Commune. This polycentric division between political and planning authorities resulted in divergences between the various players involved in the programming of the ZAC between 2015 and 2017. Faced with major difficulties in financing public facilities (schools and urban farms), the local elected representatives refused access to the development project to the inhabitants and associations. Thus, a local technician clarified the position of the elected representatives with regard to the inhabitants:

> I know that there are populations living there, that are a little marginal... But, if you want, we are not in a constituted sector, where there are already inhabitants. You see, it is true that the question of the inhabitants has always been complicated to discuss with the elected representatives. But we have always consulted, at least, with the people in the neighborhoods around [...]. But it's true that there hasn't been a political will to have a big meeting with the inhabitants.[6] [Interview with Sarah, technical officer at the Territory of Plaine Commune, April 19, 2016].

[6] Je sais qu'il y a des populations qui vivent là, qui sont un peu marginales... Mais, si vous voulez, on n'est pas sur un secteur constitué, où il y a déjà des habitants. Vous voyez, c'est vrai que la question des habitants a toujours été compliquée à aborder avec les élus. On a quand même fait de la concertation, on va dire a minima, avec les gens qui sont dans les quartiers autour [...]. Mais,

On the one hand, this interview excerpt testifies that inhabitants are marginalized in decision-making processes. On the other hand, it shows the representations of the elected representatives about them. For the authorities, these inhabitants have no legitimacy to be taken into account. As far as the position of elected representatives is concerned, it is confirmed by the severe budgetary restrictions to which the municipalities are subject. Accentuated by strong opposition between the elected representatives who do not agree on the importance to be given to this development project in the three municipalities and on the sources of funding, this phase in the programming of the ZAC ended in deadlock. In 2016, this zone of the Tartres was not only on the edge of the three municipalities for the elected representatives, it was also on the margins of the centralities of the three cities and its interest appeared to them to be secondary.

4.2 A Concession for the Management of Open Spaces for Inhabitants' Uses

The technicians of the territory then decided to set up a landscape coordination mission for the realization of a management plan for the "green heart," in order to ensure an overall coherence to the development of the ZAC and to avoid each elected representative concentrating only on his share. The other objective of using project management assistance was to minimize costs in the design, construction and future management of open spaces by pooling the use of inhabitants and associations. The landscaping agency selected in 2017 drew up a plan in the following months and worked on developing a management plan for the open spaces under concession. Faced with the many economic and political constraints and the many players who are taking part or wish to take part in the future development of the ZAC, the landscaping agency developed a project based on two main ideas: a revaluation of the heritage forms of market gardening and an economic rebalancing of the programs by distributing funds by reducing the public spaces of the "park" to "conceded" areas. The conceded areas are spaces for public use, but under private management, which are divided between market gardening plots allocated to an agricultural enterprise, allotment and garden areas and ecological compensation areas planned within the green core. This reorganization of the program of public spaces into open spaces also made it possible to insert the future allotment gardens in a local food loop[7] (BAL) on the eastern part of the ZAC des Tartres.

---

c'est vrai qu'il n'y a pas eu de volonté jusqu'à présent en tout cas, politique d'avoir un grand raout avec les habitants.

[7] The local food loop is an innovative public policy financed by the French State for economic and ecological local development.

In this project of development through conceded spaces, the gardeners occupying the Tartres zone in 2018 were able to take part in the BAL's circular economy operation, including a contributory farm, a composter and the running of an agricultural network. Martine, an agent of the Territory of Plaine Commune, explained the role that the gardeners would be able to play in this project for the local elected officials:

> They would like the growers, those in the gardens, the gardeners, they would like to be able to sign up, if there was a food loop… Knowing that, even so, the contract on the gardens is that they are not places of sale… On the other hand, there could be a barter, there could be seed exchanges. There are quite a few gardens in the garden city as well… There's something to be done.[8] [Interview with Martine, technical officer for the Territory of Plaine Commune, 12 December 2018].

In this way, the uses of the garden's inhabitants were seen as part of the development of future green spaces. The community gardens were already being promised a continuation in 2018 thanks to their assumed role as managers of green spaces. Like Mehdi, who agreed to take on the role of a caretaker for the deconstruction company, the gardeners must agree to be managers of the "green heart" in order to hope to remain in the Tartres zone.

In addition to the implementation of this circular economy loop, other criteria of municipal management policy governed the acceptability of certain uses. The very first criterion depended on the regularity of the uses and their desirability. Particularly in Stains, the policy of "reconquering allotment gardens" confirmed the important distinction made between garden use and housing use by the public authorities:

> That's the first thing: supporting families to get out of precarious housing. And the second question for the reconquest, is to clean up the management…[9] [Interview with Martine, technical officer for the Territory of Plaine Commune, 12 December 2018].

Martine specified that this measure was intended not only to reintegrate the existing gardens into associations and the allotment garden federation, but also to ensure the way in which the cultivated plots were: re-housing the people living in the gardens, reducing the use of plant protection products, redrawing the boundaries of the plots, and opening the gates of the gardens during public events. These criteria for regulating and maintaining the use of wasteland by inhabitants are evidence of the standardization processes that are taking place in the development of wasteland. These criteria meet the objective of control and monitoring sought by local authority decision-makers and technicians, but also the objective of taking advantage of them in the future development of these brownfield areas.

---

[8] Ils aimeraient que les producteurs, ceux qui sont dans les jardins, les jardiniers, ils puissent s'inscrire, s'il y avait une boucle alimentaire… Sachant que, quand même le contrat sur les jardins, c'est que ce ne sont pas des lieux de vente… Par contre, il pourrait y avoir un troc, il pourrait y avoir des échanges de semences. Il y a pas mal de jardins à la cité-jardin aussi… Voilà, il y a un truc à faire.

[9] Le premier truc, c'est ça: c'est l'accompagnement des familles pour sortir de l'habitat précaire. Et la seconde question pour la reconquête, c'est assainir la gestion…

5 Making Wasteland Desirable: Translating Resources into Figures for a Territorial Dynamic

5.1 *Desirable Uses of Wastelands*

Making a space or the characteristics of a space desirable is an essential part of a development project and, more generally, of any form of anticipation. This "idealization" (Boutinet 2012, p. 5) is a form of projection toward a desirable future, in a demonstrative and coherent approach that is guided by a search for control (Soubeyran 2014). Thus, brownfield landscaping projects consist in translating the data of the project site into a set of desirable figures. This translation takes place on the basis of the components present and, in the case of a landscape project, the natural resources (plant cover and cultivable soil) of the wasteland form initial data suitable for projection. However, there is a series of components or characters that must be transformed and translated to reach the project figure. In the case of the Tartres zone, the closed character of the fallow spaces appears to be a problem for the decision-makers, as it contributes to the image of uncontrollable spaces. In the same way, the lack of clarity and determination of uses between housing and gardens contributes to the image of social and sanitary precariousness of these living spaces and this is fought against by political decision-makers. Also, the apparent anarchy in the organization of certain gardens, even though they are mostly located on public land, contributes to making them undesirable and this is due, among other things, to the presence of vegetation and the indeterminate nature it produces.

However, the nature of the wasteland is appreciated by the public authorities and developers. In particular, the figure of the open, shared garden meets the interests of developers, firstly because it produces the image of a public, open space which can be used for environmentalist property marketing (Fig. 6).

One of the roles of the landscaping project management in the programming of the ZAC des Tartres was precisely to make the inhabitants' uses desirable for the political decision-makers. By relying not only on the many constraints in terms of regulations, but also on the strong economic restrictions of the local authorities, the planners indeed see the uses of the inhabitants on the wastelands as desirable supports for the organization of the future spaces. Nevertheless, this translation of garden use into a project figure does not go without the process of standardization and diversion which gardeners must adapt to. Indeed, gardeners are called upon to accept the frameworks of open space programming by becoming guardians of public spaces (Mattoug and Bélec 2020). These processes not only modify the forms of these living spaces, but also question the maintenance of users over time. For planners and political decision-makers, the forms of nature present on wasteland appear as a set of malleable resources, making it possible to undertake a revaluation that goes beyond the framework of the ZAC, but which is deployed throughout the territory.

Fig. 6 Real estate development for the sale of housing lots on the Avenue de Stalingrad in Saint-Denis © C. Mattoug 2019

5.2 Urban Margins as an Object for Metropolitan Upgrade

Beyond the development perimeter of the ZAC des Tartres and the integration of garden uses, several uses of the wastelands are envisaged as being inhabited uses conducive to the development of open spaces. Indeed, the landscape plan proposed in 2018 by the landscape architect contracting authority envisages the maintenance of 4 acres of "traversable" vegetable-growing agricultural land. The idea is that access to this land should remain public, thus obliging the farmer to plan his production despite the risks of theft. In the same way, two temporary occupations are envisaged by the planners during the realization of the ZAC whose role is to undertake different forms of "inhabitant contribution." Firstly, a household waste collector in Stains should enable the production of methane in a compost bin. This energy that is planned will heat greenhouses for vegetable growing and will enable inhabitants to come and participate in this production. These projects are part of the continuity of a temporary landscape architect occupation that took place from 2017 to 2019 on the Tartres, the aim of which was to liven up a vacant plot of land by introducing schoolchildren to the practice of gardening. A few hundred meters away, the creation of the Saint-Denis urban farm offers various cultural and social activities related to vegetable growing, the history of agriculture, food and ecology… A whole territorial dynamic is emerging in a craze for this domesticated and productive vegetation.

The aim of this engineering is thus to make vegetable growing a tool for urban and territorial development that goes far beyond the uses of the inhabitants. The presence of domesticated vegetation on wasteland becomes the driving force behind a productive ecologization which is manifested by the development of open spaces. This trend is evidence of a reconsideration of spaces on the margins of cities. These wastelands, which have remained unbuilt over the years and which have long been the object of marginalization, are now becoming the driving force for a renewal of social and professional practices, thanks to a renewal of representations. The tension between the uses of wastelands in this new metropolitan development illustrates the capacity of capitalist cities to reintegrate marginal phenomena as soon as they develop a new market value (Lefebvre 2000). From a space on the fringes to an economic resource of changing territories, the different forms of wasteland vegetation thus take on value with the evolution of representations.

6 Conclusion

The results of this double survey have therefore made it possible to grasp the extent to which the resources specific to wastelands are used not only for inhabitants but also for landscaping projects. By supporting a wide variety of uses and allowing such a large number of relationships, these resources also appear to be the object and theater of power relations between stakeholders. The horizon of democratization implied by the evolution of professional practices does not mean, therefore, that the processes of marginalization have disappeared: local uses are being reconsidered and are integrated into a management plan for the projected spaces, but this does not correspond to an actual distribution of decision-making power. Thus, these uses are taken into account, but the users are not necessarily and must themselves undergo processes of normalization.

However, it is necessary to nuance what we are saying. Relying mainly on an anthropological survey, we have grasped some modulations in the forms of living in these wasteland spaces which are anchored in social reality. On the other hand, the monitoring of a development project that runs until 2027 also has several limitations. Indeed, we have grasped the broad outlines of its programming at a given time, upstream of its operational phase, and there is nothing to suggest that the project will remain the same in the years to come. In this sense, the capture of social and professional practices on wastelands by an anthropological survey mechanism is similar to an ethnographic core sampling (Scapino 2016): we are inserted in a given time in a place of complexity but, above all, in the multiple reality of many points of view. Finally, we can emphasize that the ambivalent character of these resources of urban wasteland is based above all on the representations that they provoke in the inhabitants and developers. The discrepancy that we note therefore depends on the different reasons for the struggle for living spaces: living places today, anchored in the memory of places in the past or objects of desire awaiting transformation... These resources of wastelands are conducive to the creation of many images and figures for those who express the desire for them.

References

Blanc N (2016) Les friches urbaines, de la marge à la production d'espace: la trame verte urbaine [Urban wastelands, from the margins to the production of space: the urban green grid]. In: Grésillon E, Frédéric A, Bertrand S (eds) La France des marges, Armand Colin, pp 188–197

Boutinet J-P (2012) Anthropologie du projet [Project anthropology]. PUF, Paris

Brun M, Di Pietro F, Martouzet D (2019) Les délaissés urbains: supports de nouvelles pratiques et représentations de la nature spontanée? comparaison des représentations des gestionnaires et des habitants [The urban neglected: supports for new practices and representations of spontaneous nature? comparison of the representations of managers and inhabitants]. Nouv Perspect En Sci Soc 14:153–184

De Biase A (2014) Hériter de la ville. Pour une anthropologie de la transformation urbaine [Inheriting the city. For an anthropology of urban transformation, Paris, Editions donner lieu.

Demailly KE (2015) Jardiner les vacants. Fabrique, gouvernance et dynamiques sociales des vacants urbains jardinés du nord-est de l'Île-de-France [Gardening wastelands. Factory, governance and social dynamics of urban gardened vacancies in northeastern Paris Region]. PhD thesis, Université Paris 1 Panthéon-Sorbonne

Desjardins X (2008) Gouverner la ville diffuse : la planification territoriale à l'épreuve [Governing the diffuse city: territorial planning put to the test], PhD thesis, Université Paris 1 Panthéon-Sorbonne

Dorso F (2016) Tours et détours de la mauvaise ville. Paradoxes des informalités et de l'action sur l'informalité en contextes Nord et Sud » [Tours and detours of the bad city. Paradoxes of informality and action on informality in North and South contexts]. Study seminar : Pratiquer l'informalité aux marges de la précarité, CAUE 77, 14 avril 2016

Dorso F (2008) La part d'ombre : transactions et conflits entre les usages informels et les opérations de rénovation de la muraille de Théodose II à Istanbul [The shadowy side: transactions and conflicts between informal uses and the renovation operations of the Theodosius II wall in Istanbul], PhD thesis, Université Marc Bloch (Strasbourg)

Dubeaux S (2017) Les utilisations intermédiaires des espaces vacants dans les villes en décroissance [Intermediate uses of vacant space in shrinking cities], PhD thesis in geography, ENS

Franchomme M, Bonnin M, Hinnewinkel C (2013) Does biodiversity influence territorial planning? Towards a Greening of Territories. Développement durable et territoires 4:1 | April 2013. https://journals.openedition.org/developpementdurable/9749. https://doi.org/10.4000/developpementdurable.9749

Gatta F (2018) (Contre) pouvoirs urbains [(Counter) urban powers]. Ed donner lieu, Paris

Hatzfeld H, Hatzfeld M, Ringart N (1998) Quand la marge est créatrice : les interstices urbains initiateurs d'emploi [when the margin is creative: urban gaps as job creators]. FeniXX, Paris

Lefebvre H (2000) Espace et politique [Space and politics], [1972]. Editions Anthropos, Paris

Lion G (2015) Incertaines demeures [Uncertain homes], Paris, Ed. Bayard

Mattoug C, Bélec M (2020) Ressource et épuisement des espaces spontanés dans la fabrique de la ville [Resource and exhaustion of spontaneous spaces in the city factory]. In: Antonioli M, Devron G, Kaufmann V, Pattaroni L (eds) Saturations. Individus, collectifs, organisations et territoires à l'épreuve, Editions Elya, Grenoble, pp 149–164

Pichon P (2007) Vivre dans la rue. Sociologie des sans domicile fixe [Living on the street. Sociology of homeless people], Paris, Aux lieux d'être, 304 pp

Ricoeur P (1998) Architecture et Narrativité [Architecture and narrativity]. Revue Urbanisme 303:44–51

Sauvadet T, Bacqué MH (2011) Working-class spatial practices. Espaces et Sociétés 144–145, 2011–1/2:7–13

Sassen S (2006) The repositioning of citizenship and alienage: emergent subjects and spaces for politics. Migration, Citizenship, Ethnos, pp 13–33

Scapino J (2016) De la friche urbaine à la biodiversité : Ethnologie d'une reconquête (La petite ceinture de Paris) [From urban wasteland to biodiversity: ethnology of a reconquest (La petite ceinture de Paris)]. PhD thesis, MNHN

Serre M (2017) Le tiers foncier : ressources, controverses et expérimentations : l'exemple de la ville de Marseille [The land third: resources, controversies and experiments: the example of the city of Marseille]. PhD thesis, Université d'Aix-Marseille

Soubeyran O (2014) Pensée aménagiste et improvisation : l'improvisation en jazz et l'écologisation de la pensée aménagiste [Planning thinking and improvization: jazz improvization and the greening of planning thinking]. EAC, Paris

Tonnelat S (2003) Interstices urbains Paris—New York : entre contrôles et mobilités, quatre espaces résiduels de l'aménagement [Urban Interstices Paris—New York: Between Controls and Mobilities, Four Residual Planning Spaces]. PhD thesis, Université de Paris-Est Créteil et City University of New York

Toublanc M, Bonin S (2019) La fabrique politique et habitante des limites urbaines : études de cas à l'île de La Réunion [The political and inhabitant factory of urban boundaries: case studies in Reunion Island], Développement durable et territoires 10:2 | Juillet 2019. https://journals.openedition.org/developpementdurable/14503

Cécile Mattoug is an Architect and Urban Planner, and a PhD student at the University of Paris 1 Panthéon-Sorbonne (France), UMR Géographie-cités. Her research focuses on the balance of power between urban wasteland uses and the processes of urban production. She is a founding member of the "Inter-friches" (inter-wastelands) project and of the nonprofit organization "Vacance Collective" (designers and builders), and studies professional practices in urban planning.

Long-Standing Wastelands

Latency, Transactions, and the Role of Green Settings on Two Urban Wastelands in Tirana (Albania) and Istanbul (Turkey)

Stela Muçi and Franck Dorso

Abstract This chapter explores a paradox. How to explain the long-term existence of inner-city wastelands in cities characterized by densification, sharp land pressure, and large-scale urban projects? In Tirana (Albania), two large areas of meadowland have been abandoned to informal uses since the end of the communist regime and the extension of the city center in the 1990s. In Istanbul (Turkey), a strip of land adjacent to the old city walls has also been left partly as a green wasteland, despite being registered as a heritage site since 1985. Drawing on data and field surveys covering several years, we propose explanatory factors that contextualize the issue of vegetation, which will be the focus of our analysis. Finally, we propose potential avenues for broader application based on the different meanings and uses of these forms of vegetation in the city, as understood by the different actors involved, and explores some contradictions of action and intervention, in operational terms.

Keywords Tirana · Istanbul · Informality · Long-standing wasteland · Latency · Production of nature

1 Introduction

This chapter explores a paradox. How to explain the long-term existence of inner-city wastelands in cities characterized by densification, sharp land pressure, and large-scale urban projects? In Tirana (Albania), two large areas of meadowland, each covering almost 62 acres, have been abandoned to informal uses since the end of the communist regime and the extension of the city center in the 1990s. In Istanbul (Turkey), a strip of land nearly 5 miles long, adjacent to the old city walls, has also been left partly as a green wasteland, given over to long-standing

S. Muçi
APPUII – Lab'Urba, Paris, France
e-mail: smuci.appuii@gmail.com

F. Dorso (✉)
UPEC – Lab'Urba – IFEA, Paris, France
e-mail: franck.dorso@u-pec.fr

F. Di Pietro and A. Robert (eds.), *Urban Wastelands*, Cities and Nature,
https://doi.org/10.1007/978-3-030-74882-1_7

informal uses, despite being registered as a heritage site since 1985. Drawing on data and field surveys covering several years, we propose some explanatory factors that contextualize the issue of vegetation, which will be the main focus of analysis in this chapter.

The indecision observed in situ reflects first an organized latency on the part of the institutional actors, rather than abandonment of the territory. Indeed, transactions are implicitly agreed between institutional and informal actors to maintain a status quo that seems to benefit all parties, because it permits practical regulation in the urban, economic, and social spheres. We undertake a more detailed exploration of the specific role of greenery and nature in this resilience. Vegetation protects informal activities by hiding them. It is also useful for informal market gardening, grazing, gathering, and even construction practices. In symbolic terms, it forms part of a continuity that can play a positive role in fostering the acceptance of rural life, or a negative role by painting that life in an unattractive light. And finally, we note that the urban operations planned for the near future retain some of these green spaces, while imposing significant changes. To summarize, we propose potential avenues for broader application based on the different meanings and uses of these forms of vegetation in the city, as understood by the different actors involved. In operational terms, these discussions constitute an exploration of the contradictions and incoherencies of intervention, regulation, and—more broadly—of action, in the urban realm.

2 Two Long-Standing Urban Wastelands

The first field location stands on the edge of Tirana's inner city, a mile and a quarter from the central square. Bounded to the south and north by two of the capital's main highways, it has good road access and a strategic geographical position. The area gets its name, Stacioni i Trenit, from its proximity to the railroad station, abandoned a decade ago. The unoccupied greenfield site consists of two large stony urban meadows, each covering almost 62 acres, located in the middle of an unplanned residential zone. Other meadows of this kind are scattered throughout the urban fabric, but these two attract attention because of their central location and the multiple informal uses to which they are put.

The ground is covered by a fairly low layer of wild vegetation and by a few lines of trees situated close to the buildings, often in fenced plots. The land is extensively used as a crossing space, with a dozen desire lines showing pathways to the houses, to the informal market, or to the new boulevard (Fig. 1).

The uses vary according to the terrain and the times of day or year. They can attract up to two or three hundred people a day, with a balance of ages and sexes. One of the areas is mainly used for informal trading, whereas the other is dedicated to domestic activities that overflow beyond the walls of the houses: cooking, sewing, gardening, etc. There are also temporary activities that take place across the area: socializing,

Fig. 1 Aspects of vegetation in Stacioni i Trenit, Tirana © Muçi and Dorso 2018–2020

petty crime, and play areas for adults or children. These practices are subject to well-organized local control measures and reveal the overlaps in status and usage between the private and the public. In winter and in rainy periods, activity declines. One of the activities that continues throughout the year, however, is grazing. Flocks of sheep and goats, and a few cows, accompanied by their herders, spend several hours a day grazing, turning the area into an urban meadow. The built-up space is also diverse. Different forms of urban production coexist within the same territory: spontaneous buildings (precarious informal housing, lightweight structures), semi-legal constructions (permanent buildings that flout legal standards), and official constructions (big real estate projects and urban infrastructures). The urban composition of this part of the city reflects the social, political, and economic changes that have marked the country as a whole in the last half-century.

Politically, Albania has shifted from a "socialist" dictatorship to the current system of parliamentary democracy. During this transition, the instability of the institutions and internal migrations has transformed the areas on the edge of the city, traditional spaces of friction between the rural and the urban. In addition, the whole country is marked by complex land conflicts: dual property titles, property rights granted to former owners dispossessed by the communist regime, and ongoing legalization procedures for existing occupants. Stacioni i Trenit is a typical example. In 1945, when the Communist Party came to power, the cultivated land in this part of the city was nationalized. The state erected greenhouses for fruit and vegetables and a big livestock market, where farmers came several times a week to sell their produce. With the fall of the socialist regime in 1990, two phenomena occurred simultaneously: the start of procedures to restore land to dispossessed former owners, and illegal land occupation by populations from the rural exodus (Triantis and Vatavali 2016). Today, almost 30 years later, the land ownership situation in this area has still not been regularized. Nonetheless, this is where the government is initiating the "New Boulevard," one of the biggest projects in the new Tirana 2030 development plan. The objective of this project is to bring about economic and spatial transformation in this district on the edge of the city center. The extension of the city's central highway (Boulevard Zogu i Pare) will be followed by the construction of eco-neighborhoods, buildings for public, commercial, or mixed use, and public spaces and squares. This large-scale intervention, in the emulation of the models of European urban production, is intended as an antidote to the informal settlements and economic stagnation of Tirana's periurban areas. Arguments about modernization and metropolitanization are used to justify the high cost of the project and the mass demolition of houses, which has already begun with the start of construction work. The development of this project now challenges the resilience of the green wasteland located within the project zone.

In Istanbul, the space consists of a strip of land almost 5 miles long, made up of three parts: the land wall constructed by Theodosius II in the fifth century, which encloses the western side of the old city; a strip outside it, thirty to three hundred feet deep, consisting of large plots without buildings, cemeteries, road intersections, and more recently a few parks; a much shallower interior strip made up of alleyways, small waste plots, gardens of informal houses, and a few buildings running up to the wall (Fig. 2).

Overall, this is municipal land, managed by the municipality (either the city or Fatih district at different periods), together with central government and within the framework of two sets of laws, the one on heritage preservation, the other on administrative decentralization. The actors concerned come from a range of scales: international (UNESCO, transnational organizations), national (Ministry of Tourism, prefecture, chamber of architects), and local (municipality, prefecture, construction and public works companies, local civil society organizations, academic experts, and the day-to-day users of the site).

The wall performed its military role until the 1850s. After this, maintenance declined and informal activities moved in: grazing, market gardening in the southern part, Rom taverns, and small trades (tanneries, glassworks, and carpentry). In the phases of urban expansion in the 1950s and then the 1970s, other practices were introduced, at a time when the wall had ceased to be a boundary within the city: informal housing, unofficial passages, shortcuts to transportation nodes, socializing, petty crime, and concealment. These practices were broadly equal, with variations from one activity to another. Over the periods of study considered, we counted between 50 and 150 residents, and from a few hundred to a few thousand day-to-day activities, depending on the season. In 1985, the city became a UNESCO World Heritage Site. A few operations were undertaken to make safe and rebuild some of the gates. However, despite UNESCO's threats in 2003, no overarching plan was

Fig. 2 Aspects of vegetation on the city walls of Istanbul © Dorso, 2007–2016

proposed for the site. The most visible housing and most of the craft workshops were removed, but they represented only part of the activities present. The municipality sporadically showed that it had the capacity to intervene but, overall, the informal practices remained in place on the land wall. Local self-regulation of the different activities in place developed over time, adapting as necessary and continuing. From the 2010s onwards, development operations on adjacent neighborhoods affected certain points, as did the establishment of municipal parks.

The green wasteland areas here are of four types. There is the waste ground in the exterior and interior strips: meadows, slopes, forgotten cemeteries, copses located at the base of the towers or in isolation. Then there is the greenery within the wall: uneven surfaces covered with grass, copses, and trees. These two landscapes have been present for more than a century. On the northern part, behind cottages built in the 1950s on the inner side of the wall, there are small gardens in various degrees of organization. Finally, the infilled ditches in the whole southern part of the space form an extensive market gardening area tolerated under semi-formal agreements with the municipality. In addition to this, at certain points there are planned and regulated green spaces: cemeteries (Muslim and a few Orthodox Christian) and municipal parks.

Surveys in Tirana and Istanbul both drew on the paradigm of the social transaction (Rémy et al. 1978) to analyze how the actors negotiate, discuss, and face off around an issue of territorial control (Rémy 2015). Some conflicts proved insoluble: the actors overcame these without resolving them by reaching practical and temporary compromises (Gibout et al. 2009). These compromises remain revocable, if tension is revived by an external event such as the arrival or departure of one of the parties to the transaction. This series of transactional sequences provides an approach to the question of the medium and long term. In the case of informal arrangements, it is also a way to tackle the question of the legitimacy of acts and actors, who must sometimes make their arrangements in the realm of the implicit and the unsaid. The methodology of the fieldwork is shown in Box 1.

Box 1: Tools and Fieldwork

The data used in this chapter were obtained in several successive phases of research. In Tirana, two field studies, in April 2018 and March–April 2019, explored explanatory factors relating to the resilience of the Stacioni i Trenit wasteland, with a third survey planned for spring 2020. In Istanbul, there were three stages of research, in 2000–2001, 2005–2008, and then 2010–2016, which studied the evolution of the landscape and the uses of the land wall.

We focused on producing first-hand data, because there are few existing data on the two field locations. In the enquiry phases, we employed standard tools (surveys, observations, in situ or institutional interviews). In Istanbul, 100 interviews were conducted with users and residents of riverside neighborhoods between 2000 and 2016, in some cases with follow-ups with regular informants. On the institutional level, around 30 interviews were conducted. Some of them were carried out at the planning and heritage protection services of the central municipality and the district municipality of Fatih, with the municipal police of Fatih, as well as at the Center of Metropolitan Planning. Actors from civil society, various associations, NGOs, architects, town planners, and experts were also interviewed. In the current investigation in Tirana, 30 interviews have been conducted so far on the wasteland and the neighboring districts with users, bazaar and informal market sellers, neighboring businesses and shops, residents of nearby neighborhoods, and residents of the district that is expected to be demolished in the north of the wasteland. Fifteen interviews were carried out on the institutional level (municipality officers, lawyers, town planners, journalists, and associations).

Quite soon in our study, however, some particularities of the two sites prompted us to develop specific tools. On an institutional level, access to formal documents and agents' speeches was rather complicated in Tirana. This led us to conduct formal and informal interviews and to consult official and confidential documents—sometimes making it impossible to quote real names in order to protect our sources. Similar situations were encountered in Istanbul at the start of the investigations and during 2007–2008. In terms of in situ investigations, the semi-legal or illegal nature of certain situations, distrust of institutions, and the search for discretion or practical difficulties (illiteracy, difficulties with framed interactions) have led to the development of mixed tools (photo questionnaires, toolkits), to organize collective walks, and to introduce stages of long-term immersion, in the style of participant observation, over different timescales (day, week, seasons, and year).

3 Forms of Latency: Between Organized Immobilism and Institutional Abandonment

In Tirana, the context of political and state action provides a first analytical framework through which to explain the long-term persistence of these wastelands. The (non)-action of the state in the spontaneous production of the city over the last two decades can be divided into two periods:

- The first one was marked by major political transformations and state tolerance of informality. In the early 1990s, the transition between a hyper-centralized state system and a parliamentary democracy was initially characterized by an institutional void, which gave rise—in urban terms—to spontaneous urban production on a significant scale: over-occupancy and illegal occupancy of space and large-scale informal construction (Lubonja 2015). One of the reasons behind this phenomenon was the strong desire of Albanians to become homeowners, in the aftermath of the "communist" era (Pojani 2011). However, city production was also the result of a significant phenomenon characteristic of this first post-communist period: the rural exodus. Small farmers and provincials, forgotten and neglected by the new liberal system, migrated to settle near the urban centers. In both cases, the newborn political parties supported, or even encouraged, mass population movements, and tolerated informality. The Democratic Party was the first to encourage the populations from northern Albania to settle in the outskirts of Tirana, in order to consolidate its power and build an activist electoral base close to the capital. The second wave of migration to the urban centers took place in 1997 following an economic and political crisis that affected the country and for a few months brought it close to civil war. The Socialist Party continued the same "laissez-faire" policy, leading to the creation of large numbers of informal neighborhoods. The spontaneous production of Albania's cities in the 1990s combined the informalities of activities and construction with political informality (Jacquot and Morelle 2018) that reflected informal arrangements with and within legal institutions. These informal processes, in their different scales, nevertheless led to the establishment of an economic and social equilibrium between the different actors: the state, vulnerable populations from the rural world, and the former landowners (Pojani 2013).
- The second period was characterized by the recentralization of power and greater state control. Paradoxically, however, it is this phase that explains the situation of latency in Stacioni i Trenit. Toward the end of the 1990s, the state wished to re-establish control over the bottom-up production of the city, by proposing new land laws and legalization policies aimed in particular at disadvantaged social groups. The creation of the new body charged with running the regularization process (ALUIZNI[1]) marked a new era characterized by bureaucratization, along

[1] *Agjencia e Legalizimit, Urbanizimit dhe Integrimit the Zonave dhe Ndertimeve Informale:* Agency for Legalization, Urbanization, and Integration of Informal Areas and Buildings (author's translation).

with corruption and clientelism (Aliaj 2009). Political informality shifted: corrupt practices did not disappear but were concentrated in the relations between the state and private construction firms. However, these measures were not applied uniformly across the whole territory. In the area we have been studying, in fact, land conflict remains unresolved. The state let the regularization processes drag on. This prevented the development of the territory and for several years made the district unattractive to private investors, despite its position in a city in the throes of a real estate boom. Today, the construction of a big infrastructure has considerably increased land values here and is opening the way to private investors and developers.

In Istanbul, the municipality succeeded in 1985 in getting the city registered as a UNESCO World Heritage Site, for four major components, including Theodosius's wall. This process took place at a time when the military were restoring power to the civilians, after the 1980 coup d'état. The government embarked upon a neoliberal-inspired policy of national economic development. For its part, the city of Istanbul hoped that the UNESCO label would help develop the tourist industry. Restoration work began in 1986 and 1987, but the municipality was subsequently content with partially removing the informal settlements most visible from the beltway that runs alongside the structure, and with reconstruction projects that attracted criticism from specialists in archaeological conservation. There was no global conservation plan. The city has a difficult relationship with its ramparts. In the 1960s, and sporadically afterwards, plans were made to demolish the wall and thereby gain more land for construction. The only moment of actual investment takes place one day year, on May 29, to commemorate the conquest of the city by the Ottomans, a warlike celebration. In terms of the historiography of the city, the site is a reminder of an earlier non-Turkish period that remains a delicate issue (Pérouse 2003). The seesawing between the obvious neglect of the structure and this brief and fervent celebration highlights the sharp tensions that still exist within a complex process of identity construction.

At the end of the 2000s and in the next decade, a few measures temporarily interrupted this neglect. In the run-up to "Istanbul cultural capital of Europe" year in 2010, proposals were made for the development of high-rise hotels or prestigious stores. They stayed on the drawing board. At this time, however, the municipality gave a more active turn to its "urban regeneration" policies, which have been interpreted as forced population change operations designed to favor social categories assumed to be close or favorable to the AKP Party (Montabone 2013).[2] The demolition and reconstruction of the Sulukule district wrought profound alterations (Erdi Lelandais 2014), with collateral impact on the uses of the wall, though it did not affect the landscape of grassy slopes running along the bottom of the ramparts. The same process then took place in the Ayvansaray district, right in the north. In the last couple of decades, Istanbul municipality has also pursued the development of urban parks, among other things, with the aim of promulgating a way of life and family uses consistent with the conservative values preached by the AKP (Karaman 2017). A few

---

[2] *Adalet ve Kalkınma Partisi*, Justice and Development Party.

parks of this kind have been developed on the outside of the wall, including Topkapi Park, home to the Panorama Museum (which also celebrates the capture of the city on May 29, 1453). However, these actions are part of an overarching project to develop or preserve the wall. Their approach to it is utilitarian, exploiting its capacity to bring added value to a neighboring project. Up to 2016, recent projects had not modified this institutional neglect, which has permitted the long-term existence of large green areas around and within the land wall.

4 Implicit Transactions and Practical Compromises: Mutual Interest in Maintaining the Status Quo

Let us remain in Istanbul to explore the deeper reasons for this neglect, and for the status quo that it maintains. By cross-referencing data on the spatial configurations, identifying practices, and conducting interviews with the actors (institutional and informal), we were able to reveal mechanisms of two kinds.

First, the informal practices that take place along the wall perform, in their diversity, a regulatory role in several respects. We identified five: economic regulation (by hosting informal activities that represent all or part of household incomes); urban regulation (by offering an integrative halfway house in the geographical and social mobilities of rural or foreign migrants); housing regulation (by offering the possibility of squats or informal housing, a function that is currently in decline); psychosocial regulation (by offering places where people can escape from or transgress the social and moral order of the surrounding traditional working-class neighborhoods); and finally symbolic regulation (by offering a medium for the construction of a normative discourse on the good and the bad city). The form of the area, in particular the vegetation, plays an important role in containing these acts and these functions, a point we will return to later.

On the other hand, the actors are aware of these processes to different degrees. The development of informalities has been analyzed as a form of structural regulation which helps governments that cannot or will not implement public economic urban policies of significant scale (Brown and McGranaham 2016). The tolerance and maintenance of these regulations by the informal sector are thus not always the result of incompetence (often an assumption made by experts and developers), but rather local forms of arrangement between the actors (which also maintain existing inequalities and balances of power). In our two decades of study on the site, we thus saw practical compromises established and negotiated over different timescales. Nonetheless, such arrangements create an underlying situation that is problematic for the actors concerned. They remain illegal and illegitimate, especially in a city where municipal policy is aligned with the international standards of world cities. The problematic nature of these compromises applies right down to individuals in municipal departments, who risk their professional legitimacy. That is why protective mechanisms are developed to mask these compromises and prolong the status quo. At the micro level, the arrangements remain in the realm of the implicit and the unsaid (that's why it is sometimes useful to be able to conduct an off-the-record survey alongside the main survey). In some cases, the institutional actors also choose not to know what is happening on the wall (Dorso 2019). At the meso level, a transaction of secondary importance (heritage preservation) conceals a fundamental—but unspoken—transaction, relating to social harmony (Dorso 2018).

In Tirana, the status quo on the green wasteland of Stacioni i Trenit is largely attributable to the problem of the site's land ownership status. The first transactional level developed at the institutional level, between the public authorities and the plot owners. This transaction stretches back a long time, from collectivization through to re-privatization in a context of urban transformation and expansion (INSTAT 2010). The land issue has been settled over different timescales and with different results in the rural areas, the tourist areas in the south, and the denser urban zones. In the interviews conducted at the municipality, the operational personnel claim that the government could have regularized the issues around the central plots in Tirana as quickly as elsewhere. The communist state had kept precise records, and neither officials nor residents are convinced by the argument of technical difficulty. Especially beyond the beltway most of the comparable sites have been regularized. In Stacioni i Trenit, however, the process has remained frozen. The government and the municipality talk about difficulties in order to maintain these areas in a state of indecision that ultimately benefits their plans—in particular the Tirana 2030 plan, which includes big development projects largely allocated to private companies (Grimshaw in our sector). Indeed, as long as the regularization processes are not completed, the legal compensation mechanisms for expropriation are less costly.

In parallel with the global transactional level, where the situation has long been frozen, another level has emerged at the local scale. In our field surveys, we were able to see that the green wasteland and the presence of informal activities were in no way an indication of neglect or abandonment. The tract owners tolerate the informal occupants in exchange for a protective role: the activities occupy the land and prevent other informal—more permanent or uncontrolled—uses. This is what we learn from Mirela,[3] an urban planner at the city council of Tirana and owner of a plot of land awaiting regularization in Stacioni i Trenit:

> I know the history of this wasteland […]. My parents were landlords of one plot. We are still waiting for the documents of ownership, procedures have been dragging on for twenty years […]. This is an area of great potential for urban development. The government is dragging legalization procedures because there are interests behind. In the meantime people have found tacit agreements, although conflicts have been present over the years. My brothers had to defend our land by force. Others preferred to rent their plot to shepherds who exploit it for grazing sheep or cows, while fending against possible foreign occupants.

An experiment conducted in April 2018 around the co-development of a soccer field for local children revealed these arrangements: the tract owners were afraid that permanent occupants might settle and prevented the project. So far, these local compromises have greatly contributed to the persistence of the status quo.

[3] The first names have been changed.

5 The Positive Roles of Vegetation in the Long-Term Resilience of the Wastelands

A comparison of the two spaces shows convergent mechanisms in terms of latency and transactions between actors. These convergences allow us to contextualize the explanatory factors in terms of vegetation and nature in the city, elements that we will now explore. The approach to nature, to greenery, the dynamics of life and biology, will be related here to their interactions with human and social uses.

Protection

The wall in Istanbul hosts multiple forms of vegetation. Shrubland and small woodland copses, as well as high grass on the uneven tracts, provide hiding places for practices that are prohibited or seen as immoral (trafficking, meetings, drinking, sexual activities, housing, unauthorized settlements, isolation, and concealment). The vegetation provides protection from prying eyes (from the internal streets and the exterior beltway), as well as from the interior itself, by maintaining separation between activities that largely take place in rituals of mutual avoidance and coexist in a broadly effective self-management. It is in the interest of the people involved to maintain this discretion. The informal practitioners let the greenery grow or control it in ways that favor their activities. It is also in the interests of the institutional actors that these implicitly tolerated practices should not be too visible. For the market garden plots, the municipality has even imposed a regulation requiring the installations (greenhouses) to be invisible. The interviews conducted with operational officials in the municipal departments show that these invisibilities also protect the compromises, the institutions, and the actors themselves: if no one sees them, no one is supposed to know, and one is less vulnerable to criticism or legal proceedings for incompetence or illegality (Dorso 2019). Over the period of study, therefore, the actors tended to either maintain the vegetation or let it spread freely.

Protecting by concealing is a process that is less developed in Stacioni i Trenit, in Tirana. Most of the vegetation consists of large meadowlands, stony in places. The height of the grass varies from season to season and, to a lesser degree, from area to area. A stand of poplar trees marks the edge of the tract, at the end of the strip of houses built without permits, which divides the area into two parts. However, this stand of trees is surrounded by a fence and, during the period of our observation, seemed largely unoccupied.

Utility

In Stacioni i Trenit, the vegetation is used as grazing for flocks of sheep and goats, as well as a few cows. The herders, originally from villages north of the capital, move their flocks from one area to another.

> We live further away, in the districts across the river. We come here because we don't find such a large meadow elsewhere. It's good for the sheep, the grass is tall, the space is large [...] We make a living on that, if there is no meadow we do not know what will become of us. (Dardan, shepherd on the wasteland, near the market, 50 years old).

Henhouses and dog kennels are scattered around the houses in the central strip. In the mornings, an informal junk and salvage market is attached to the bazaar located in the south of the area, where the urban highways meet. The traders spread sheets or blankets on the grass, along a track that crosses the tract. In the afternoon, collectors spread across the field, salvaging remains from the market that have been caught on the bushes. A few groups living in the Rom neighborhood further to the north move in to sort through the salvaged items, which will be reinjected into this endless circular economy. Children come to play, some accompanied by adults, some on their own, of all ages and in relative parity.

> Our children have been playing here for years. For us it's a good thing, we can sit next to them or watch them from our windows. We feel reassured [...]. The land belongs to the neighbor, it is a private property. When the grass grows, he brings his cows here... so children have to find some place else to play. (Zana, mother of a young boy, living in one of the houses built in the middle of the wasteland, 35 years old).

In the northern sector, older men have formed the blocks of stone into tables where they play dominoes in good weather.

The users say that they would like to retain these places for their various forms of utility: commercial, economic, and recreational extension to the domestic space. The lot owners themselves, whether or not they live in situ, also favor these activities, which, as we have seen above, serve a guarding function, provided that they are known and managed. For all this, the green wasteland provides favorable conditions, which the people concerned wish to maintain.

Similar forms of utility are found in Istanbul. The grassy slopes are used for grazing sheep (especially before the festival of Eid) and, more rarely these days, for horses. On the most neglected lots, older women and a few men pick plants (salad, wild berries, and medicinal herbs).

> Here it's quiet, I'm eating my bread and cheese. Do you want some? And you can pick wild arugula. It's paradise here, right? (Ahmet, user, 45 years old).

In some areas, soil is extracted for sale, extending the utility to the whole field. Wood from the trees is used for informal buildings (internal housing, stables, and sheds) or burnt in *mangals* (portable barbecues) or the bonfires where men gather at night to talk and drink.

> Look, here I am fine, I come alone, I put my radio on, I listen, I look at the road below, this is where I am most at peace. (Bülent, user installed in front of Edirnekapı, 55 years old).
>
> We don't cause any damage, we take care of things. (Elias, occupant of the Silivrikapı crypt, 40 years old).
>
> There have always been bostans (market gardens) but the interior of the walls, the interior pass of the walls, is of great importance in Istanbul folklore. (Ezra, town planner, heritage specialist, university professor, 50 years old).

Here again, the interests of the people concerned are to maintain these resources and conditions. The market gardeners offer a particular illustration: while the municipality frequently threatens their facilities (on the pretext of clearing, securing, or "cleansing" the area), in interviews the market gardeners claim to be the real developers of the zone.

> We sell the products to the people who come. But the land is not good here. It's embankment, look! There is dirt, rubbish, lots of stones, bottles of wine… Yep! With the gardens, it's green, it's clean. Otherwise it would always be the discharge. So we maintain it, and in addition it is us who pay the rent! (Mert, market gardener near Silivrikapı, 40 years old).
>
> The municipality threatens to evict us, they say that we dirty, that we degrade. But look, we are the ones who maintain the premises here. And we do it better than the construction companies, they come to do work, and after two years everything falls to the ground… (Mehmet, market gardener near Mevlânakapı, 45 years old).

In parallel to these complaints, the market gardeners have in the last decade or so made attempts to extend their land sections and the processes they use to protect the plots (fences and large dogs).

Continuity

In this next section, we move on from actions to representations—those of the actors involved, and those of the broader society within which they operate. In Tirana, grazing presents no local problems: it is consistent with the agricultural history of the area which, on either side of the old railway, was home to greenhouses and the livestock market. The inhabitants say that they felt no sense of invasion in the 1990s, when the shepherds from the north first brought their flocks for grazing. They also remember the area as a place without buildings, as evidenced by old photographs (Gegprifti 1990).

> Before the '90s there were no dwellings here. At first [in the '50s] there were private fields, then the state took everything to build fruit and vegetable greenhouses. There was also a large cattle market, which was easily accessible by the villagers since the train station was close by. (Mira, saleswoman of a kiosk in the middle of the wasteland, 60 years old).

In Istanbul, one can go back to nineteenth-century engravings and watercolors depicting the wall to illustrate the taste for green areas, gardens, and nature in Ottoman and Turkish culture—a taste found in all social categories. Today, every fragment of green space in the city is still invaded at weekends by family picnics, *mangals* of all ages, sexes, and conditions, and by strollers. Engravings or photographs from the 1960s (during the big phases of urban expansion in Istanbul) also show scenes of grazing, although—as we will see further on—there may be less consensus over these historical testimonies and representations of the agricultural world than those associated with strolling in the countryside.

The presence of vegetation in the heart of the city, and the uses associated with it, reflect a continuity that is perceived locally as normal or positive. There are no objections to it and in some cases, it is perceived as more broadly positive. This symbolism is found in the language: both languages use a related term to refer to greenery, *jeshil* (in Albanian) and *yeşilik* (in Turkish), both of which often have positive connotations.

6 Ambiguities of Urban Vegetation: Negative or Ambivalent Factors

6.1 A Continuity that Is Problematic with Respect to the Values Promulgated in Certain Sections of Society

This ongoing positive representation is not unanimously shared by all categories of actors. In Tirana, the culture and aspirations that are now dominant in the media, in the middle classes, and in some sections of the working classes are characterized by an attraction to the West. Despite the disappointments and resentments expressed in surveys (notably toward France), linked with complex issues around migration, the urban lifestyles of richer European countries are highly valued. For this reason, some interviewees identify waste ground and grazing with a backward rural world, in opposition to the coveted urban modernity of Germany, France, or the United States.

Until the early 1990s, Albania was 70% rural, a rurality that was not so much the result of backwardness as of political choices associated with the mode of development and territorial planning pursued during the communist era (Jarne 2018). In the surveys, the interviewees also refer to disorder and dirt, associated in particular with the stalls of the informal traders, the scattered remnants of the morning market, or else the holes or foundations left by forgotten or uncompleted buildings.

> The pasture is not a problem, although animals are perceived as dirty in the capital. What makes this place dirty is market waste… and the fact that people don't clean, don't take care of their environment. (Astrit, 40 years old, living in new buildings north of the wasteland).

In a transforming country like Albania, opinions and social categories are undergoing rapid restructuring and the generalization of the results of the surveys in progress remains complex on these subjects. Within the scope of the study, we can see different trends emerging: rich and middle classes are rather unfavorable to informal and agricultural uses of wastelands, while working classes and local residents close to the places are rather favorable. However, defining boundaries or categories is a complex task in Tirana. Representations are affected by the personal backgrounds of individuals. Thus, family memories or, from another perspective, nostalgia for the communist regime can generate—like a mirror-image—a negative representation of the Western world.

> For poor people, it was better under Enver Hoxha. (Mira, Saleswoman on the wasteland, 60 years old).

The same ambivalent structure of representations around the green wasteland was observed on the land walls of Istanbul. However, these images run in parallel with a negative storyline about cutthroats that is widespread at all levels of society. The development, dissemination, and reproduction of this fiction reflects a process of normative construction: evoking the bad city through the image of the wall as a wasteland riddled with forbidden or immoral practices is a way to define a fundamental otherness, against which the standards of the good city and of good behaviors are constructed and refreshed.

> So there, animals, vehicles, carts and squats are prohibited! Before there were one hundred and fifty meters of gecekondu around the wall. Now it's razed to the ground, everything is prohibited! But they live there, anyway, no one cares! In the evening, beware, it is dangerous: there is alcohol, rakı, wine, and also heroin. (Bahar and Can, sitting in the courtyard of the Mihrimah mosque, 60 years old).
>
> During the day it is quiet, but as soon as the sun goes down there are thieves, they drink, they piss, they take your money, they stab you. (Mehmet, sitting on old steps, 60 years old).
>
> This is a good place, here, but there are people who come to smash everything, who litter, you see these bottles … And there was a fire too. (Ömer, walker on the wall, 50 years old).

It is in the interests of everyone, including neighbors and users of the wall, to maintain this narrative, which positions the speaker (necessarily as a member of the better categories) and performs a certain protective function (the cutthroat storyline keeps the place relatively safe). That is why a single individual may simultaneously hold a critical attitude to the wasteland and be a regular user of it, as evidenced by these two scenes:

> "Don't walk around here, it's dangerous!" Say Cem and Mehmet, both in their 40 s, as they enter the wall, and whom I later see inside.
>
> "You shouldn't walk on the wall, there are dangers, there are sniffers, you can get yourself stabbed…" said Ozan, trader in household linen, 45 years old, whom I also met later on the site.

In Tirana, as in Istanbul, these negative or ambivalent narratives and representations around the wasteland are not always without consequences: they represent a reservoir of arguments that institutions or companies can use to justify their plans to develop the sites.

6.2 Tenuous Links Between Environmental Concerns and the Conservation of Green Spaces

At this point, we wondered about the role of ecological and environmental factors in the two situations. We might have expected a convergence between the different views on vegetation described above and actions or discourses associated with environmental protection and the resistance to climate change.

In Tirana, and to a lesser degree in Istanbul, the environmental concerns found in Western Europe are little, if at all, present in the local populations. Nature is used for utilitarian purposes often inherited from a rural tradition. People are motivated by economic factors (often involving multiple activities) which play a big part in day-to-day life. Environmental protection is not a matter of urgency, nor is it very present in representations or practices, as evidenced by the garbage scattered in the streets in both city and countryside. However, this is not necessarily an indication of entropy and waste: there are other ways of sorting and recycling. A whole informal or semi-formal economy of waste recovery is in place in both Tirana and Istanbul, especially on our two study sites (often practiced by semi-sedentary Rom families). The idea of

Fig. 3 Waste and recovering in Tirana and Istanbul © Muçi 2019 and Dorso 2013

waste reduction at source, on the other hand, is neither spoken about nor practiced, given that the dominant paradigm is about catching up with the model of the classic Western consumer society. In fact, it might be argued that there is a temporarily irreconcilable division between popular aspirations and an environmentalist credo found only among the most comfortable sections of local society (Fig. 3).

Environmentalist attitudes to the preservation of green areas are more present in civil society organizations or activist groups, though their impact is limited. The 2013 Gezi Park episode in Istanbul cast light on these movements which, after a period of repression, sought to spread through projects for shared or guerilla gardening (Fautras 2016). On the walls, attempts at a rapprochement between the civil society sector and the market gardeners date back to the late 1990s, but they remain focused on urban farming (Robin 2011). Their impact remains limited and they have not prevented a recent hardening of the municipality's position on farming-related structures.

Finally, it is questionable whether environmental factors relating to the two spaces have been taken up by the institutional actors and project developers. In both Tirana and Istanbul, the institutions refer to the environment in their communication as a strategic tool of internationalization, with the aim of promoting the compatibility of environmental responsibility with developmental policy. However, there are a number of contradictions in this discourse. As a potential obstacle to the extension of big projects (Fautras et al. 2020)—as in the deforestation that followed the construction of the third airport—or to the exploitation of natural resources necessary for this development, nature (specifically in the city) in Istanbul is envisaged purely as an artifact (Pérouse 2014). This position is not favorable to the maintenance of the wastelands, as evidenced by interviews with institutions:

> We could not do any work [survey or project] about these people [the users]. But, around the walls, on the map, we have green spaces, we have protected the walls, and a completely green area (Aslı, urban planning service, Central Municipality of Istanbul, 40 years old).
>
> First there was a protection plan, which provided for the establishment of a green strip around the walls. And then actions on specific points. For example [...] Sulukule, there is a lot of

degeneration there (Zeynep, service Müzekent (City-Museum) of the Metropolitan Planning Center, 50 years old).

— In fact, some [of the gardeners] will leave, some of them will stay. Some will leave, and there will be protection for open green spaces. And some will stay, those who have been there for a long time, and who continue to work…

— Are they aware, those who will stay, and those who will have to leave?

— No, they don't know yet. (Fadime, urban planning service, Central Municipality of Istanbul, 40 years old).

The municipality's actions relating to nature around the wall have been concerned with creating parks primarily for sociopolitical ends (Karaman 2017). In Albania, the Tirana 2030 project includes an inventory of the city's green spaces. We were able to consult documents which are partly confidential and not yet made public, in which only official green spaces are counted. Green wasteland and the surroundings of watercourses (for example, the Tirana River, north of the center) are not included in this count, although they are integrated into the future development projects. At Stacioni i Trenit, this question has been delegated to the company that will be developing the site. Here again, the environment remains subordinate to a construction project that will greatly increase density in the area. The park planned for part of the current wasteland area is presented, in the documents we were able to see, as a recreational amenity associated with the construction of a tetragon of high-rise towers (Fig. 4).

In the center of the boulevard we will have a central park, similar to New York's Central Park, with services and leisure facilities. It will contribute by improving the living environment of citizens, and above all by revaluing the land throughout the territory. Said Erjon Veliaj, mayor of Tirana in October 2015.[4]

Fig. 4 General view of the site (© Milliard 2019) and a banner of the central park project displayed along the wasteland (© Muçi and Dorso 2019)

[4] Video extract on the national media Ora News, October 25, 2015 https://www.youtube.com/watch?v=VbNqV5PvUko.

7 When the Green Zone Survives the Urban Plan

The final paradox reveals that the two sites will probably remain green zones in the coming years, precisely as a result of projects designed to increase urban density that do not support the current endemic uses of the two wastelands. Is this, once again, a form of resilience?

The contract for the Tirana 2030 urban transformation project, initiated by the creation of the new Bulevardi i Ri road north of the city center, has been awarded to the firm Grimshaw. The plan in the "Tirana Boulevard, Central Park and River Project" is to construct apartment blocks and towers (medium to high prestige) along the whole length of the road, as far as the River Tirana. The total perimeter of the project includes spaces with and without buildings. In April 2019, we were able to follow the expulsions and house demolitions in the northernmost part, near the river. The organized latency of the public authorities and the failure to settle questions of land and real estate ownership are directly illustrated in this episode: without the completion of the regularization process for (and between) ownership of the land and of the buildings, the inhabitants have no formal legal right to oppose the demolitions. Central government and the municipality chose an English private firm, Grimshaw, to develop this large urban area, and it will gradually acquire ownership of all the plots to implement the project.

On the Stacioni i Trenit tetragon, the central strip of houses will also be razed, but the project provides for the retention of a green zone. The masterplan we were able to see shows the construction of a ring of 10-storey towers surrounding an urban park. The green space thus seems to be retained, but several changes are apparent. The future "Central Park" will be landscaped, designed as a place of recreation primarily intended for the residents of the towers. Finally, the status of the land will change, becoming a public amenity managed by Grimshaw. The total situation is reversed: this will be a public space managed by a private company, whereas at present there is an area of private lots with public uses.

If the project comes to fruition, it will maintain the green zone in the future, but only in formal terms: the appearance and types of use will change profoundly, as will the way the site is managed. The questions raised here are about appropriation, power, and control. As we heard from the interviewees in situ, and as we have seen elsewhere with the tolerance of the landowners (who prefer informal uses to buildings):

> Nothing is abandoned today. (Linda, architect-urban planner at the municipality of Tirana, 30 years old).

The retention of the green zone will simply entail a change of register in control of the place. With subsequent research, it will be possible to assess whether (and how) landscaping signals the end of the green wasteland, rather than its resilience.

This question has a much longer history in Istanbul. In the contemporary period, initial landscaping efforts on the wall date back to the interwar period. Between 1935 and 1951, the French urbanist Henri Prost worked on the masterplan for Istanbul and designated a *non-œdificandi* zone for the wall 85–110 feet in depth on the inner side, and 550 feet on the outer side. These measurements decreased over time, but the general principle was respected, even after the construction of the external beltway that runs alongside the wall. They continue to apply to a good proportion of the grassy slopes that extend below the uneven and wooded heights of the rampart.

Although no formal development plan for the wall as a whole was subsequently established, two types of projects are currently changing the configurations of these green spaces. Urban regeneration operations on adjacent districts (Sulukule, Ayvansaray) have driven out the traditional uses and, with them, some of the endogenous modes of regulation (copresence, avoidance, distance, and mutual aid), making these spaces less safe. The specific landscape of the land wall is sometimes associated with or "captured" by these projects, for example, in publicity documents or on the big fences surrounding the construction sites: they show a "cleaned up" wall and an ordered landscape.

Moreover, the landscape parks extend up the grassy slopes to the foot of the wall, occupying what remains of Prost's *non-ædificandi* zone. The big Topkapı Park built between 2004 and 2009, or the older IBB Soğanlı Bitkiler Parkı, located right in the south, is highly controlled in their layout and rules of use, which are also, as we have said, socially targeted: recreational family uses on a traditional and conservative model (Karaman 2017). Here again, while the map still shows green, the forms that this vegetation takes and the social practices that produce it or use it have changed. Landscaping is thus always an operation with goals, resources, actors, and interests. It is a social construct like any other, in which nature is simultaneously context, resource, and instrument of an economic, urban, social, and political project.

8 Conclusion

This journey across the fields has provided us with elements of explanation for the long-standing survival of the two green wastelands. The presence of nature on these central spaces is the product of a history longer than that of our observations alone—several centuries for Istanbul, at least half a century for Tirana. Recent changes (urban expansion, densification, population movements, political and economic developments) have prompted the different actors to take hold of these territories, each according to their interests and strategies. In Tirana as in Istanbul, the frictions or conflicts associated with land ownership and use have found temporary forms of resolution in an organized latency. This is in no way an abandonment. The tensions around adjacent projects and operations show clearly that control of these territories remains an important issue, with winners and losers. These dynamics are found in many urban situations but often over shorter periods. One of the factors that helps to explain the long timescale here relates, in part, to the place and particular role of vegetation in the two situations.

The practical and protective aspects of nature perform important services, and the actors try to preserve the conditions of their existence. Vegetation and its uses also form part of a continuity that encourages positive, or at least normal, representations of the situation. This cumulative process provides conditions (or indeed arguments) that favor maintaining the sites as they are. However, local changes in lifestyles and aspirations can prompt certain population categories to associate these socio-spatial situations with negative images (rural world, backwardness) or feed into normative discourses and narratives that contrast the figure of the lowlife with that of the good city. The positive representation of vegetation seems to have won the day so far, but for how long?

Recent urban projects show that these zones will remain partially green. However, this is an ambiguous form of resilience. Environmental measures are tilted in the direction of intensive urban projects. These operations seek neither to recognize nor to draw upon the existing fabric, which is usually reduced to a glance at the green areas on the urban map. In Tirana's Central Park, as in the parks of Istanbul, there are going to be profound changes in the composition of the vegetation, in the status and management of the spaces, and in the regulation of practices. The old compromises are under threat. And the vegetation in the city is considered here in different conceptions according to the categories and the position of the actors in the contexts. The landscaping arrangements proposed by developers and elected officials reveal measures of territorial management and control. The users' conceptions are oriented toward the availability and the usefulness of a space which could, in certain situations, be questioned, in the continuation of the surveys, from the perspective of the commons (Le Roy 2016).

These changes might prompt us to question the urban act and, perhaps, the creative act at work behind it. The two green wastelands have survived until today because, though they are not abandoned, the institutional actors (public or private) have intervened little on the ground, leaving the inhabitants and users in relative control of the spaces, by means of more or less explicit compromises. In other words, a form of non-action (withdrawal, restraint) has favored the presence and resilience of this urban vegetation, whose positive aspects in biological, spatial, or social terms are taken into consideration elsewhere and in several analyses in this book (although the connection between those aspects proves complex).

It is both a stimulating and sensitive issue to envisage this non-action as a possible form of action. That form of action can be considered unorthodox and viable in a different way, and still remains ambiguous: weak actors may possess a degree of leeway, but in a system of social relations that does not change the general order of positions and resources. Nonetheless, what these two sites show us is that a relative withdrawal of the strong actors (albeit tactical and temporary) enables the weaker players to act and develop skills, and that it is these conditions which explain, beyond ethical judgments, the continued existence of these large areas of urban vegetation over long timescales.

Future debate could therefore begin with the question of the taboo of non-action, which opens up at least three avenues of discussion. First, from a sociological perspective, the question of power: making a form of neglect official and removing it from the realm of the implicit signifies a sharing of power. Under what circumstances might a strong player consider it? This is a question that could be directed not only at decision-makers, but also at researchers: analyzing and writing about these situations has its own impact; we ourselves are stakeholders in these power relations, by our presence, by our publications, by our actions. Should we always pursue our research, and if so under what conditions (why and for whom)? In anthropological terms, the taboo relates to its reciprocal term: the impulse to act and to create. This can be explored with respect to the professional actors (professions are constituted on the basis of an act, a gesture, an intervention, and *out* of the processes of recognition, of legitimization, and of preservation that accompany them), and more broadly than in general human terms. How should this power to act be directed, particularly in the configurations that we have just described? Thirdly, one might return to the concrete conditions of the operational sphere: actors and actions are conditioned by the processes and pressures associated with the economic imperatives of the development business (landscape, forest, urban). People act because their firms have to function and because, as individuals, they need to maintain their own conditions of existence. All of them are dependent on a given mode of social production. Exploring these processes leads one to make close links between the analysis of nature in the city and the analysis of development firms. In other words, after focusing on demand (practices, consumption), they prompt us to turn our attention to supply, toward an anthropology of the production of nature.

References

Aliaj B (2009) Hapësira qytetare drejt fundit të tranzicionit?!: Këndvështrime filozofike dhe teknike [Civic space towards the end of the transition?!: Philosophical and technical perspectives]. Albania: Botime Afrojdit, Tirana

Brown D, McGranaham G (2016) The urban informal economy, local inclusion and achieving a global green transformation. Habitat Int 53:97–105

Dorso F (2019) L'hypothèse de l'inconnaissance dans les dynamiques d'(in)visibilisation de la ville informelle [The hypothesis of unknowledge in the dynamics of (in)visibility of the informal city]. Territoire en mouvement 43. Du visible et de l'invisible dans la fabrique de la Ville et les études urbaines [The visible and the invisible in the City Factory and Urban Studies]. https://journals.openedition.org/tem/6201

Dorso F (2018) Un récit de la bonne et de la mauvaise ville. Travail de l'altérité dans le processus de construction normative [A narrative of the good and the bad city. The work of alterity in the process of normative construction]. Recherches Sociologiques et Anthropologiques 49(2):135–156

Erdi Lelandais G (2014) Right to the city as an urban Utopia? Practices of everyday resistance in a Romani neighbourhood in Istanbul. In: Erdi Lelandais G (ed) Understanding the city: Henri Lefebvre and urban studies. Cambridge Scholars Publishing, Cambridge, pp 69–90

Fautras A (2016) Les nouveaux bostan d'Istanbul : quelle pérennisation pour les jardins de la contestation ? [Istanbul's new bostans: what sustainability for the gardens of protest?] Eur J Turk Stud 23. https://journals.openedition.org/ejts/5400

Fautras A, Hervet F, Perouse JF (2020) Le nouvel Istanbul: dimensions, dynamiques et vulnérabilités [The new Istanbul: dimensions, dynamics and vulnerabilities]. Historiens Et Géographes 450:62–69

Gegprifti L et al (1990) Tirana [Tirana]. Polygraphischen Kombinat, Neue Druckerei, Tirana

Gibout C, Blanc M, Foucart J (2009) Transactions et sciences de l'homme et de la société [Transactions and human and social sciences]. Pensée plurielle 20(1). de Boeck Editeur, Bruxelles

Instat (2010) Socio-demographic statistics in Albania: selected topics and future developments. Shtëpia Botuese & Shtypshkronja Pegi, Tirana

Jacquot S, Morelle M (2018) Comment penser l'informalité dans les villes "du Nord", à partir des théories urbaines "du Sud"? [How to think about informality in "northern" cities, based on "southern" urban theories?]. Métropoles 22. https://journals.openedition.org/metropoles/5601

Jarne A (2018) La fragmentation de l'espace urbain postsocialiste en Albanie [The fragmentation of post-socialist urban space in Albania]. Belgeo 4. https://journals.openedition.org/belgeo/30849

Karaman H (2017) Le Topkapı Kültür Parkı : fabriquer un parc public à Istanbul [Topkapı Kültür Parkı: building a public park in Istanbul]. Eur J Turk Stud. https://journals.openedition.org/ejts/5389

Le Roy E (2016) Des Communs « à double révolution » ["Double revolution" Commons]. Droit Et Société 94(3):603–624

Lubonja F (2015) Libri i deshtuar i arkitektures shqiptare [The failed book of Albanian architecture]. Perpjekja, Tirana

Montabone B (2013) Droit à la ville et contestation de l'ordre moral urbain en Turquie [Right to the city and challenge to urban moral order in Turkey]. EchoGéo. https://echogeo.revues.org/13567

Pojani D (2013) From squatter settlement to suburb: the transformation of Bathore, Albania. Hous Stud 28(6):805–821

Pojani D (2011) Urban and suburban retail development in Albania's capital after socialism. Land Use Policy 28(4):836–845. Elsevier

Pérouse JF (2003) La muraille terrestre d'Istanbul ou l'impossible mémoire urbaine [Istanbul's terrestrial wall or the impossible urban memory]. Rives Nord-Méditerranéennes 16:27–44

Pérouse JF (2014) Le "mouvement de Gezi" ou le choc des systèmes de valeurs environnementales dans la Turquie en croissance [The "Gezi movement" or the shock of environmental value systems in growing Turkey]. Méditerranée 123:49–56

Rémy J (2015) L'espace, un objet central de la sociologie [Space, a central object of sociology]. Erès, Paris

Rémy J, Voyé L, Servais E (1978) Produire ou reproduire. Tome 1: conflits et transactions sociales [Produce or reproduce. Volume 1: conflicts and social transactions]. Vie ouvrière, Bruxelles

Robin C (2011) La possibilité d'une agriculture : les maraîchers des murailles à Istanbul [The possibility of an agriculture: market gardeners in the city walls of Istanbul]. https://fr.calameo.com/books/002193621ec70e961166d. Accessed 4 June 2020

Triantis L, Vatavali F (2016) Informality and land development in Albania: land reforms and socioeconomic dynamics in a coastal settlement. Geografiska Annaler (ser B Hum Geogr) 98(4):289–303

Stela Muçi is an Architect-Urban Planner and Lecturer at the Paris Urban Planning School, Lab'Urba (France). She has worked in the participation and co-production schemes implemented with the residents of Seine-Saint-Denis and in Hérault (France) with the nonprofit organization APPUII. Her research focuses on abandoned lands, urban wastelands, and the informal practices of residents, in France and Albania.

Franck Dorso Anthropologist and sociologist **Franck Dorso** has worked in local development and training, in France and Turkey. He is now an Associate Professor at the Paris Urban Planning School, Lab'Urba (France) and at the French Institute of Anatolian Studies, where he runs the Master's Degree course *Urban alternatives and experimental approaches - Public spaces.* His research focuses on informality, the logics of holding back and transgression, and the notion of urban valve.

Let It Grow? Social Representations of Nature on Contaminated Brownfields

Marjorie Tendero and Cécile Bazart

Abstract This study focuses on social representations of nature in the case of contaminated brownfield sites in France. Data were collected using an open-ended questionnaire in a French national cross-sectional survey administered to people living in municipalities impacted by contaminated brownfields. We use lexical methods to examine how respondents perceive nature in the case of contaminated brownfields. Three complementary software for text analyses were used (IRaMuTeq, Hyperbase, and Tropes). This strategy allows us to identify and understand, in-depth, different dimensions related to nature in the case of contaminated brownfields. The results show that respondents associate different forms of nature, from wild gardens to more domesticated forms such as vegetable gardens and crop fields, with contaminated brownfields. Knowledge of these types of nature that are distinguished by respondents is important in designing brownfield redevelopment projects and in improving the management of urban green spaces such as brownfields, which are an important source of biodiversity in the city.

Keywords Contaminated brownfields · Nature · Social representations · Temporary nature · Textual analysis

1 Introduction

At the EU-28 level in 2018, there were more than 650,000 contaminated sites (Payá Pérez and Rodríguez Eugenio 2018). Brownfields represent a substantial proportion of contaminated sites in many European countries (Dixon et al. 2011). Brownfields refer to "any land or premises which have previously been used or developed and are not currently fully in use, although it may be partially occupied or utilized. It may also be vacant, derelict or contaminated" (Alker et al. 2000, p. 49). In France, at the

M. Tendero (✉)
ESSCA School of Management, SMART-LERECO, Angers, France
e-mail: Marjorie.TENDERO@essca.fr

C. Bazart
CEE-M, Montpellier, France
e-mail: cecile.bazart@umontpellier.fr

F. Di Pietro and A. Robert (eds.), *Urban Wastelands*, Cities and Nature,
https://doi.org/10.1007/978-3-030-74882-1_8

beginning of the year 2020, there were more than 7 250 soil-contaminated sites[1] and soil-contaminated brownfields[2] represent between 138,000 and 154,000 ha in urban areas. Moreover, 66,000 to 77,000 ha of soil-contaminated brownfields are located near existing transportation and utility infrastructure (ADEME and QuelleVille? 2015). Thus, the redevelopment of contaminated brownfields has gained widespread political support. Indeed, their redevelopment is recognized as a key instrument to fight urban sprawl, and its negative impacts, such as soil sealing, loss of agricultural land, air pollution, traffic fatalities, and jams (Dorsey 2003). The 2014 Access to Housing and Urban Renovation law (ALUR in French) and the 2018 Evolution of Housing, Development and Digital technology law (ELAN in French) are the most recent attempts in France to remove the legal and financial barriers to the reuse of brownfield land.

Furthermore, long considered as antithetical (Kowarik 2018; Raymond and Simon 2012), the concept of urban nature is now deemed to be an important element on the urban and political agenda in achieving sustainable development. In France, the 2015 Paris Climate Agreement and the 2016 Energy Transition for Green Growth law aim to reclaim biodiversity, nature, and landscape and to promote environmental preservation. We also observe a growing interest and a high desire for nature by inhabitants of Western societies (Bailly and Bourdeau-Lepage 2011; Bauer 2005; Bontefeu 2009; Bourdeau-Lepage and Vidal 2013). With a bit more than 80% of the French population living in urban areas in 2020 (Statista 2019), *homo urbanus* is also a strict *homo qualitus*[3] according to the expression used by Bailly and Bourdeau-Lepage (2011; Bourdeau-Lepage 2018).

There exist different forms of urban nature, for instance, urban green spaces, such as parks and community gardens, planted trees or vegetation along current or former transport routes such as the "Petite ceinture" in Paris (Scapino 2016), urban farms and beehives, green roofs, flower beds or flowerpots on balconies, donkey "school buses",[4] and lawnmowing sheep (Lelay et al. 2019). Informal urban green spaces such as derelict spaces and brownfields with the growth of wilderness or green features are another example of urban ordinary nature (Loures and Vaz 2018; Rupprecht and Byrne 2014). Ordinary nature corresponds to everyday nature (Kaplan et al. 1998) or the wider countryside (Adams et al. 1994) and is generally opposed to the extraordinary nature, referring to heritage and remarkable species and sites (Godet 2010; Pellegrin et al. 2018). This ordinary urban nature provides a plethora of benefits. Environmental and health benefits are well documented in the literature in the case of formal greenspaces such as forests, parks, and gardens, for instance (Capaldi et al. 2015; Franco et al. 2017; Laille et al. 2014). However, it is less the case for urban

[1] https://www.georisques.gouv.fr/risques/sites-et-sols-pollues/donnees#/type=instructions.

[2] Contaminated brownfields hereafter.

[3] Antoine Bailly and Lise Bourdeau-Lepage to refer to people who consider nature and the preservation of the environment as important components of their well-being used the expression *homo qualitus*. This expression is a pun on the expression *homo economics* used to describe rational economic agents in economics.

[4] The cities of "Louviers" and "Val de Reuil" in France, for example, are pioneers in testing the donkey school buses thanks to the "Le chemin du halage" organization.

wastelands and brownfields (Botzat et al. 2016; Brun et al. 2018). These spaces are considered as an important source of biodiversity for temporary nature (Kattwinkel et al. 2011; Triplet 2020). According to the definition given by Triplet (2020), temporary nature corresponds to the "development of pioneering habitats, fauna, and flora on areas waiting to be redeveloped" such as brownfields (Triplet 2020, p. 818). Therefore, the ecological benefits and in particular the biodiversity of urban wastelands and brownfields are well documented in the literature (Bonthoux et al. 2014; Gardiner et al. 2013; Harrison and Davies 2002; Hunter 2014; Macadam and Bairner 2012; Mathey and Rink 2010; Muratet et al. 2007; Westphal et al. 2004), including in the case of contaminated brownfields (ADEME 2014). At the same time, other environmental benefits, such as the impact of brownfields on the microclimate and urban heat islands, and also on social interactions are less documented (Hou et al. 2018; Keniger et al. 2013; Koch et al. 2018). Besides, as people prefer domesticated nature (Hofmann et al. 2012; Kim and Miller 2017; Lafortezza et al. 2008), brownfields may be negatively perceived, especially in the case of contaminated brownfields, which raises concerns about health risks and safety issues (Bambra 2018; Bambra et al. 2015; Gilderbloom et al. 2014; Maurel et al. 2010). Nature-based solutions such as phytoremediation to redevelop contaminated brownfields are preferred by people (Faivre et al. 2017). It is thus important to understand what the perceptions of nature on brownfield sites are.

To the best of our knowledge, a few attempts have been made to assess how urban brownfields are perceived by inhabitants, and, in particular, if whether or not they value urban nature on such sites (Brun et al. 2019, 2018; Mathey et al. 2018). In their study in the cities of Dresden and Leipzig (Germany), Mathey et al. (2018) show that brownfield redevelopment with spontaneous vegetation should be linked to aesthetic motives to avoid the impression of neglect and therefore foster the reuse of the sites by local inhabitants. Focusing on the conurbations of Tours and Blois (France), Brun et al. (2018) show that perceptions and valuation of wastelands depend on the plant successional stage and that the intermediate grassland stage is better and positively perceived by residents. They also show that wastelands represent freedom from the controlled built environment.

Previous studies have been focused on urban wastelands or on brownfields that are not contaminated (Brun et al. 2019, 2018; Mathey et al. 2018). Contaminated brownfields can remain in a transitional state for decades because soil and water contamination entails high remediation costs, complex administrative procedures, and long administrative delays (Adams et al. 2001; Rosato et al. 2010). During this transitional state, diverse vegetation structures are exhibited.

We investigated social representations regarding contaminated brownfields. French psychologist Serge Moscovici defined social representations as "socially elaborated systems of values, ideas, and practices that define an object for a social group" (Moscovici 2003). People to understand and communicate about their environment use social representations. We aimed to understand how people living in a municipality impacted by contaminated brownfields perceived them.

Moreover, we assumed that people perceived a form of urban nature on contaminated brownfield sites, examined how people perceive nature in the case of contaminated brownfields, and tested whether their perception is positive or negative.

The chapter is organized as follows. Section 2 explains the selection of the study area and respondents. It also presents research methods and how the data were collected and analyzed using computer-based textual analyses. Section 3 focuses on the results. They show that people associate different forms of nature: from wilderness to the more domesticated forms. These results are discussed in the light of existing literature regarding the typologies of nature in Sect. 4. We also conclude and give some suggestions for future research.

2 Methodology

Data were collected from people living in municipalities impacted by at least one contaminated brownfield using an online questionnaire (Box 1). Figure 1 shows the different criteria of inclusion to obtain our geographically targeted panel (Box 2).

Box 1: How to Identify Contaminated Brownfields in France?

In France, the ministry in charge of the environment has kept a national register of sites that have been reported by the local authorities, among them the Regional Directorate of Environment, Planning, and Housing (DREAL in French), as being actually or potentially contaminated sites and in need of specific government action. These sites are listed in a database named BASOL and among them are brownfield sites. We have extracted contaminated sites that were listed in BASOL on 27th October 2015. At that time, there were 6294 contaminated sites (7256 at the beginning of 2020). We focused our study on brownfields located in Metropolitan France because brownfields are a relatively scarce phenomenon in French overseas territories (in 2015, they represented only 2.01%, 1.80% in 2020, of the total contaminated sites that were listed in BASOL and this proportion is stable over time). We identified brownfield sites using the site status variable which distinguishes between industrial sites still used, reused sites, and industrial brownfields. According to this variable, among the 6168 contaminated sites in Metropolitan France, 2133 (33.88%) were brownfield sites. These 2133 brownfield sites were crossed with data from the National Institute of Statistics and Economics Studies (INSEE in French) to identify impacted municipalities. The 2133 brownfields were in 1375 municipalities. This represented only 3.75% of the total number of municipalities that existed in France in 2015.

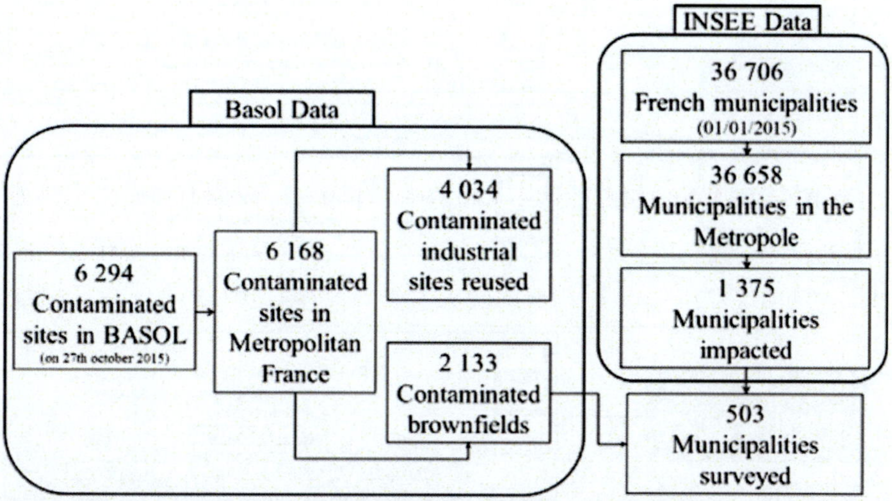

Fig. 1 Inclusion criteria for the generation of the targeted panel using data from BASOL and INSEE

Box 2: Sampling

To obtain a geographically targeted panel, a quota on the number of people to be surveyed in each municipality impacted by a contaminated brownfield was set up using data from the 2015 National Institute of Statistics and Economic Studies (INSEE) Census. A professional panelist (Survey Sampling International) was in charge of quotas on age, gender, and municipalities to be surveyed considering the number of contaminated brownfield sites. The online questionnaire was sent using email addresses.

The data were collected in December 2015 by means of an online cross-sectional questionnaire using Limesurvey[5] (Schmitz 2015). The survey was designed to take approximately 20 min to be completed. The aim of this questionnaire was fourfold: (i) collect representations, perceptions, and knowledge regarding soil contamination and (ii) contaminated brownfields, (iii) identify people's preferences and expectations regarding brownfield redevelopment, and (iv) examine the level of trust given to different stakeholders involved during a brownfield redevelopment project, such as elected representatives, urban developers, and environmental organizations. Our analysis focuses here on the first two aims.

To retrieve social representations regarding contaminated brownfields, we used the following open-ended question: "What words or phrases pop into your mind when thinking of urban brownfields?" Free association is a technique frequently used for the collection of social representations (Dany et al. 2015). This technique is advantageous: it is easy to use, to implement in a survey, and to understand by the participants. To analyze the answers given to this question, we used textual analysis (Box 3).

[5] https://www.limesurvey.org/.

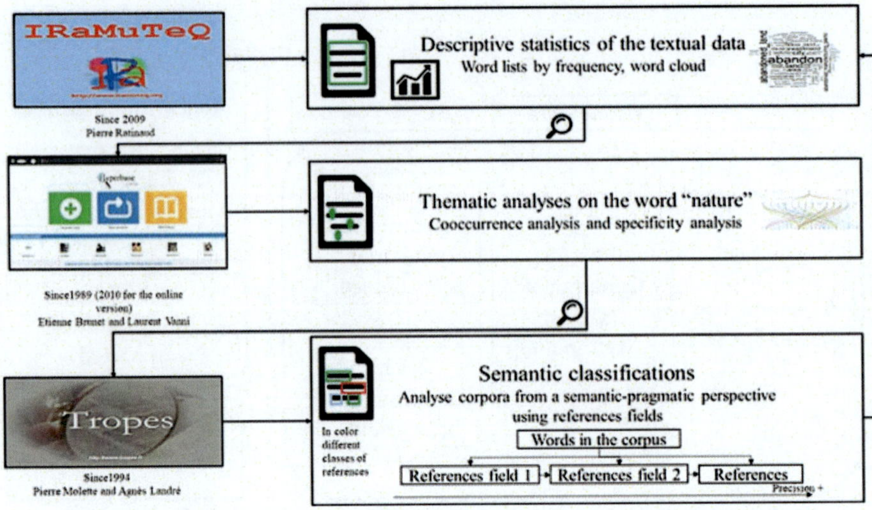

Fig. 2 Textual data analyses conducted using IRaMuTeQ, Hyperbase Web, and Tropes

Box 3: What is Textual Analysis?

Textual analysis aims to examine the content, the structure, and the meaning of texts (Given 2008). The methods of textual analysis have been developed by researchers since the 1960s in several disciplines: linguistics, discourse analysis, statistics, information technology and computer science, and socioeconomic survey analysis (Pêcheux et al. 1982, p. 19). In our study, textual data come from the answer to an open-ended question ("What words or phrases pop into your mind when thinking of urban brownfields?"). One of the main advantages of textual analysis is that it uses responses to open-ended questions in the form that they are collected. Thus, it avoids the bias introduced in the stage of thematic post-codification (Lebart et al. 1998).

We also asked our respondents whether they were aware of the existence of a brownfield site near their home, the number of brownfields they perceived near their home, as well as whether they visit one of them. We also collected information about the characteristics of the brownfields in the questionnaire. Sociodemographic characteristics were also collected at the end of the survey.

All questions in our survey are required so that respondents must answer them before submitting their questionnaire. Open-ended questions were also mandatory. This kind of question requires more effort and time to respond. As a result, some respondents typed "noise" responses such as "n/a," "I do not know," or "no." For this reason, 264 respondents (32.87%) failed to give even a single word or expression in response to that question. These 264 answers were excluded from the analysis; therefore, our sample contains 539 answers distributed among 503 municipalities.

Faced with a large amount of textual data, we required computer assistance to organize, describe, and compare texts (Lebart et al. 1998). Depending on the nature of the corpus and the aim of the study, different software can be used (Pincemin 2018). We used IRaMuTteQ, Tropes, and Hyperbase Web. Figure 2 describes the textual analyses conducted and Box 4 briefly presents these software packages.

Box 4: Textual Software used

IRaMuTeQ stands for the interface of R for multi-dimensional text and questionnaire analysis, and is a free and open-source software created in 2009 (Ratinaud and Déjean 2009). This software provides users with different text analyses such as basic word frequency, similarity analysis, descending hierarchical classification, and correspondence factor analysis (Souza et al. 2018). Besides, it is possible to use a tailored dictionary to perform more precise, accurate, and relevant analyses regarding a research interest, which is one of the strengths of this software compared to other textual statistical approaches. We used IRaMuTeQ to perform textual statistics using a tailored dictionary. To construct this dictionary, we began with preliminary descriptive statistics to identify the words and expressions that were not recognized by the software. Then, according to their word frequencies and cooccurrences, we added these unrecognized words or expressions to a tailored dictionary considering the meaning of the words in context.

Cooccurrence analysis shows the existing links between words used in a text. It counts paired data within a collection unit such as a text corpus (Buzydlowski 2015). Specificity analysis determines the typical words in a corpus. Calculation of specificities enables the detection of words or expressions that appear frequently compared to their frequency in the whole text (Lafon 1980). This preliminary investigation yields insights into the words that may be linked to ordinary nature, such as green space, tall grass, or wild plants.

Hyperbase **Web** is the online version of the free and open-source software Hyperbase, created in 1989 (Brunet and Luong 1989). Because of the authors' academic background in literacy studies, Hyperbase was initially set up to explore corpora of literature including whole books. Hyperbase Web has been developed since 2010 and provides a toolbox supplying a variety of text statistics, such as cooccurrences, distances, k-means, and z-tests, for corpus exploration. Hyperbase also allows a good interconnection with other textual analysis software such as IRaMuTeQ or Tropes (Kasparian 2014; Mayaffre and Ben Hamed 2014).

Tropes is a free and open-source semantic analysis software which was created in 1994 (Molette et al. 2013). Based on a pre-established lexicon that contains 20,000 equivalent classes divided into broad semantic categories, the software identifies the themes tackled in the text and shows how these themes are linked to one another using concept maps or constellations graphs. In our study, Tropes was a powerful tool for extracting relevant information by detecting contexts and isolating the main themes. Three levels of classifications are used to perform the content analysis. The most fine-grained analysis identifies the "references" of the words used in the text. References gather closely related words in the text. The reference fields enable Tropes to elaborate a representation of the context using the semantic equivalents dictionary of Tropes. This dictionary is composed of three different classification levels. The references are at the more precise level of classification; the software displays only significant fields. These references provide a detailed overview of the words in the corpus. These references are grouped together into reference fields 2 which, in turn, are merged into the broader reference fields 1, corresponding to the most coarse-grained level of analysis (Piolat and Bannour 2009).

Thanks to the quotas used for the proportions of people to survey in each municipality impacted by contaminated brownfields, the distribution by regions of residence reflects the regional distribution of contaminated brownfields. Table 1 gives an overview of the socioeconomic characteristics of the respondents. We have a gender-balanced sample with 50.65% of men and 49.35% of women as well as an age-balanced sample thanks to the quotas that were used for these two variables. Besides, the proportions by occupational categories, educational qualification, and family structure in our sample follow the distribution that was observed in 2016 in the French population according to INSEE data.[6] This was also the case for the distribution of people according to city size: 79.78% of the people were living in a city of less than 10,000 inhabitants. This proportion is close to those observed in census data at that period. Therefore, our sample is representative of the French population.

3 Results

3.1 *The Perception of Contaminated Brownfields*

Even though all respondents were living in a municipality impacted by contaminated brownfields, only 21.52% of our respondents were aware of the existence of such sites near their place of residence (Table 2). The perception of contaminated brownfields is low.

Table 3 shows that, according to the residents, brownfields are most often former industrial sites (73.28%) or former public facilities such as abandoned schools or hospitals (22.41%). These percentages reflect the French industrial heritage. Brownfield sites are mainly located in urban periphery areas (55.17%) or in the city center (37.93%). According to our sample, redevelopment of 59.48% of brownfields has been pending for at least five years.

3.2 *General Representations of Contaminated Brownfields*

We performed textual statistics using word frequency to investigate the words most often used. The results of these textual statistics are presented in Table 4.

Table 1 Sociodemographic and geographical characteristics of the respondents

Demographic characteristics—N = 539	Frequency	Percentage
Gender		
Male	273	50.65
Female	266	49.35
Age (mean = 45.84; Std. Dev. = 13.11)		
18–24	32	5.94

(continued)

[6] See https://www.insee.fr/fr/statistiques/2011101?geo=FRANCE-1.

Table 1 (continued)

Demographic characteristics—N = 539	Frequency	Percentage
25–39	142	26.35
40–59	266	49.35
64	99	18.37
Educational qualification		
None/primary school	26	4.82
Secondary school/vocational education	94	17.44
High school diploma	126	23.38
Associate degree (2-year diploma)	128	23.75
Undergraduate/Graduate degree (3-year or more diploma)	165	30.61
Occupational categories		
Farmer	2	0.37
Self-employed, small shopkeepers, business owners	14	2.60
Senior managers and higher managerial, administrative and professional positions	74	13.73
Intermediate managerial, administrative positions	95	17.63
Services and sales workers, clerical support workers	146	27.09
Manual workers	38	7.05
Inactive	170	31.54
Unemployed	29	5.38
Retired	75	13.91
Students	20	3.71
Other	46	8.53
Marital status		
Married or in a civil partnership	343	63.64
Single, divorced, widowed	196	36.36
Political orientation The question about political orientation was the only optional question in the survey		
Right-wing	136	25.23
Centre	66	12.24
Left-wing	143	26.54
Ecologist party	40	7.42
Did not answer	154	28.57
City size		
10,000 inhabitants	430	79.78
10,000	109	20.22

Table 2 Self-reported awareness and perceptions of brownfield sites

Location characteristics—N = 539	Frequency	Percentage
Aware of the existence of a brownfield near the place of residence		
Yes	116	21.52
No	423	78.48
Number of brownfield sites perceived near the place of residence (Mean = 1.90; Std. dev = 1.28)		
1	59	50.87
2	34	29.31
3	23	19.82
Visits to the brownfield site		
Yes	38	32.76
No	78	67.24

Table 3 Main characteristics of the brownfield sites. *The percentage may exceed 100 as the brownfield types and their locations refer to the overall brownfields that are near the place of residence of the survey respondents. These percentages are not specific to a brownfield site

Brownfield characteristics—N = 116	Frequency	Percentage
*Types of brownfield**		
Industrial brownfield	85	73.28
Military brownfield	10	8.62
Harbor brownfield	4	3.45
Railway brownfield	22	18.97
Former public infrastructures (e.g. schools, hospitals)	26	22.41
*Brownfield location**		
In the city center	44	37.93
In urban periphery area	64	55.17
In rural area	26	22.41
Duration of indeterminate status of the brownfield		
5 years	47	40.52
5 years	69	59.48

Brownfields were considered as abandoned land in urban areas that are unused. The use of the word "empty" underlines this phenomenon: brownfields are uninhabited, unoccupied, or uncultivated areas. Respondents referred to former industrial sites or manufacturing plants, which may explain why they perceived these spaces as contaminated. They mentioned, for instance, pollution and the presence of garbage. Besides, respondents also referred to the etymological sense of brownfield by using the expression "fallow land," meaning that they considered brownfields to be land uncultivated for years.

Table 4 Frequency of the active forms (>10) using IRaMuTeQ; adj: adjective

Active forms	Freq	Type
Abandon	89	Noun
Abandoned land	61	Noun
Land	56	Noun
City	48	Noun
Area	40	Noun
Wastelands	36	Noun
Fallow land	32	Noun
Abandoned manufacturing plants	29	Noun
Pollution	25	Noun
Industrial	25	Noun
Site	22	Noun
Unmaintained	20	Adj
House	19	Noun
Build	18	Verb
Contaminate	17	Verb
Manufacturing plants	16	Noun
Garden	16	Noun
Building	16	Noun
Unbuilt	15	Adj
Abandoned building	15	Noun
Soil	14	Noun
Garbage	14	Noun
Empty	14	Adj
Wild plants	13	Noun
Urban area	13	Noun
Uninhabited	13	Adj
Dump	13	Noun
Waste	12	Noun
Urban	12	Adj
Uncultivated	12	Adj
Nature	12	Noun
Unused	11	Adj
Space	11	Noun
Ruin	11	Noun
Abandoned areas	11	Noun

3.3 *The Relative Importance of Nature in the Case of Contaminated Brownfields*

We use a word cloud in Fig. 3 to highlight the most frequent words used in the corpus. We underline in green the words and expressions that may refer to nature in the case of brownfields. In this word cloud, we observed that nature appeared as a peripheral element: the proportion of words that referred to nature is relatively small.

To examine more, in-depth, the perception of nature, we performed a cooccurrence analysis and a cluster analysis using Hyperbase Web. Figure 4 shows the existing links between the word "nature" and the other words used in the corpus. This graph shows the importance of the links between the word nature, at the top of the graph, and its first-order cooccurrences are shown in the middle of the graph. Cooccurrences are reiterated to identify second-order cooccurrence at the bottom of the graph. The cooccurrence analysis shows that the word "nature" is used with the expression "wild plants" and with sentences and words meaning that nature "reclaims its right" and is "unmaintained" because "vegetation grows."

We can also observe the word "free," which is linked with the fact that brownfields remaining unoccupied and unmaintained represent an opportunity for future redevelopment projects. The problem of waste disposal was also mentioned.

The results of the specificity analysis, which determines the typical words used with the word nature, are given in Table 5.

This specificity analysis showed which words are used in combination with the word "nature" in the answers that were given by our respondents. The specificity

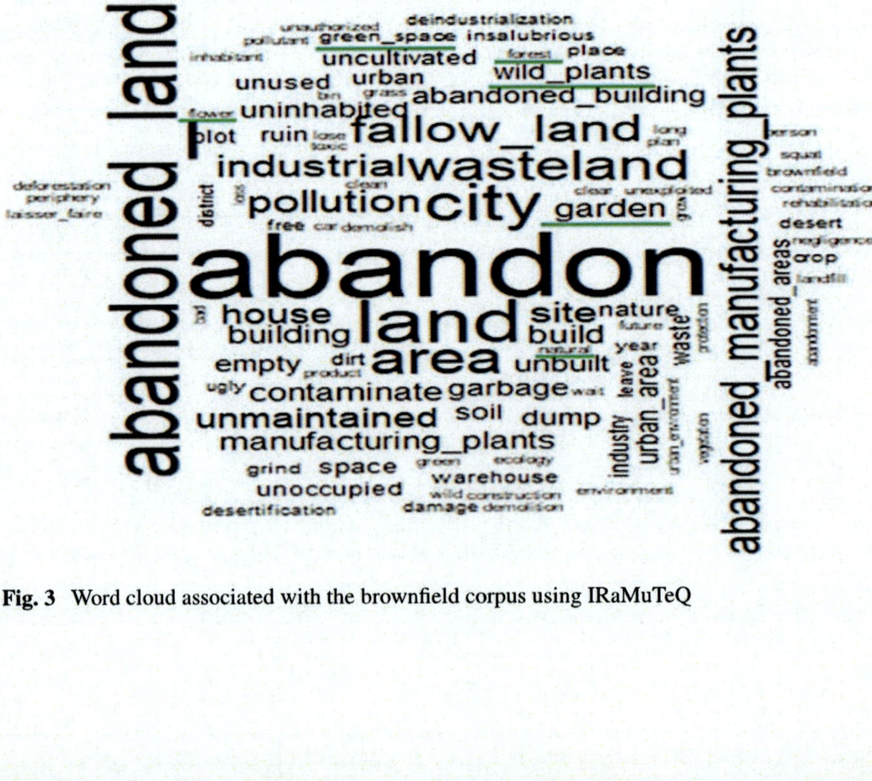

Fig. 3 Word cloud associated with the brownfield corpus using IRaMuTeQ

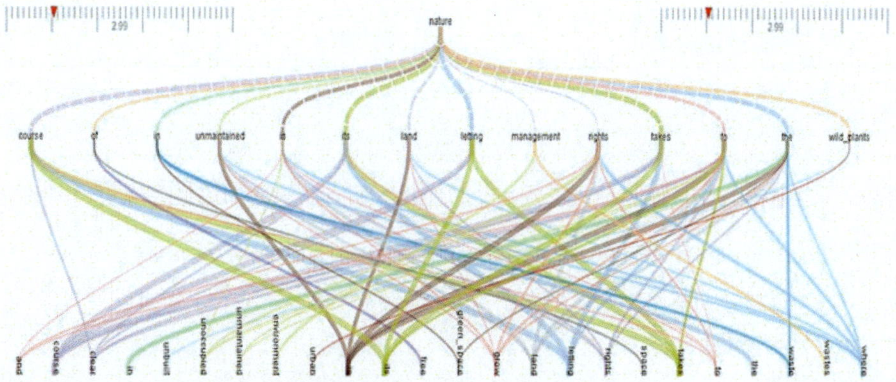

Fig. 4 Cooccurrence analysis of the word "nature." The thicker the line, the stronger the link is with a specific word

Table 5 Specificity analyses of the word "nature"

Words	Specificity	Frequency
Wild plants	3.8	13
Unmaintained	3.59	20
Rights	3.01	2
Management	3.01	2
Natural	2.89	3
Clear	2.89	3
Environment	2.89	3
Grow	2.8	4
Free	2.73	5
Waste	2.73	5
Green space	2.67	6
Sites	2.62	7
Wastes	2.62	7
Space	2.51	10
Unoccupied	2.51	10
Urban	2.44	12
Unbuilt	2.36	15
Industrial	2.17	25

analysis results showed that nature is often used with the terms "wild plants" and "unmaintained." Figure 5 displays the results of the factorial correspondence analysis for the two first axes of the 200 most frequent words in the whole corpus.

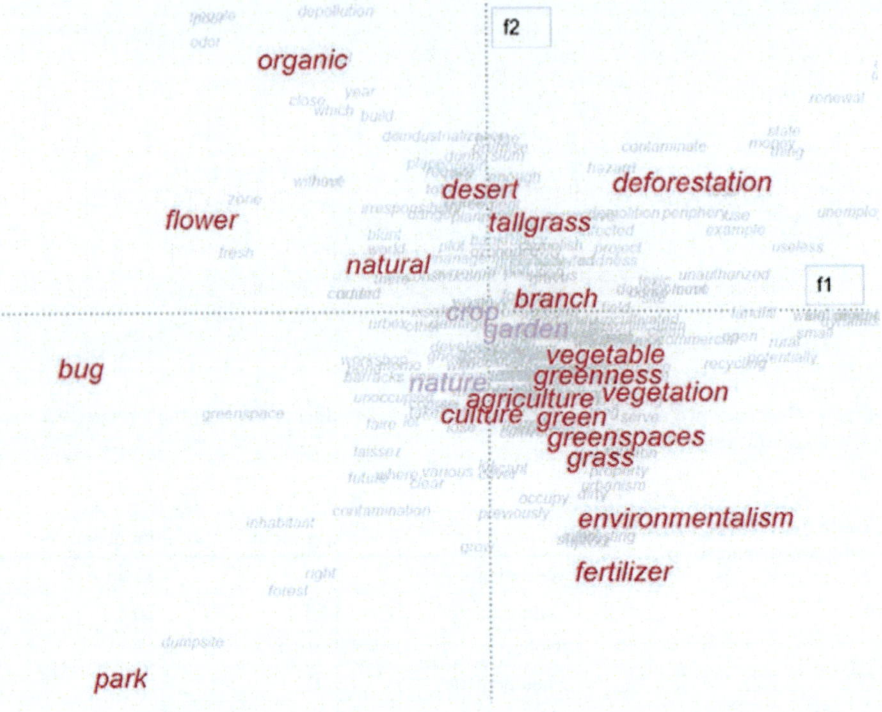

Fig. 5 Correspondence analysis using Hyperbase Web. The words related to nature are highlighted in pink

We have highlighted the words related to nature in pink. The results of the factorial correspondence analysis reveal that respondents perceived a form of biodiversity (bug, flower, tallgrass, for example) and refer to different elements of nature (tree, greenspace, park, forest). Different kinds of gardening practices are mentioned (organic garden, fertilizers, for instance) as well as agricultural practices related to fallow land: uncultivated soil, grazing.

We carried out semantic analyses using Tropes. References gather closely related words in the text. The reference fields enable Tropes to elaborate a representation of the context using the semantic equivalents dictionary of Tropes. This dictionary is composed of three different classification levels. The references are at the more precise level of classification. Table 6 displays the references of the words contained in the corpus in decreasing frequency. Only significant fields are displayed by the software. The frequency counts are the number of words in a reference. We chose to specify the complete expressions or cooccurrences to gain in precision. These references provide a detailed overview of the words in the corpus.

This semantic classification confirms that brownfields are considered as abandoned and unmaintained land. Contamination is perceived by respondents, as are issues regarding waste disposal. Different types of nature are cited. One type of

Table 6 References using tropes

References	Freq	Words or expressions used
Abandoned land	61	Abandoned land
Land	52	Empty land, industrial land, uncultivated land
Area	51	Industrial area, uninhabited area, abandoned area
City	49	City
Wasteland	36	Wasteland
Fallow land	32	Fallow land
Abandoned manufacturing plant	28	Abandoned manufacturing plants
Pollution	25	Pollution
Building	22	Former industrial buildings, insalubrious buildings, unused building
Unmaintained	20	Unmaintained, unmaintained garden, nature unmaintained, crops unmaintained, unmaintained land
House	17	Abandoned house, without house, house in ruins
Site	15	Landfill site, contaminated site, abandoned site
Abandoned building	15	Abandoned building
Unbuilt	15	Unbuilt area, unbuilt plots
Factory	14	Manufacturing plants, former manufacturing plants
Urban area	13	Urban area
Garbage	12	Garbage, polluting garbage, garbage on the ground
Wild plant	12	Wild plants, let wild plants grow
Nature	12	Urban nature, nature, let nature take its course, nature
Garden	10	Vegetable garden, green garden organic garden, garden poorly maintained, urban garden, wild garden
Greenspace	6	Greenspaces
Landfill	6	Landfill site, unauthorized landfill
Crop	6	Crops, without crops, uncultivated crops
Graminaceous plants	5	Tallgrass, grass
Ecology	4	Ecology
Brownfield	3	Brownfield
Flower	3	Flowers
Vegetation	3	Vegetation
Plant	3	Plant, plants
Deforestation	3	Deforestation

nature is linked to agricultural activities and especially fallow land: pieces of land that are left with no crops for a while to recover soil fertility. This type of nature is also mentioned in the land references with the expression “uncultivated land” and in the specific references related to crops. The second form of nature associated with the concept of wilderness is wild plants, and the nature references, for example, are associated with the expression “let nature take its course.” A more domesticated nature is also cited with the garden and the green spaces references.

We determined the proportion of respondents who expressed a vision related to nature to examine its relative importance in our sample. To do so, we created an associated variable corresponding to a global vision of nature. This variable comprises the following words (in alphabetical order): “ecology,” “environment,” “flowers,” “forest,” “garden,” “grass,” “greenness,” “green space,” “leaves,” “nature,” “tallgrass,” “tree,” “vegetation.”. This variable was constructed as a thematic post-codification after the textual data analysis of the corpus to avoid biases (Guerin-Pace 1998; Lebart et al. 1998). According to this variable, we obtained the proportion of respondents who expressed a vision of nature in a broad sense.

Only 14.1% of respondents associated a form of nature with contaminated brownfields in a broad sense of the term nature. Indeed, 14.1% of respondents cited at least one element, which is related to a vision of nature in the case of contaminated brownfields. Thus, a small proportion of respondents referred to nature, 8.72% of them expressed a vision of wild nature, and 5.38% expressed a vision, which is more related to agricultural activities (fallow land, cultivated soil, and crops). We conclude that the wilderness form of nature and the agricultural form of nature are the two main visions of the nature of contaminated brownfields that respondents expressed in our sample.

Moreover, among the respondents who expressed a representation of contaminated brownfields related to nature, 30 were women (63.83%) and 17 (36.17%) were men. Table 7 provides some descriptive statistics of this subsample. Those respondents also considered the spontaneous vegetation on brownfield sites using the expression “wild plants,” “garden,” and “nature.”

However, in our survey, 63% of the respondents preferred green space redevelopment projects (Tendero 2018).

4 Discussion and Conclusion

The aim of this study was to focus on social representations of contaminated brownfield sites in France and to analyze the importance given to nature in these representations. Using an open-ended question in a French national cross-sectional survey administered to people living in municipalities impacted by contaminated brownfields, we analyzed the representations of contaminated brownfields and the elements that may be related to nature in the case of contaminated sites. We performed textual statistics using IRaMuTeq to carry out an exploratory analysis of the corpus. We used the cooccurrence analysis using Hyperbase Web and a semantic classification using

Table 7 Sociodemographic characteristics of the subsample (n = 47)

Variables	Freq	%
Gender		
Male	17	36.17
Female	30	63.83
Age		
18–24	3	6.38
25–39	11	23.40
40–59	29	61.70
64	4	8.51
Educational qualification		
None/primary school		
Secondary school/vocational education		
High school diploma		
Associate degree (2-year diploma)		
Undergraduate/Graduate degree (3-year or more diploma)		
Occupational categories		
Farmer	0	0
Self-employed, small shopkeepers, business owners	2	4.26
Senior managers and higher managerial, administrative, and professional positions	6	12.77
Intermediate managerial, administrative positions	7	14.89
Services and sales workers, clerical support workers	13	27.66
Manual workers	3	6.38
Inactive	16	34.04
Marital status		
Married or in a civil partnership	31	65.96
Single, divorced, widowed	16	34.04
Region of residence		
Île-de-France	8	17.02
Hauts-de-France	6	12.77
Grand Est	7	14.89
Auvergne-Rhône-Alpes	3	6.38
Nouvelle-Aquitaine	3	6.38
Centre-Val-de Loire	2	4.26
Normandy	5	10.64
Occitanie	1	2.13
Provence-Alpes-Côte d'Azur	3	6.38
Bourgogne-Franche-Comté	2	4.26

(continued)

Table 7 (continued)

Variables	Freq	%
Brittany	5	10.64
Pays de la Loire	2	4.26
Corse	0	0

Tropes. This strategy allowed us to identify and understand more, in-depth, different dimensions related to nature in the case of contaminated brownfields (Fig. 6).

Our results showed that only 14.1% of respondents associated contaminated brownfields with a form of nature. Indeed, 14.1% of respondents cited at least one element which is related to a vision of nature in the case of contaminated brownfields. A small proportion of respondents referred to different forms of nature. Figure 7 summarizes the results in the light of the existing typologies of nature in the literature (Buijs and Volker 1997; Hunt 2000; Kowarik 1992). This figure is adapted from the typologies developed by Buijs and Volker (1997), Hunt (2000), and Kowarik et al. (1992) and shows the links between each category in these three typologies and our results. People distinguish different forms of nature. The first form of nature corresponds to a wild garden with wild plants and tall grass. This vision of nature is negatively perceived from an aesthetic point of view because this space is considered as abandoned and unmaintained. The second form of nature corresponds to agricultural activities and refers more precisely to fallow land, which corresponds to the etymological sense of the word brownfield or *friche* (wasteland) in French. The third form of brownfield corresponds to vegetable gardens or organic gardens and green spaces that could represent potential temporary uses or future redevelopment projects. The last form refers to the state of brownfield in itself: a vacant land.

Contaminated brownfields are a form of temporary ordinary nature. As contaminated brownfields remain in a transitional state for decades, spontaneous nature vegetation grows. However, this nature is negatively perceived because people perceive such sites as contaminated and unmaintained. Similar results have been found using

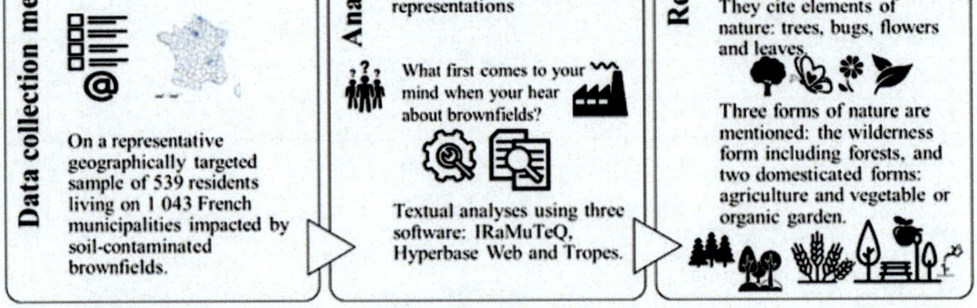

Fig. 6 Graphical synthesis of the research process and the results

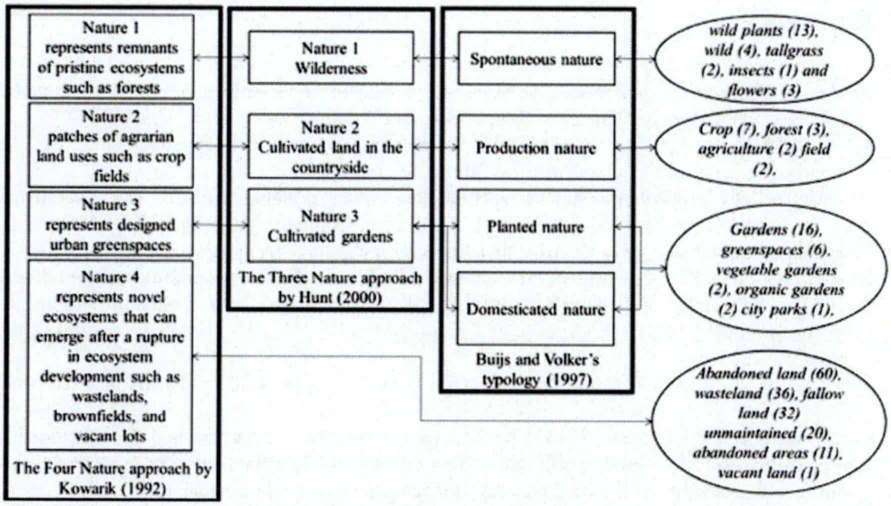

Fig. 7 Summary of the results within existing typologies on nature. In italics are the words that are cited in the corpus and their frequencies are presented in italics in parentheses

focus groups among inhabitants of the city of "La Possonnière" ("Maine et Loire" department) in France when the focus groups described a contaminated brownfield located in this city which was characterized by uncontrolled vegetation growth (Tendero and Plottu 2019). A minority of people also consider that such sites may offer an opportunity for productive nature linked with the possibility of common gardening practices and agricultural activities. It is interesting to note that this type of answer is closely linked to the preferences expressed by our respondents regarding possible brownfield redevelopment projects. In our survey, most respondents (63%) preferred green space redevelopment projects (Tendero 2018). Other studies in Italy (Turvani et al. 2010), the United Sates (Greenberg and Lewis 2000), and China (Li et al. 2016) have found similar results regarding preferences for the redevelopment of contaminated brownfields.

Our sample does not provide enough information regarding the perceptions of the respondents who said they visited a brownfield site (only 38 answers). However, this factor could have an influence on their representations (Brun et al. 2018, 2019). An interesting research avenue would be to develop further the representations of people who have visited a contaminated brownfield to analyze the potential differences of representations between the spaces perceived and experienced.

Acknowledgements The University of Montpellier, the SRUM 2015, and the CEE-M (Center for Environmental Economics—Montpellier; formerly Montpellier Laboratory of Theoretical and Applied Economics) supported the implementation of this survey. We would like to thank Jean-Marc Rousselle (CEE-M) who administered the survey. We thank participants of the international symposium "Urban wastelands: a form of urban nature" which was held in May 2019 for valuable comments and suggestions. We also thank Francesca di Pietro and Amelie Robert for their remarks on a previous version of this chapter.

References

Adams D, Disberry A, Hutchison N, Munjoma T (2001) Ownership constraints to brownfield redevelopment. Environ Plan Econ Space 33:453–477. https://doi.org/10.1068/a33200

Adams WM, Hodge ID, Bourn NAD (1994) Nature conservation and the management of the wider countryside in eastern England. J Rural Stud 10(2):147–157

ADEME (2014) Biodiversité et reconversion des friches urbaines polluées [Biodiversity and contaminated brownfield redevelopment] - Collection Connaître pour Agir - ADEME (Agence de l'Environnement et de la Maîtrise de l'Énergie)

ADEME, QuelleVille? (2015) Estimation du gisement foncier des friches urbaines potentiellement polluées [Estimation of the land base of potentially contaminated urban wastelands, a summary] - Note de synthèse. ADEME (Agence de l'Environnement et de la Maîtrise de l'Énergie) - Quelle Ville?

Alker S, Joy V, Roberts P, Smith N (2000) The definition of brownfield. J Environ Plan Manag 43:49–69

Bailly A, Bourdeau-Lepage L (2011) Concilier désir de nature et préservation de l'environnement: vers une urbanisation durable en France [Toward sustainable urbanization in France: desire for nature and protection of the environment]. Géographie Économie Société 13:27–43

Bambra C (2018) Placing health inequalities: where you live can kill you. In: Crooks VA, Andrews GJ, Pearce J (eds) Routledge handbook of health geography. Routledge Handbooks, Routledge, p 386

Bambra C, Cairns JM, Kasim A, Smith J, Robertson S, Copeland A, Johnson K (2015) This divided land: an examination of regional inequalities in exposure to brownfield land and the association with morbidity and mortality in England. Health Place 34:257–269. https://doi.org/10.1016/j.healthplace.2015.05.010

Bauer N (2005) Attitudes towards wilderness and public demands on wilderness areas. In: Kowarik I, Körner S (eds) Wild urban woodlands new perspectives for urban forestry. Springer Berlin Heidelberg, Berlin, Heidelberg, pp 47–66

Bontefeu E (2009) La demande sociale de nature en ville. Enquête auprès des habitants de l'agglomération lyonnaise [Social demand for nature in the city. Survey of residents of the Lyon metropolitan area]. Urbia 8:21–38

Bonthoux S, Brun M, Di Pietro F, Greulich S, Bouché-Pillon S (2014) How can wastelands promote biodiversity in cities? A Review. Landsc Urban Plan 132:79–88. https://doi.org/10.1016/j.landurbplan.2014.08.010

Botzat A, Fischer LK, Kowarik I (2016) Unexploited opportunities in understanding liveable and biodiverse cities. A review on urban biodiversity perception and valuation. Glob Environ Change 39:220–233. https://doi.org/10.1016/j.gloenvcha.2016.04.008

Bourdeau-Lepage L (2018) Nature and well-being in the French City: desire and Homo Qualitus. Built Environ 44:339–353. https://doi.org/10.2148/benv.44.3.339

Bourdeau-Lepage L, Vidal R (2013) Nature urbaine en débat: à quelle demande sociale répond la nature en ville? [Urban nature in debate: what social demand does nature in the city respond to?] Déméter 2013 195–210

Brun M, Di Pietro F, Bonthoux S (2018) Residents' perceptions and valuations of urban wastelands are influenced by vegetation structure. Urban For Urban Green 29:393–403. https://doi.org/10.1016/j.ufug.2017.01.005

Brun M, Di Pietro F, Martouzet D (2019) Le délaissés urbains: supports de nouvelles pratiques et représentations de la nature spontanée? Comparaison des représentations des gestionnaires et des habitants [Can they support new uses and perceptions of spontaneous nature? Comparison of residents and planners' perceptions]. Nouvelles Perspectives En Sciences Sociales 14:153–184. https://doi.org/10.7202/1062509ar

Brunet É, Luong X (1989) Computer processing and quantitative text analysis: HYPERBASE, an interactive software for large corpora. In: Data analysis, learning symbolic and numeric knowledge. Nova Science Publishers, United Kingdom, pp 207–214

Buijs AE, et Volker CM (1997) « Publiek draagvlak voor natuur en natuurbeleid ». SC-DLO Report 546. The Public Basis of Nature and Nature Policy. Wageningen University, Wageningen, The Netherlands

Buzydlowski JW (2015) Co-occurrence analysis as a framework for data mining. J Technol Res 6:1–19

Capaldi C, Passmore H-A, Nisbet E, Zelenski J, Dopko R (2015) Flourishing in nature: a review of the benefits of connecting with nature and its application as a wellbeing intervention. Int J Wellbeing 5:1–16. https://doi.org/10.5502/ijw.v5i4.449

Dany L, Urdapilleta I, Monaco GL (2015) Free associations and social representations: some reflections on rank-frequency and importance-frequency methods. Qual Quant 49:489–507. https://doi.org/10.1007/s11135-014-0005-z

Dixon T, Otsuka N, Abe H (2011) Critical success factors in urban brownfield regeneration: an analysis of 'hardcore' sites in Manchester and Osaka during the economic recession (2009–10). Environ Plan Econ Space 43:961–980. https://doi.org/10.1068/a43468

Dorsey JW (2003) Brownfields and greenfields: the intersection of sustainable development and environmental stewardship. Environ Pract 5. https://doi.org/10.1017/S1466046603030187

Faivre N, Fritz M, Freitas T, de Boissezon B, Vandewoestijne S (2017) Nature-based solutions in the EU: innovating with nature to address social, economic and environmental challenges. Environ Res 159:509–518. https://doi.org/10.1016/j.envres.2017.08.032

Franco L, Shanahan DF, Fuller R (2017) A review of the benefits of nature experiences: more than meets the eye. Int J Environ Res Public Health 14

Gardiner MM, Burkman CE, Prajzner SP (2013) The value of urban vacant land to support arthropod biodiversity and ecosystem services. Environ Entomol 42:1123–1136. https://doi.org/10.1603/EN12275

Gilderbloom JI, Meares WL, Riggs W (2014) How brownfield sites kill places and people: an examination of neighborhood housing values, foreclosures, and lifespan. J Urban Int Res Placemaking Urban Sustain 1–18. https://doi.org/10.1080/17549175.2014.905488

Given L (Ed) (2008) Textual analysis. In: The SAGE encyclopedia of qualitative research methods. SAGE, 2455 Teller Road, Thousand Oaks California 91320 United States, pp 865–866

Godet L (2010) La «nature ordinaire» dans le monde occidental ["Ordinary Nature" in the Western World]. Espace Géographique 39:295. https://doi.org/10.3917/eg.394.0295

Greenberg M, Lewis MJ (2000) Brownfields redevelopment, preferences and public involvement: a case study of an ethnically mixed neighbourhood. Urban Stud 37:2501–2514

Guerin-Pace F (1998) Textual statistics. An exploratory tool for the social sciences. Popul Engl Sel, New Methodological Approaches in the Social Sciences 10:73–95

Harrison C, Davies G (2002) Conserving biodiversity that matters: practitioners' perspectives on brownfield development and urban nature conservation in London. J Environ Manage 65:95–108. https://doi.org/10.1006/jema.2002.0539

Hofmann M, Westermann JR, Kowarik I, van der Meer E (2012) Perceptions of parks and urban derelict land by landscape planners and residents. Urban For Urban Green 11:303–312

Hou D, Song Y, Zhang J, Hou M, O'Connor D, Harclerode M (2018) Climate change mitigation potential of contaminated land redevelopment: a city-level assessment method. J Clean Prod 171:1396–1406. https://doi.org/10.1016/j.jclepro.2017.10.071

Hunt JD (2000) Greater perfections: the practice of garden theory. If gardening. Penn studies in landscape architecture. University of Pennsylvania, University of Pennsylvania Press, USA

Hunter P (2014) Brown is the new green: Brownfield sites often harbour a surprisingly large amount of biodiversity. EMBO Rep 15:1238–1242. https://doi.org/10.15252/embr.201439736

Kaplan R, Kaplan S, Ryan R (1998) With people in mind: design and management for everyday nature. Island Press, Washington, USA

Kasparian S (2014) Les chiffres et les lettres peuvent-ils se marier? 15 ans de recherches au LADT [Can letters and numbers get married? 15 years of research at LADT]. Rev Univ Monct 45:1. https://doi.org/10.7202/1038898ar

Kattwinkel M, Biedermann R, Kleyer M (2011) Temporary conservation for urban biodiversity. Biol Conserv 144:2335–2343. https://doi.org/10.1016/j.biocon.2011.06.012

Keniger L, Gaston K, Irvine K, Fuller R (2013) What are the Benefits of Interacting with Nature? Int J Environ Res Public Health 10:913–935. https://doi.org/10.3390/ijerph10030913

Kim EJ, Miller P (2017) Residents' perception of local brownfields in rail corridor area in the City of Roanoke: the effect of people's preconception and health concerns factors. J Environ Plan Manag 60:862–882. https://doi.org/10.1080/09640568.2016.1182898

Koch F, Bilke L, Helbig C, Schlink U (2018) Compact or cool? The impact of brownfield redevelopment on inner-city micro climate. Sustain Cities Soc 38:31–41. https://doi.org/10.1016/j.scs.2017.11.021

Kowarik I (1992) « Das Besondere der städtischen Flora und Vegetation ». Schriftenreihe des deutschen Rates für Landespflege 61(5):33–47

Kowarik I (2018) Urban wilderness: Supply, demand, and access. Urban For Urban Green 29:336–347. https://doi.org/10.1016/j.ufug.2017.05.017

Lafon P (1980) Sur la variabilité de la fréquence des formes dans un corpus [On the variability of the frequency of forms in a corpus]. Mots 1:127–165

Lafortezza R, Corry RC, Sanesi G, Brown RD (2008) Visual preference and ecological assessments for designed alternative brownfield rehabilitations. J Environ Manage 89:257–269

Laille P, Provendier D, Colson F, Salanié J (2014) Les bienfaits du végétal en ville. Étude des travaux scientifiques et méthode d'analyse [The benefits of plants in the city. Study of scientific work and method of analysis] (Rapport Plante & Cité). Plante & Cité

Lebart L, Salem A, Berry L (1998) Exploring textual data, text, speech, and language technology. Kluwer Academic, Dordrecht, Boston

Lelay M, Dechaume-Moncharmont C, Vion M, Douchet M (2019) L'éco-pâturage pratiqué par SNCF Réseau [Eco-grazing practiced by the French rail network]. Sci Eaux Territ Numéro 27:86. https://doi.org/10.3917/set.027.0086

Li X, Jiao W, Xiao R, Chen W, Bai Y (2016) Regional variations of public perception on contaminated industrial sites in china and its influencing factors. Int J Environ Res Public Health 13:410. https://doi.org/10.3390/ijerph13040410

Loures L, Vaz E (2018) Exploring expert perception towards brownfield redevelopment benefits according to their typology. Habitat Int 72:66–76. https://doi.org/10.1016/j.habitatint.2016.11.003

Macadam C, Bairner S (2012) Urban biodiversity: successes and challanges: brownfields: oases of urban biodiversity. Glasg Nat 25

Mathey J, Arndt T, Banse J, Rink D (2018) Public perception of spontaneous vegetation on brownfields in urban areas—results from surveys in Dresden and Leipzig (Germany). Urban For Urban Green 29:384–392. https://doi.org/10.1016/j.ufug.2016.10.007

Mathey J, Rink D (2010) Urban wastelands—a chance for biodiversities in cities? Ecological aspects, social perceptions and acceptance of wilderness by residents. In: Urban biodiversity and design, conservation science and practice. Müller Norvert, Werner Peter & Kelcey John, pp 406–424

Maurel N, Salmon S, Ponge J-F, Machon N, Moret J, Muratet A (2010) Does the invasive species Reynoutria japonica have an impact on soil and flora in urban wastelands? Biol Invasions 12:1709–1719

Mayaffre D, Ben Hamed M (2014) Récits de mort et souvenir traumatique. Trames et traces lexicales des témoignages sur la Shoah [Stories of death and traumatic memory. Wefts and lexical traces of testimonies on the Shoah]. Argum Anal Discours. https://doi.org/10.4000/aad.1836

Molette P, Landré A, Ghiglione R (2013) Tropes. Version 8.4. Manuel de référence [Tropes Version 8.4. Users manual]

Moscovici S (2003) Des représentations collectives aux représentations sociales: éléments pour une histoire [From collective representations to social representations: elements for a story]. In: Les représentations sociales. Presses Universitaires de France, p 79

Muratet A, Machon N, Jiguet F, Moret J, Porcher E (2007) The role of urban structures in the distribution of wasteland flora in the Greater Paris Area, France. Ecosystems 10:661–671

Payá Pérez A, Rodríguez Eugenio N (2018) Status of local soil contamination in Europe (Joint Research Centre Technical Reports No. EUR 29124 EN), Revision of the indicator "Progress in the management contaminated sites in Europe." European Commission - Joint Research Centre (JRC)

Pêcheux M, Léon J, Bonnafous S, Marandin J-M (1982) Présentation de l'analyse automatique du discours (AAD69): théories, procédures, résultats, perspectives [Presentation of automatic speech analysis (AAD69): theories, procedures, results, perspectives]. Mots 4:95–123. https://doi.org/10.3406/mots.1982.1053

Pellegrin C, Sabatier R, Napoléone C, Dutoit T (2018) Une définition opérationnelle de la nature ordinaire adaptée à la compensation écologique. Le cas contrasté des régions Centre, Champagne-Ardenne et Paca [A working definition of "ordinary nature" adjusted to ecological offset policies tested in three French regions]. Nat Sci Sociétés 26:170–188. https://doi.org/10.1051/nss/2018034

Pincemin B (2018) Sept logiciels de textométrie [Seven textometry software] (Working paper). Laboratoire IHRIM- Université de Lyon, Lyon, France

Piolat A, Bannour R (2009) An example of text analysis software (EMOTAIX-Tropes) use: the influence of anxiety on expressive writing. Curr Psychol Lett 25

Ratinaud P, Déjean S (2009) IRaMuTeQ: implémentation de la méthode ALCESTE d'analyse de texte dans un logiciel libre [Iimplementation of the ALCESTE method of text analysis in free software]. Modélisation Appliquée Aux Sci Hum Soc MASHS 8–9

Raymond R, Simon L (2012) Biodiversité : les services écosystémiques et la nature en ville [Biodiversity: ecosystem services and nature in cities]. Rev For Fr. https://doi.org/10.4267/2042/48442

Rosato P, Alberini A, Zanatta V, Breil M (2010) Redeveloping derelict and underused historic city areas: evidence from a survey of real estate developers. J Environ Plan Manag 53:257–281. https://doi.org/10.1080/09640560903529550

Rupprecht CD, Byrne JA (2014) Informal urban greenspace: a typology and trilingual systematic review of its role for urban residents and trends in the literature. Urban For Urban Green 13:597–611

Scapino J (2016) De la friche urbaine à la biodiversité. Ethnologie d'une reconquête: La petite ceinture de Paris [From urban wasteland to biodiversity. Ethnology of a reconquest: the small belt of Paris] (PhD thesis in architecture, space planning). Muséum National d'Histoire Naturelle (MNHN), Paris

Schmitz C (2015) Limesurvey: an open source survey tool

Souza MAR de, Wall ML, Thuler AC de MC, Lowen IMV, Peres AM (2018) O uso do software IRAMUTEQ na análise de dados em pesquisas qualitativas. Rev. Esc. Enferm. USP 52. https://doi.org/10.1590/s1980-220x2017015003353

Statista (2019) La démographie en France [The French demography] (Dossier Statista sur la démographie en France No. did-31527-1), Dossier Statista. Statista

Tendero M (2018) Reconversion et aménagement durable des friches urbaines polluées - Élaboration d'une méthode participative d'aide multicritère à la décision [Sustainable contaminated urban brownfield redevelopment—development of a participatory method of multi-criteria decision support (PhD thesis in economic sciences). Agrocampus Ouest, Angers, France

Tendero M, Plottu B (2019) A participatory decision support system for contaminated brownfield redevelopment: a case study from France. J Environ Plan Manag 62:1736–1760. https://doi.org/10.1080/09640568.2018.1512476

Triplet P (2020) Dictionnaire encyclopédique de la diversité biologique et de la conservation de la nature [Encyclopedic dictionary of biological diversity and nature conservation], 6ème. ed

Turvani M, Tonin S, Alberini A (2010) Stakeholder preferences for contaminated land clean-up and redevelopment programmes in Italy. J Urban Regen Renew 3:315–323

Westphal L, Levengood J, Wali A, Soucek D, Stotz D (2004) Brownfield redevelopment: a hidden opportunity for conservation biology. Bengston David, Columbia University, New York

Marjorie Tendero is an Assistant Professor of Economics at the ESSCA School of Management, Angers (France). In 2018, she defended her PhD thesis on the redevelopment of contaminated brownfields. Her research looks at environmental problems and land use issues, such as the redevelopment of brownfields, the uses of community gardens, and vacant land. She uses economic psychology and urban and economic geography methodologies to analyze the distribution of degraded spaces, including vacant land.

Cécile Bazart is an Assistant Professor at the University of Montpellier (France). Her research interests focus on incentives and environments supporting compliance. She uses economic psychology and behavioral as well as experimental methodologies to explore the wide domain of pro-fiscal, pro-social, and pro-environmental behaviors, including issues regarding the acceptance of brownfields.

Getting to Know Urban Wasteland—A Look at Vacant Lands as Urban Green Space in Japan

Minseo Kim and Christoph D. D. Rupprecht

Abstract Urban spaces are dotted with various interstitial spatial areas from very narrow spaces between buildings or structures to huge spaces between parcels. These in-between spaces are filled with plants that represent the surrounding nature, partly or entirely. In urban areas with past human interference, can we thus consider or recognize this quasi-nature as urban green space? This research begins on the premise that the role of urban green space is important in supporting the combined well-being of urban residents. We thus review the potential of vacant lands as urban wastelands in the context of the state of affairs in Japan, which is undergoing a paradigm shift in urban green policy. We surveyed Ichikawa City (Japan) as an example of a city that has already been or is currently being urbanized. The survey combined field surveys and perception surveys to identify vacant lands and to understand residents' perceptions. The quantity of vacant lands observed corresponds to about 1.43% of Ichikawa City. Residents with higher exposure to traditional green space in their daily lives were more aware of the existence of vacant lands. In addition, respondents who see vacant lands as an urban green space show a positive and active attitude toward existing urban green space and urban nature. Moreover, they respond more strongly to the issue of the non-sustainability of vacant lands than to the issue of private property. As a result, vacant land may have a high tendency to be perceived by residents as an intimate, local space, suggesting usability. We finally highlight that vacant lands can serve an alternative or supplementary role in cities with limitations to creating new urban green space.

Keywords Vacant land · Urban wasteland · Green space policy · Informal green space · Participatory management

M. Kim (✉)
Department of Academic Research and Industry-Government Collaboration, Planning and Promotion Division, Hiroshima University, Hiroshima, Japan
e-mail: hellominseo@gmail.com

C. D. D. Rupprecht
FEAST Project, Research Institute for Humanity and Nature, Kyoto, Japan
e-mail: crupprecht@chikyu.ac.jp

F. Di Pietro and A. Robert (eds.), *Urban Wastelands*, Cities and Nature,
https://doi.org/10.1007/978-3-030-74882-1_9

1 Introduction

Urbanization, which changes the living structure and lifestyle of residents, has become a global phenomenon (WHO Regional Office for Europe 2016). The urban environment is influenced by population growth as rural land uses and vegetated lands are removed to provide facilities for accommodating human needs. Urbanization, if promoted without regulation, eliminates more vegetation and increases an overtly man-made environment, thereby depriving residents of the opportunity to encounter nature. Therefore, many scholars have focused their attention on urban green space (UGS) as being important to improve not only the quality of life of residents, but also the urban environment (Kabisch et al. 2015). In cities where urbanization is underway or has already plateaued, the role of UGS has been increasingly emphasized in terms of environment, economic, and social importance (Wright Wendel et al. 2012). Although the definition and classification of UGS may differ depending on the context, UGS embraces "natural surface" or includes "nature settings," and also includes "blue space" such as water elements (WHO Regional Office for Europe 2017). Accepting this contextual definition, many researchers have studied urban parks, forests, gardens, etc. to improve the quality of life of urban dwellers and the urban environment.

Despite the issues around the function and benefits of UGSs, an overall difficulty lies in creating new green spaces in dense urban environments. These difficulties are pronounced for cities that have already undergone or have intensified urbanization processes, imposing financial and spatial burdens. The state of affairs related to public infrastructure projects is particularly noticeable in countries such as Japan, where economic and population growth has peaked, and cities have begun to decline (Yokohari et al. 2010; Rupprecht 2017). Dense cities there have already been challenged by the lack of spaces for new UGS. For some Asian countries, it has been pointed out that in cities where urbanization has progressed, dense infrastructure such as cables and pipes underground could increase the difficulty of developing new green spaces (Tian and Jim 2012). Expectations and demands of people toward green space today are also influenced by various social changes. Further challenges imposed on UGS today are not only spatial limitations and rapidly changing social needs, but also low financial support to local government. UGS often has low policy priorities at national or regional levels across the world. This eventually leads to a vicious circle of concerns about UGS, as low policy priorities and budget serve as a barrier to meeting social needs and improving low quality. These challenges may lead us to ask whether we should explore a progressive paradigm for UGS.

In shrinking cities, there is an increasing interest in integrating wasteland, which would be defined as unmanaged and neglected areas, into green infrastructure (Nejman 2018). Wasteland occurs in response to dramatic economic, social, and policy changes in urban structure due to the declining population in the process of de-industrialization (Martinez-Fernandez et al. 2012). In Europe, the reorganization of cities that has been repeated over the decades has led to the formation of land that was left unused. In Japan, population decreases have been recorded since 2008, and the number of vacant lands (VLs) and empty houses has risen (Mizuno et al. 2016). Wasteland generally tends to be perceived as a signal of a downturn by local people (Corbin 2003). Thus, it may also be perceived as functional exhaustion and criminal association of urban degeneration (Maurer et al. 2000). Nevertheless, in more recent studies, wastelands are attracting attention as an opportunity to integrate UGSs in the context of urban decline.

The rate of VL is increasing in the majority of cities in Japan, which face a population-declining society. According to the land survey conducted by Japan's Ministry of Land, Infrastructure, and Transport, the proportion of empty houses and VL increased 1.4 times over the decade from 2003 to 2013 (Hiroshi 2018). Lands, which had risen in price during the development of the city, are often left unmanaged due to the decline in use-value and loss of asset-value as the city enters an age of decline. The owners of the neglected VLs do not perform the necessary maintenance due to decreased desire to own and use them. As a result, the spaces begin to turn green through spontaneous vegetation.

Urban spaces go through cycles of planning and (re)development regularly and repeatedly, which could result in spatial by-products, such as wastelands, arable lands, brownfields, as well as VLs, which could be perceived as leftover spaces. These are created not as a result of destruction and degradation, but as a result of differences in time as spatial by-products of policy action. In our previous research, we investigated these spatial by-products, which we called informal green space in Ichikawa City in Japan, and classified nine types: VLs, street verges, water verges, gaps, brownfield, unimproved lands, parking lot verges, railroad verges, and overgrown structures (Kim et al. 2018). This chapter addresses only VL as an urban wasteland from the nine types of informal green space, based on Japan's UGS policy, which is considering VL to supplement the supply of UGS in a daily living environment close to residents.

2 Turning Point of Urban Green Space Policy in Japan

A characteristic of VL in declining regional cities in Japan is that it generally occurs in the gradual process of population shrinkage. It is often pointed out that the developers' misguided plans or land speculation following the collapse of the bubble economy period may be a cause of VL occurrence, but this is not the dominant factor (Yamada et al. 2016). A decline in residential attractiveness and abandonment of management due to the aging of landowners may result in the inability to determine landownership. Therefore, VL remains neglected for a long time until it is reused for a new purpose. The Japanese government has been trying to tackle national-level policies for VL and empty houses. However, the countermeasures for VL by the administration have remained at the level of individual buildings or sites, so they have not reached the point of discussion of what status such sites should have throughout the city structures (Mizuno et al. 2016).

The national budget in Japan for promoting public infrastructures, which includes the maintenance and creation of UGS, has been steadily decreasing since it peaked in 1997 (Ministry of Finance 2017). Japan's population grew after the period of high economic growth, but after reaching its peak in 2008, it began to shrink. The Japanese government intends to establish and review urban policies that take into account social issues raised by changes in the population structure, such as low birthrate and aging population. Therefore, today, Japan's urban park and green space policies place more weight on maintaining and utilizing its current inventories effectively than building new facilities to cope with the changing times. The shift in the stance of these policies is interpreted as a response to the aging of social overhead capital facilities that was intensively invested during the high economic growth period. Recently, Japan has been experiencing a multiplication of needs for green spaces and open spaces, including urban parks, against the background of a declining population, and financial and administrative restrictions by local government (City Bureau of MLIT 2018). Under these circumstances, the Urban Green Space Conservation Act[1] was partially amended to actively create facility green spaces and conserve appropriate UGSs while maximizing private cooperation in 2017. From the 1970s, when the economy was booming, Japan's per capita park area exceeded the national average of 10.0 m^2 as a result of continuous investment in social overhead capital.[2] However, this figure represents only the national average, and still falls short of the level recommended by the Urban Park Act in many local cities. Therefore, a private-participatory green space system, "Citizen Green Space Certification System," was established to take into account connections with the private sector in order to secure green spaces in a familiar place of daily life for urban residents and to promote practical activities therein. This system provides privately owned land for residents to use, and the person in charge of installing and managing it as a green space establishes the plan and uses it for a certain period after obtaining recognition from the city's decision-making authority, such as the mayor.[3] Recent policy trends for VL in Japan seem to be moving toward detecting social changes and establishing and systematically reorganizing them as a system not only for individuals but also for cities as a whole.

3 Ichikawa City as Research Site

Ichikawa City in Japan could be regarded as an example of a city with limited budget and spatial room to create new UGSs, although the municipal government wants a qualitative as well as a quantitative increase of UGSs. The city was formed and developed by the functional expansion of Tokyo during the high economic growth period in Japan, and most of the residents came from outside of the city. Thus, it has been pointed out that the local community attachment is relatively low compared to

[1] Japan enacted the Urban Green Space Conservation Act (都市緑地保全法) in 1973 and revised the name to the Urban Green Space Act (都市緑地法) in 2004, but the English name is maintained as the Urban Green Space Conservation Act.

[2] The Urban Park Act in Japan recommends an urban park area of 10 m^2 per capita.

[3] Article 60 of the revised Urban Green Space Conservation Act.

other regions around it (Ichikawa City Urban Planning Division 2017). From the late 1950s to the late 1990s, Ichikawa's population soared, urban development sprawled, and land prices increased exponentially. Therefore, the municipal government has been planning to provide qualitative and quantitative UGSs as a means to enhance community attachment and improve the quality of life for residents. According to the Green Master Plan of Ichikawa City, the government is planning in three phases[4] of green space boost projects from 2003 to 2025, with green space area per person (m^2/person) as an indicator (Ichikawa City Urban Planning Division 2004). Despite the long-term plan of over 20 years, the target quantity has not been achieved at each phase, and even if the final target is completed, it is far below the 10.0 m^2 recommended by Japan's Urban Park Act. In addition, the limitations of budget and space to create and maintain a new urban green space are one of the causes of failure of the plan.

4 Collecting Data

Many types of green spaces, such as urban parks and conservation forests, are officially registered, managed, and included in urban planning or governance. However, the urban wasteland is not classified formally or data-driven by most governments (Feltynowski et al. 2018). Therefore, this research obtained data through sampling fieldwork and a mail-back survey to identify urban wastelands and to understand residents' perceptions (Rupprecht et al. 2015; Kim et al. 2018).

We surveyed and identified informal green spaces distributed in Ichikawa City and classified them into nine types (Kim et al. 2018, 2020). Here, only VL, considered to be highly available as supplementary urban green space in terms of human use, will be covered.[5]

Box 1: Methodologies used in this Chapter

Sampling fieldwork

The sampling method in our research was applied by referring to "Regular" and "Equal-stratified" in soil science. 229 sample sites were placed on the intersection between 500 m grids on the study site (Fig. 1). The size of each sample is 0.25 ha (a square with sides of 50 m), and the total combined area is 57.25 ha, which is equivalent to about 1% of the whole city. The size and placement interval of the samples being set up in our research take into account the installation standard of "city block parks," the smallest unit in Japan's urban park installation standard.

Mail-back survey

[4] The indicator at the time when the municipal government began to declare the plan was 2.70 m^2, and the next goal was set at 3.85 m^2 for 2015 before the final goal of 4.73 m^2 per person by 2020.

[5] To propose the supplementary availability of urban green spaces, we identified and classified nine types of informal green spaces: vacant lots, street verges, water verges, gaps, brownfields, unimproved lands, parking lot verges, railroad verges, and overgrown structures (Kim et al. 2018), referring to prior research (Rupprecht and Byrne 2014a, b).

Questionnaire data were collected by means of a mail-back survey that respondents could complete at any time at home. The questionnaire kit, which consists of a question sheet, a statement of the research purpose, and a prepaid envelope for respondents to reply, was distributed around each sample site. We collected 517 significant responses from the 3,700 mail-back survey kits we distributed to 185 sample sites (Response rate: 13.97%) (Fig. 2[6]). The composition of the questionnaire was developed from the results of the pilot workshop conducted with 70 undergraduate students before the main survey.

Data collection for fieldwork and residents' perception was carried out in 2017. For fieldwork, we placed sample sites using a geographic information system (ArcMap 10.5), extracted the coordinates of each site, and conducted a direct visit. To analyze statistically and chart the perception survey, Excel 2016 and IBM SPSS (version 26) were used.

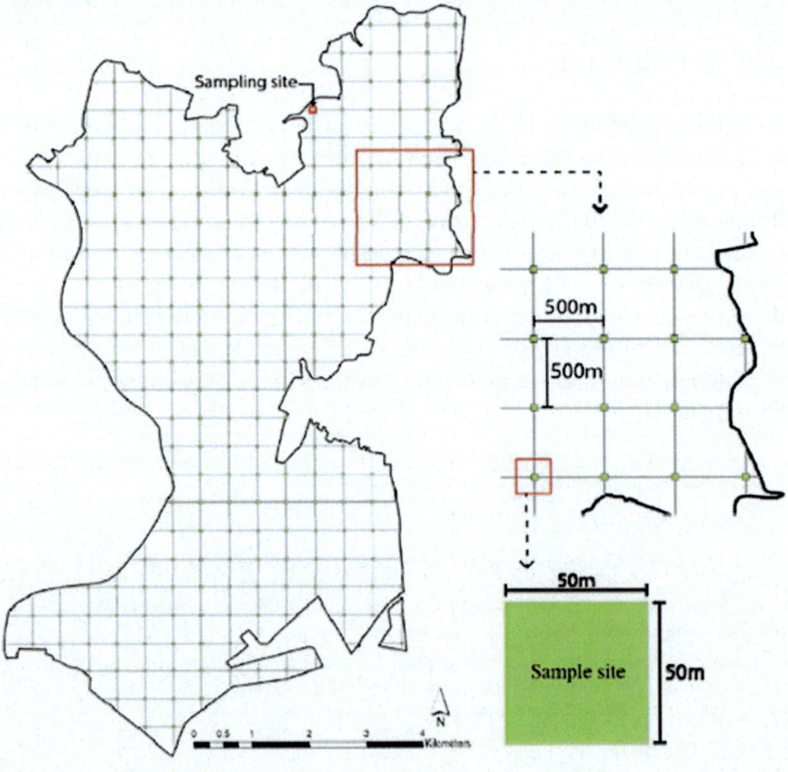

Fig. 1 Distribution of sample sites across Ichikawa City

[6] Figure 1 is a re-use of Fig. 2 in the journal Land (Residents' Perception of Informal Green Space—A Case Study of Ichikawa City, Japan, by Kim et al. 2018).

Fig. 2 Number of responses per sample site

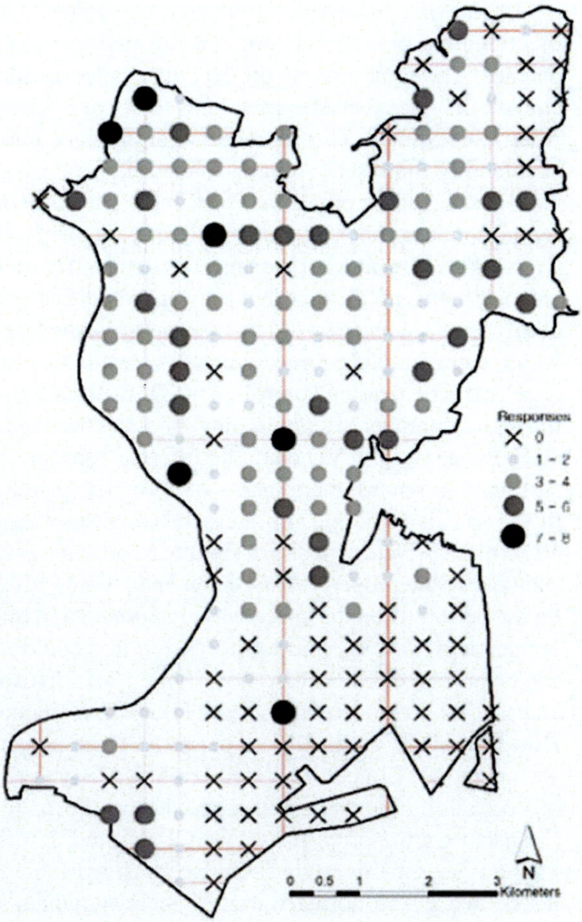

5 Vacant Lands in Ichikawa City

VL identified in this study is space temporarily left unattended after the previous function has ended. Since human activity has been involved more than once and re-intervention is likely to occur, such space is different from the original state of nature, such as ancient forests. But as humans rarely intervene in VL after abandonment, it is slowly covered with herbaceous plants such as weeds. Vegetation inside these spaces is thus not designed or intended, but spontaneous, a mix of various species that produce and reproduce without human care or intent (Del Tredici 2010). The less human care there is, the more spontaneous vegetation is produced inside the space, the more quasi-nature its forms, and the more greenery it provides. However, since the greener VL without human intervention may be considered as housing sites left unattended in urban areas, it can be associated with negative perceptions from residents because of unauthorized dumping of garbage.

A total of 229 sample sites were designated for VL information collection in this research. In the fieldwork, VL information was collected from 44 sites with 47 patches. The total area of the VL distributed in this study site was 0.819 ha. The amount of VL gathered corresponds to about 1.43% of the total sample site quantity when extrapolated. This represents a significant amount, considering that 364 UGSs as a facility that users are able to use, such as urban parks, account for only about 2.46% of the entire city area. The sample sites for which VL is not identified include sections that cannot be accessed directly: centers of construction sites, center of the river, highway, centers of industrial spaces, and conservation area. In this research, the characteristics of VL distribution were identified according to land use patterns. To clarify the classification of land use patterns, the land was divided into 15 categories, which were simplified into four types: residential area, commercial area, industrial area, and urbanization control area.[7] Of the total sample sites, land use patterns with the highest number of sample sites allocated were residential areas, with 125 samples.

Previous uses of VL could be inferred through its location, association with the surrounding environment, and structures left inside and outside the space. Most of the VL consisted of housing sites, and sometimes large structures were present. Since VL is often private property, a signboard or fence is frequently erected to exercise the right of private property (Rupprecht and Byrne 2014a). Although differences depend on the period when the space was abandoned and the degree of intervention through human activities, vegetation succession that could represent the surrounding nature can be observed. Given that most of the VL is distributed within the residential area, and that its previous purpose was housing, residents have good access to the sites. However, since VL is private property, there is a limitation to access to the space within and for direct use. VL in Ichikawa City is classified according to the phases of vegetation succession and access to the site, as follows (Fig. 3).

Because a majority of patches were found in residential areas, most of the VL's previous uses in this study site were housing (Table 1). Identified VL was frequently distributed in the northern part of the city, which has a relatively low population density and high aging index. Besides, the proportion of large-sized patches was higher in the northern part. This study site clearly differs in population density and aging index between the northern and southern parts. The northern part of Ichikawa has a high distribution ratio of low-rise houses or low-rise multiplex houses, while the southern part has a high distribution ratio of mid- and high-rise mansions with industrial and commercial areas.

[7] Currently, Ichikawa City is composed of more than 70% urbanized areas, including residential, commercial, and industrial sectors, and about 30% of the urbanization control area.

Fig. 3 Photographs of vacant land

Table 1 Identified vacant lands in sample sites by land use patterns

Land use patterns	Counts of sample sites	Size
Residential area	30 (68.2%)	0.4267 ha (57.04%)
Commercial area	2 (4.5%)	0.0281 ha (3.44%)
Industrial area	1 (2.3%)	0.207 ha (25.17%)
Urbanization control area	11 (25.0%)	0.117 ha (14.35%)
Total	44 (100.0%)	0.819 ha (100.0%)

6 Perception of Vacant Land as Urban Green Space

Of the 229 located distribution sites, 44 were excluded because they contained either no residences or were inaccessible. We distributed 3,700 kits to 185 sites and received 567 responses from 157 sites. 517 valid samples were used as statistical data.

Residents were aware that VLs were located around their living environment. More than 70% of all respondents answered that they knew that VLs were scattered around their residence. The mean number of VLs they perceived was 4.84. Despite the high recognition of the existence of VLs, experience of using them was very low at 23.4% (Fig. 4). The experience of using VL included direct experiences, such as exercise and relaxation inside, and indirect experiences, such as observation of animals and plants.

Gardening activities can be psychologically and aesthetically therapeutic for gardeners and promote self-sufficient food consumption to enhance satisfaction in life (Clatworthy et al. 2013). In this study, gardening activities are collectively referred to as horticultural activities performed in private areas, such as home gardens and verandas, to improve personal living conditions. They are often performed in residential spaces, the most familiar place for participants in green activities. Therefore, we conducted a chi-square test to know whether the frequency of exposure in residential life to the green environment and the frequency of experience of green activities in the most familiar place were significantly linked to VL and to awareness of its existence.

In Table 2, significant differences are visible in the frequency and distribution of each variable. The awareness of the existence of VL was significantly different

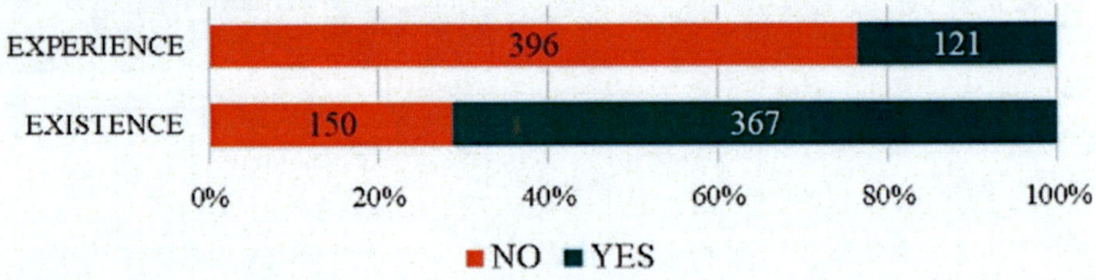

Fig. 4 Experience and recognition of the existence of vacant lands (n = 517)

Table 2 Chi-square test between the awareness of the existence of VL and the composition of residents

Exposure to green environment in ordinary life			Awareness of existence		Total
			No	Yes	
Housing type	No	Count	49 (43.8%)	63 (56.3%)	112
		% of total	9.5%	12.2%	21.7%
	Shared green	Count	36 (34.3%)	69 (65.7%)	105
		% of total	7.0%	13.3%	20.3%
	Home garden	Count	65 (21.7%)	235 (78.3%)	300
		% of total	12.6%	45.5%	58.0%
Chi-square = 22.089, df = 2, Sig(p) = 0.000					
Gardening activity	No	Count	78 (43.1%)	103 (56.9%)	181
		% of total	15.1%	19.9%	35.0%
	Yes	Count	72 (21.4%)	264 (78.6%)	336
		% of total	13.9%	51.1%	65.0%
Chi-square = 26.809, df = 1, Sig(p) = 0.000					

depending on what types of green space respondents owned in their residence. In the case of respondents who did not have green space in their house, there was no significant difference in recognizing the existence of VL in their living surroundings. However, there was a marked difference in respondents who owned a "shared green space" or "home garden." In particular, residents who own a private home garden perceived the existence of VL around their living environment about three times more than those who did not. Likewise, the difference in awareness of the existence of VL was more pronounced in respondents who experienced gardening than in those who had never done gardening before. The chi-test results show that the distribution of the perception of the existence of VL varies depending on the frequency of exposure to a green environment.

Do residents perceive VL as an urban green space? Or is there any room for perception? The three questions asked to understand whether residents recognize VL as green space were as follows: are you willing to use VL? [USE]; are you affected favorably by VL in your daily life? [DAILY LIFE]; can you feel nature, whether it is direct or indirect? [NATURE]. Results reveal that a majority of respondents disagreed with the three variables. About 30% of respondents agreed with [USE] and [NATURE], while only about half that percentage agreed with [DAILY LIFE] (Fig. 5). Therefore, [DAILY LIFE] could be interpreted as showing that the residents do not perceive VL as contributing favorably to their everyday lives.

We extracted cases as negative respondents [Neg.] and positive respondents [Pos.] to variables [USE], [DAILY LIFE], and [NATURE] for a more definitive comparison of preference trends toward VL (Table 3). [Neg.] all answered "strongly disagree" and "disagree" for variables [USE], [DAILY LIFE], and [NATURE], totaling 161 (31.14%) out of 517 samples. In contrast, 44 [Pos.] answered both "strongly agree"

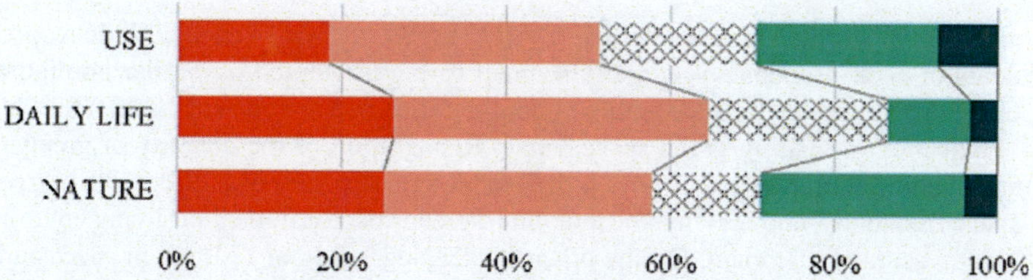

	NATURE	DAILY LIFE	USE
s.disagree	25.1	26.3	18.4
disagree	32.7	38.3	32.9
neither	13.3	21.9	19.1
agree	24.8	10.1	22.2
s.agree	4.1	3.5	7.4

s.disagree disagree neither agree s.agree

Fig. 5 Attitude for vacant land as green space

Table 3 Comparison of sample composition between Neg. and Pos. respondents

Composition of respondents		Neg N = 161	Pos N = 44
Gender	Male	69 (42.9%)	20 (45.5%)
	Female	92 (57.1%)	24 (54.5%)
Age	20s	8 (4.9%)	2 (4.5%)
	30s	15 (9.3%)	4 (9.1%)
	40s	27 (16.8%)	10 (22.7%)
	50s	30 (18.6%)	11 (25.0%)
	60s	37 (23.0%)	5 (11.4%)
	Over 70	44 (27.3%)	12 (27.3%)
Children family	No	126 (78.3%)	28 (63.6%)
	Yes	35 (21.7%)	16 (36.4%)
Employment status	Yes	81 (50.3%)	18 (40.9%)
	No	80 (49.7%)	26 (59.1%)
Public experience	No	139 (86.3%)	37 (84.1%)
	Yes	22 (13.7%)	7 (15.9%)
Frequency of visiting green space	Never	62 (38.5%)	5 (11.4%)
	Occasionally	53 (32.9%)	15 (34.1%)
	Frequently	46 (28.6%)	24 (54.5%)
Recognition of the quantity of greenery surrounding	Lack	50 (31.1%)	7 (15.9%)
	Moderate	38 (23.6%)	10 (22.7%)
	Plentiful	73 (45.3%)	27 (61.4%)

and "agree" for the variables [USE], [DAILY LIFE], and [NATURE]. While respondents of each propensity generally seemed to be distributed in similar configurations, differences in the distribution of configurations appeared in "Children family," "Frequency of visiting green space," and "Recognition of the quantity of greenery surrounding." In the case of [Pos.], they tended to take care of children, visit green space frequently, and feel that the amount of green space around their living environment was more abundant. On the other hand, [Neg.] did not visit green spaces and felt that there was not enough green space around their living environment. In addition, [Neg.] had a higher proportion of employment status than [Pos.]. The largest proportion of respondents was among the over-70s for both tendencies, followed by those in their 50s for [Pos.], and those in their 60s for [Neg.].

The response frequency between [Neg.] and [Pos.] to five variables[8] describing why they are reluctant to use VL is as follows (Fig. 6). [Pos.] tend to be more

[8] Description of variables: I'm concerned about the conflict with the landowner of the site [OWNER]; Signs or fences make it difficult to get into the site [ENTER]; It seems to be polluted and dirty [DIRTY]; It is too small or narrow to use [SMALL]; It may be either developed or disappear someday [DEVELOP].

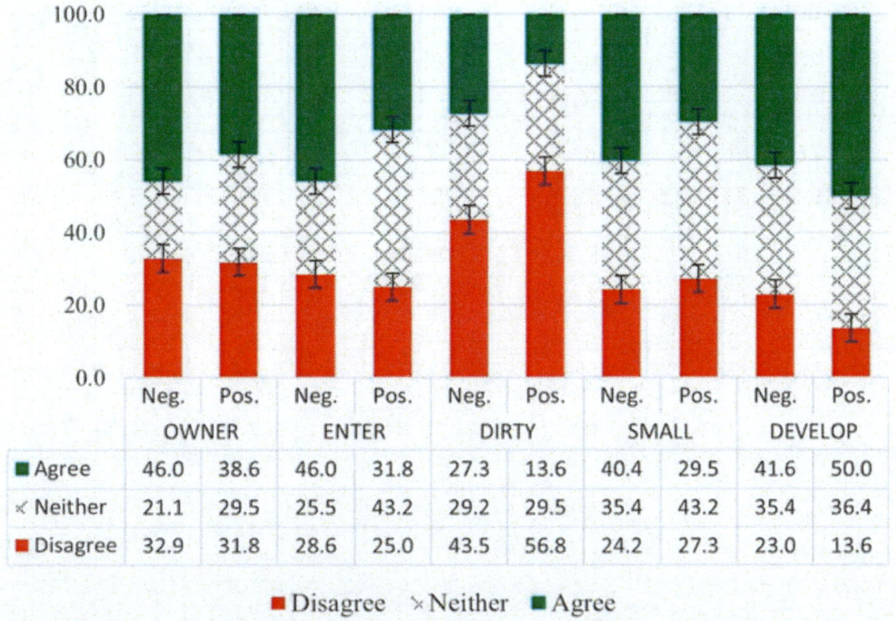

	OWNER Neg.	OWNER Pos.	ENTER Neg.	ENTER Pos.	DIRTY Neg.	DIRTY Pos.	SMALL Neg.	SMALL Pos.	DEVELOP Neg.	DEVELOP Pos.
Agree	46.0	38.6	46.0	31.8	27.3	13.6	40.4	29.5	41.6	50.0
Neither	21.1	29.5	25.5	43.2	29.2	29.5	35.4	43.2	35.4	36.4
Disagree	32.9	31.8	28.6	25.0	43.5	56.8	24.2	27.3	23.0	13.6

Fig. 6 Reasons for reluctance to use vacant land between Neg. and Pos. respondents

proactive about the use of VL than [Neg.]. In particular, [Neg.] strongly agreed on all reasons for reluctance, but [Pos.] strongly agreed on the variable [DEVELOP]. In other words, the transience of VL can affect motivation to use it. The variables for [OWNER] differed little between [Neg.] and [Pos.]. In addition, it could be suggested, through the response distribution of the variable [DIRTY], that neither [Neg.] nor [Pos.] respondents had a hostile attitude toward VL.

We hypothesize that VL is not an institutionally recognized green space such as an urban park, but that this liminal space composed of spontaneous vegetation can provide users with the UGS-like potential to meet aesthetic or recreational needs. Therefore, we examined the general attitude toward UGS and urban nature,[9] depending on whether VL is recognized as a green space (Fig. 7).

[Pos.], the group that perceives VL as green space, has a more intimate attitude toward UGS and urban nature than [Neg.], the group that does not. In particular, referring to the variables [VOLUNTEER] and [TIME], [Pos.] show a more active willingness to engage in nature conservation activities, and they learn about the

[9] Description of variables: It is important to coexist with plants, animals, and humans in an urban environment [COEXIST]; I'm willing to participate as a volunteer to conserve nature [VOLUNTEER]; I'm willing to arrange a time for conserving nature [TIME]; I'm willing to pay some money to conserve nature [MONEY]; I've known plants, animals, and insects that are often observed in or near my area [LOCAL]; I can feel the community attachment from plants, animals, and insects that are often observed in or near my area [ATTACHMENT]; The neighborhood green space should be managed [GREEN].

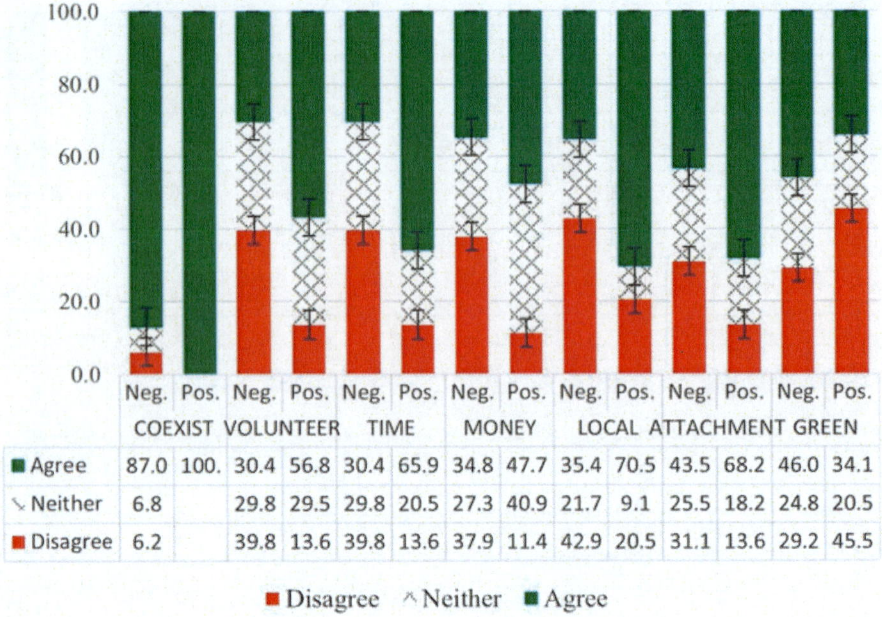

	Neg.	Pos.	Neg.	Pos.	Neg.	Pos.	Neg.	Pos.	Neg.	Pos.	Neg.	Pos.	Neg.	Pos.
	COEXIST		VOLUNTEER		TIME		MONEY		LOCAL		ATTACHMENT		GREEN	
Agree	87.0	100.	30.4	56.8	30.4	65.9	34.8	47.7	35.4	70.5	43.5	68.2	46.0	34.1
Neither	6.8		29.8	29.5	29.8	20.5	27.3	40.9	21.7	9.1	25.5	18.2	24.8	20.5
Disagree	6.2		39.8	13.6	39.8	13.6	37.9	11.4	42.9	20.5	31.1	13.6	29.2	45.5

Fig. 7 Attitude to UGS between Neg. and Pos. respondents

region through the nature of their neighbors. Also noteworthy is that the proportion at which [Pos.] perceive the surrounding green space as something that must be managed is lower than that of [Neg.]. In other words, the stance toward green spaces of respondents who do not recognize VL as green space could be inferred as space where management should be performed. VL is neither a natural resource nor an officially or institutionally recognized green space. Therefore, it is important to tap the potential as a green space that the VL offers, but it is also necessary to arouse interest from direct and indirect users of the VL.

7 Discussion and Conclusion

In this chapter, we have examined urban wasteland as a supplementary urban green space in response to the financial and spatial constraints in green provisioning in contemporary urban areas. Recently, Japan revised a part of the Urban Green Conservation Act to provide residents with friendly green space and open space in their living environment and to make improvements in the urban environment. This targets several local cities with limited budgets, workforce, and space constraints to create new places. In addition to the issues of budgets, human resources, and spaces that challenge the construction of UGS, the revisions reflect changing trends of users toward UGS. Today, as cities decline, the number of VLs in Japan is increasing due to various social factors, such as the aging of landowners and the abandonment of the management of land price depreciation. Therefore, the "Citizen Green Space Certification System" was established in 2017 to publicly share private lands and provide administrative benefits to the landowners who provided the land (City Bureau of MLIT 2018). However, since the law was revised not long ago and Japanese citizens' awareness of private land is conservative, the system has yet to be established. Therefore, VL has not been officially recognized and designated as official UGS by users or by the government or landowners. As a result, it may be difficult for residents to perceive VL as equivalent to existing UGS, such as urban parks. In this chapter, we have investigated the distribution characteristics and perception toward VL from the point of exposure to familiar UGS in an attempt to deepen understanding of these issues.

UGSs have a role as a part of day-to-day life in the nature of the city and the implications of how they are integrated into urban planning are wide ranging. It is therefore important to explore the answer to the question of whether VL can serve as a supplementary green space (Kim et al. 2020). This study proposes to explore the ultimate value VL can have, through its distribution characteristics and the perception of residents. Considering that institutionalized UGS accounted for 2.46% of the Ichikawa area, the finding that the amount of VL as urban wasteland present in the sites accounts for 1.43% of all sample sites supports the view that VL has the potential to supplement UGS in Ichikawa. In addition, VL is frequently distributed throughout people's living environment, so the merit of VL is that it is "easy to reach" for people who are familiar with and exposed to traditional UGS. VL is not a nature reserve. This is a space where human activity has previously ended and which has been completely or irregularly neglected. Therefore, VL should be evaluated as space available for use by humans or non-humans. Both could be not only users of space, but also indirect beneficiaries.

Green space management by the local community encourages user participation by improving the production of ecosystem services as well as biodiversity (Dennis and James 2016). With regard to community space, appropriate participatory management through on-site management can promote physical activities to improve participants' health (Hynes and Howe 2004; Alaimo et al. 2008). In the revised Urban Green Space Conservation Act of Japan, the government aims to create a shared community by encouraging private sector participation. Therefore, we suggest that it is important to provide the groundwork for participation in VL management with consideration of the structural complexity and intensity of management. The management system of the participatory green space can lead to a more positive perception of urban wasteland, and the importance of various experiences related to green space in everyday life should be emphasized in order to promote the willingness to participate in an urban wasteland. A good starting point for participatory VL management would be the eight planning principles outlined previously (Rupprecht 2017), with slight adaptations to cater to the specifics of VL in comparison with informal green spaces as a whole.

This study is limited to Japan as a representative of countries with cities that are undergoing shrinking and aging. However, after experiencing rapid urban growth, more advanced study of the availability of urban wasteland could be considered by comparing other Asian cities sharing similar demographic and social issues. In addition, the role of the government and stakeholders in how urban wasteland is integrated into urban policy should be studied.

Acknowledgements We would like to thank Katsunori Furuya for constructive comments on the previous version of this manuscript. We are very grateful to all respondents for participating in this study.

References

Alaimo K, Packnett E, Miles RA, Kruger DJ (2008) Fruit and vegetable intake among urban community gardeners. J Nutr Educ Behav 40:94–101. https://doi.org/10.1016/j.jneb.2006.12.003

City Bureau of MLIT (2018) The abstract and the aim of the act for Partial Amendment of the Urban Green Space Conservation Act, etc.

Clatworthy J, Hinds J, Camic PM (2013) Gardening as a mental health intervention: a review. Ment Heal Rev J 18:214–225. https://doi.org/10.1108/MHRJ-02-2013-0007

Corbin CI (2003) Vacancy and the landscape: cultural context and design response. Landsc J 22:12–24. https://doi.org/10.3368/lj.22.1.12

Del Tredici P (2010) Spontaneous urban vegetation: reflections of change in a globalized world. Nat + Cult 5:299–315. https://doi.org/10.3167/nc.2010.050305

Dennis M, James P (2016) Site-specific factors in the production of local urban ecosystem services: a case study of community-managed green space. Ecosyst Serv 17:208–216. https://doi.org/10.1016/j.ecoser.2016.01.003

Feltynowski M, Kronenberg J, Bergier T et al (2018) Challenges of urban green space management in the face of using inadequate data. Urban For Urban Green 31:56–66. https://doi.org/10.1016/j.ufug.2017.12.003

Hiroshi O (2018) Utilization of vacant lots in urban areas

Ichikawa City Urban Planning Division (2004) Ichikawa green master plan 2004

Ichikawa City Urban Planning Division (2017) Urban infra of Ichikawa based on data 2017

Kabisch N, Qureshi S, Haase D (2015) Human-environment interactions in urban green spaces—a systematic review of contemporary issues and prospects for future research. Environ Impact Assess Rev 50:25–34. https://doi.org/10.1016/j.eiar.2014.08.007

Kim M, Rupprecht C, Furuya K (2018) Residents' perception of informal green space—a case study of Ichikawa City, Japan. Land 7:102. https://doi.org/10.3390/land7030102

Kim M, Rupprecht CDD, Furuya K (2020) Typology and perception of informal green space in urban interstices. Int Rev Spat Plan Sustain Dev 8:4–20. https://doi.org/10.14246/irspsd.8.1_4

Martinez-Fernandez C, Audirac I, Fol S, Cunningham-Sabot E (2012) Shrinking cities: urban challenges of globalization. Int J Urban Reg Res 36:213–225. https://doi.org/10.1111/j.1468-2427.2011.01092.x

Maurer U, Peschel T, Schmitz S (2000) The flora of selected urban land-use types in Berlin and Potsdam with regard to nature conservation in cities. Landsc Urban Plan 46:209–215. https://doi.org/10.1016/S0169-2046(99)00066-3

Ministry of Finance (2017) Budget for public work projects

Mizuno A, Takehito U, Hirofumi A (2016) Status of countermeasures for vacant houses and lots in Japan and a cooperation method with policies for compact city. J City Plan Inst Jpn 51:1101–1108

Nejman R (2018) The right to wild. Green urban wasteland in the context of urban planning. Urban Develop Issues 59:43–53. https://doi.org/10.2478/udi-2018-0027

Hynes HP, Howe G (2004) Urban horticulture in the contemporary United States: personal and community benefits. Acta Hortic 643:171–181. https://doi.org/10.17660/ActaHortic.2004.643.21
Rupprecht C (2017) Informal urban green space: residents' perception, use, and management preferences across four major Japanese shrinking cities. Land 6:59. https://doi.org/10.3390/land6030059
Rupprecht CDD, Byrne JA (2014a) Informal urban green-space: comparison of quantity and characteristics in Brisbane, Australia and Sapporo, Japan. PLoS ONE 9
Rupprecht CDD, Byrne JA (2014b) Informal urban greenspace: a typology and trilingual systematic review of its role for urban residents and trends in the literature. Urban For Urban Green 13:597–611. https://doi.org/10.1016/j.ufug.2014.09.002
Rupprecht CDD, Byrne JA, Ueda H, Lo AY (2015) "It's real, not fake like a park": residents' perception and use of informal urban green-space in Brisbane, Australia and Sapporo, Japan. Landsc Urban Plan 143:205–218. https://doi.org/10.1016/j.landurbplan.2015.07.003
Tian Y, Jim CY (2012) Development potential of sky gardens in the compact city of Hong Kong. Urban For Urban Green 11:223–233. https://doi.org/10.1016/j.ufug.2012.03.003
WHO Regional Office for Europe (2016) Urban green spaces and health: a review of the evidence. World Heal Organ 1:92
WHO Regional Office for Europe (2017) Urban green spaces: a brief for action 1–24
Wright Wendel HE, Zarger RK, Mihelcic JR (2012) Accessibility and usability: green space preferences, perceptions, and barriers in a rapidly urbanizing city in Latin America. Landsc Urban Plan 107:272–282. https://doi.org/10.1016/j.landurbplan.2012.06.003
Yamada C, Terada T, Tanaka T, Yokohari M (2016) Directions for vacant lot management in the outer suburbs of the Tokyo metropolitan region. Urban Reg Plan Rev 3:66–84. https://doi.org/10.14398/urpr.3.66
Yokohari M, Amati M, Bolthouse J, Kurita H (2010) Restoring urban fringe landscapes through urban agriculture: the Japanese experience. Disp 46:51–59. https://doi.org/10.1080/02513625.2010.10557086

Minseo Kim is a Landscape Planner and Project Researcher at the Hiroshima University of Japan, where she has been in charge of urban planning and the international academic exchange. Her research interests lie in improving the green space policy, ranging from the planned space to the informal space with a wasteland. Her current research looks at the wasteland's open potential as a supplementary urban green space not only for human well-being but also the urban environment.

Christoph D. D. Rupprecht is an urban geographer and Senior Researcher with the FEAST Project at the Research Institute for Humanity and Nature, Kyoto (Japan). He is interested in the potential of urban wastelands to function as informal green spaces that provide a diversity of benefits to humans and more-than-humans, in shrinking Japanese cities. His other interests include the agency of spontaneous vegetation challenging orthodox notions of urban planning, wastelands as potential sites for edible landscapes, and multispecies approaches to landscape stewardship.

Wastelands at Port-City Interfaces. The Search for Water Spaces to Evade the Constant Hustle and Bustle of City Life

Kristel Mazy

Abstract Port-city interfaces, near urban centers, are under intense land pressure within the context of increased competition for the development of these sites. Some interfaces are wastelands, induced by the remoteness of port installations from urban centers. For other interfaces, the word "wasteland" can be instrumentalized by planners to reallocate underused, yet active, port spaces. This paper aims to highlight the points of view of users, mostly inhabitants of these neighborhoods, in Brussels and Lille, based on a study of territorial representations. This study shows that these representations are clearly different from the conflictual environment of decision-making. The representation of a "quiet space" and a "breathing, natural space," for leisure and relaxation, dominates the discourse of users. These representations make sense in very densely built environments, which are landlocked by the passage of major transport infrastructures. This desire for temporary withdrawal can be related to moments allowing daily pressures to be relieved. The development of port-city interfaces as true interstitial breathing spaces within urbanized spaces could be explored for planners.

Keywords Port-city interfaces · Inland waterways · Dense built environment · Territorial representations · Feelings of nature

1 Introduction

Along the waterside, port-city interfaces offer many different types of landscapes. They can be industrial, urban, natural, hybrid, and can even be left or mutated for other purposes. This hybridization attests to a shift that tends to place these spaces in the spotlight of different private and public strategies. Deindustrialization has in fact left behind a difficult socioeconomic situation in Northwest Europe, which is crossed by a network of rivers. This is a sensitive environment which is particularly

K. Mazy (✉)
University of Mons, Mons, Belgium
e-mail: Kristel.MAZY@umons.ac.be

F. Di Pietro and A. Robert (eds.), *Urban Wastelands*, Cities and Nature,
https://doi.org/10.1007/978-3-030-74882-1_10

at risk of flooding and pollution, and which has a significant number of wastelands or brownfields.

This article aims to give an overview of the role of these designed areas from the points of view of both their creators and their users.

First, I will focus on putting the subject into context by introducing the port-city relationship from a geohistorical approach. I will then look at the process of creation of wastelands, within the conversion process of port areas. Today, occupancy rates of port-city interfaces can vary significantly, from being very active to under-occupied, or wasteland. Some port areas actually never become wasteland, which can be foreseen or anticipated by planning bodies. Through three Brussels and Lille case studies, I will outline the physical particularities of these areas, which are densely built and isolated by transport infrastructures.

Further in the analysis, I will highlight the territorial representations of these lands, which are crossed by a waterway between the city and the port. Using a qualitative research method based on interviews to study the representations, I have used thematic analysis as a method of content analysis. From the conception and decisional viewpoints, these *designed* spaces are evidence of strong competition regarding the use of the ground between urban and port stakeholders. The idea of wasteland was raised by the developers to reallocate underused port areas that were still active to speed up the conversion process. On the other hand, from the users' point of view, these areas of life are represented as peaceful and breathing sites that can stimulate the senses as well as the imagination. Even though they are far from being natural landscapes, they do share some features which closely resemble nature. The recognition of the role of these interstitial breathing spaces, sparsely dense, within urbanized spaces in dense urban environments as a secondary function for ports could open avenues of reflection for urban planning.

2 Obsolescence and Conversion of Sea-Port Areas

2.1 Origin of Port and Industrial Wastelands

Port and infrastructure wastelands are the results of the gradual disconnection between a city and its port (Bird 1963; Hoyle 1989; Hall 2012). This started in the medieval era in the oldest European cities, and this disconnection has been increasing since the Industrial Revolution.

Ports have gradually moved away from the urban center thanks to two, often simultaneous forces: technological port priorities (brought about by new property needs and proximity with new, higher performance transportation infrastructures) and/or property and real estate resulting from the urban development. These two factors involve different levels of analysis. Centripetal forces are linked with the process of urban conglomeration. Conversely, on a larger scale—in the urban area—centrifugal forces push the more common functions, like logistics or transportation activities, toward the suburbs to maintain high added-value activities in the city center. These, therefore, contribute to the dispersal of activities within the suburbs (Fremont 2011, p. 2).

From the second half of the twentieth century, this split intensified in Western cities with the instigation of technological evolutions, such as containerization, and the desire for proximity to highways, the new transportation mode providing flexible solutions to the new economic requirements of the "just-in-time" phenomenon. During the deindustrialization period, competition between emerging countries, coupled with the transfer of production plants to the suburbs and the development of service industries within city centers, led to the formation of industrial wastelands.

These rapid changes made some facilities obsolete, whereas others located farther from the city, such as container terminals, remained useful. Sometimes signs of decline, sometimes signs of unsuitability, some interfaces tended to be abandoned, and, in some cases, became wastelands under the influence of the separation of ports from cities.

From the Brussels and Lille cases, I establish the sequence of the evolution of the city-river port relationship, according to a geohistorical approach. First, through observations of historical maps, I identified successive developments that not only changed both the relationship between the city and its port but also between the city and its waterway. On this basis, I have produced synthetic maps (Mazy 2014). They show, for each period, the extension of urbanization and the main urban changes affecting the waterway and associated practices (establishment of craft trades, port facilities, etc.). Through schematic sequences, these synthetic maps enabled me to draw the evolution of the city-port relationship, based on the cases of Brussels and Lille (Fig. 1).

2.2 *Obliteration of the Wasteland Phase*

In this context of port-city disconnection, waterfronts have become spaces that are subject to strong property tensions. This is especially the case within the framework of urban projects that contributed to their conversion and to a decrease in their industrial function (Prelorenzo 1999). Conversion zones do not only concern abandoned areas.

> Riverbank areas that are in high demand do not go through the wasteland stage. Sometimes, the reuse anticipates an obsolescence that has not yet been confirmed (Chaline 1988, p. 8).

This rapid conversion of port lands, sometimes eliminating the wasteland phase, can be explained by two factors which, from the 1950s, placed waterfronts at the center of urban strategies aiming to revitalize maritime cities.

The first factor is related to the political will bring back city dwellers to damaged central districts following deindustrialization:

> This way, the reorganization of the interface comes as a means to refocus the urban area to compensate a certain splitting and dispersion of the demographic streams, due to urban sprawl. It can allow some centrality to emerge regarding new functions which better match

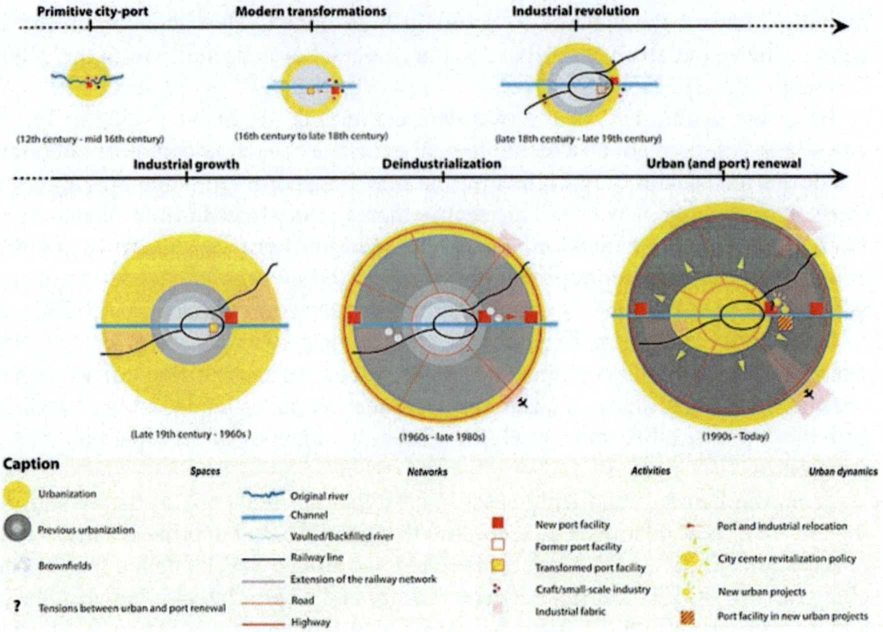

Fig. 1 Schematic evolution of the relationship between inland ports and cities (Mazy 2014)

> the new urban dimensions, since traditional city centers are no longer adapted to this role of urban conglomeration center (Boubacha et al. 1997, p. 18).

The second factor is a necessary precondition to the revitalization of coastal municipalities. This comes from the context of political decentralization that gives more initiative back to the cities.

> People from city centers that were neglected by ports suddenly emerge as major property opportunities in the politics of "urban recapture" that is part of most elected representatives' discourses. Local governments ask that "urban ports" be given back to the city. During the '80s, the will to re-establish a direct relationship with quays confiscated for a long time by the port authority was necessary to the promotion of the coastal city's image (Baudouin and Collin 1996, p. 25).

With the London experience as a flagship project, the construction concepts of the waterfronts quickly spread to fluvial port cities. From the 1980s, rivers and canals became vehicles for projects in numerous cities based on them and their estuary ports, like Bilbao, Hamburg, Nantes, and Bordeaux. This also applied to internal port cities, such as Paris, Lyon, and, more recently, Lille and Brussels. Often, and for many of these cities, the rediscovery of the river became an urban issue in a context of internationalization (Bétin and Cottet-Dumoulin 1999, p. 118). This growing interest in canals and their influence on territory planning led B. Le Sueur to invent the neologism *flurbanization*.

> In response to the fluvial exodus that took workers away from the rivers, our contemporaries tended to come back to the watersides as they lacked urban nature and leisure spaces. If we compare this move to the one that affected the rural world years ago, I could easily talk of *flurbanization* that slowly spreads in harmony. Canals then became a factor of economic balance again, carrying memory and identity (Le Sueur 1997, p. 203).

The spread of concepts of the conversion of *waterfronts* in coastal port cities, along with the growing stakes of the attractiveness of cities requiring internationalization in a context of territorial competition, is pushing fluvial cities to speed up the conversion sequences of port areas that are close to urban centers.

2.3 Port-City Interfaces: Splits in Cities?

Currently, port-city interfaces contain wastelands, industrial or port activity areas, which are sometimes underused or very active, and future urban project areas. I took a deeper look into three case studies that show the diversity of these occupation levels: the Biestebroeck and Vergote reservoirs in Brussels, and the area near the port of Lille.

Close to urban centers, the developed areas that surround the three sites show a very high density. Orthophotos (Fig. 2) reveal a very tight urban network that comes mainly from the industrial Golden Age at the end of the nineteenth century and beginning of the twentieth century. Additionally, perception of the density is strengthened by the enclaving of these areas between the transportation infrastructures that were used, or that are still being used, to transport merchandise. Varying from one case to another, the urban barriers are shaped by the canal, the railway network, the metropolitan roads, and even highways. In some cases, embankments not only isolate port activities from their visual and sound pollution, but also increase their isolation among other urban functions. In addition, the density is also exacerbated by the high mineral characteristics of these areas, which provide few open and green public spaces.

We will see that social tensions stemming from negotiations concerning the organization of these port-city areas are based on spatial objects that are subject to tensions through their own physical structure. However, the canal, the ponds, and the port fabric stand out from these dense—or even oppressive—environments, offering a breath of air and empty spaces among the constructed masses of the districts. Indeed, a common feature provided by the three case studies, and inland port docks more generally, is that they are sparsely built to receive materials for storage (mainly construction materials, oil tanks, and logistical activities), delivered by ships. Another common feature of these port spaces is that they are mainly mineral, made of concrete docks, though spontaneous vegetation grows on neglected or "waiting" space. And all of these spaces are open to the waterway. These common characteristics make these spaces singular in cities. For example, a zoom in on the Biestebroeck Reservoir shows these sparse, mineral spaces, along the waterway (Fig. 3). While the morphological characteristics of the three sites are mainly similar, some nuances appear linked to the type of materials stored, the typology and age of warehouses, the width

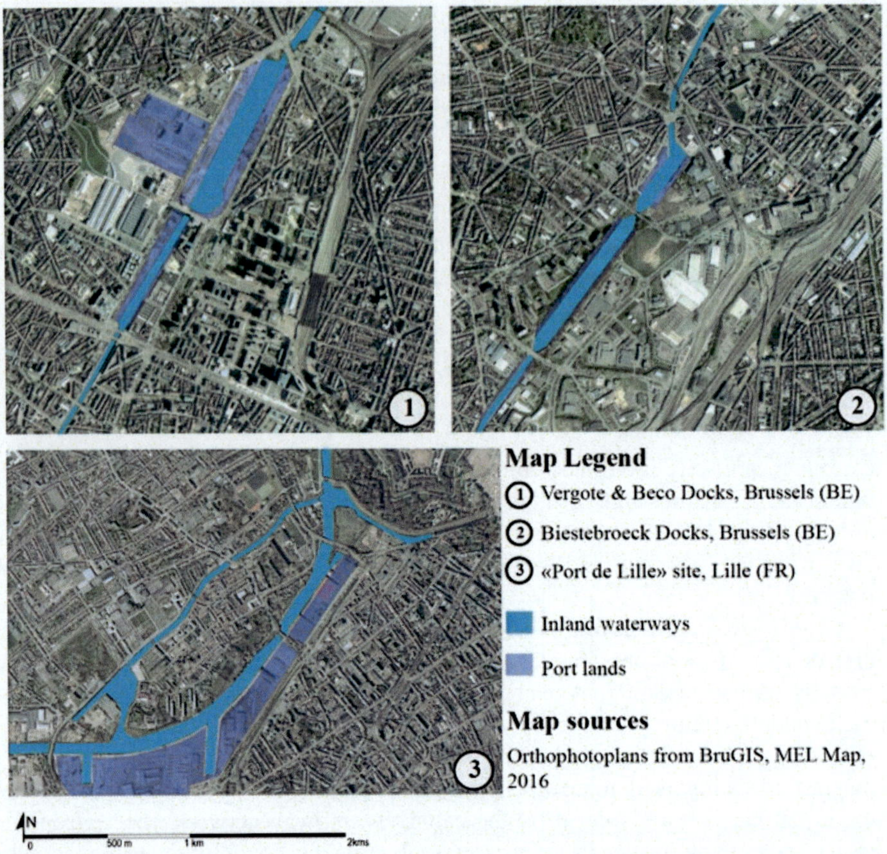

Fig. 2 Orthophotos of Vergote, Beco, and Biestebroeck Docks in Brussels, and "Port de Lille" site in Lille

of the waterway, or the presence of vegetation. However, these nuances do not influence the representations of these spaces expressed by stakeholders and users. These representations are further developed below.

In addition to their useful functions, which are linked to good transportation activities, do the singular spatial qualities of these port areas play another role in the city? Are they helping to create specific uses that are developed from their singularities?

Fig. 3 Orthophotos of Biestebroeck Docks in Brussels

3 Spatial Representations of Port-City Interfaces: Tension or Control Areas?

3.1 Tension Areas: Using the Notion of "Wasteland" to Reconsider Port Areas

The first part of the analysis considers port-city interfaces as "built territories." It highlights the highly "geopolitical" dimension (Subra 2007), or at least the balance of power that arises during their conception phase, especially through the project process.

In many cases of *waterfronts*, the urban project accelerates the disconnection between the city and the port. These disconnections are embodied by the reproduction of functional cuts of the industrial era. The creation of a "city" system encircles, and eventually excludes, freezes, or relocates the remaining port activity areas. It is very difficult to lead interdisciplinary approaches between urban and port spheres because of institutional, cultural, and economic factors. At an institutional level, the project processes struggle with an isolation that is linked to the port-city governance structure for the decisional systems. The barriers that tend to separate the city from the port culturally relate to the complexity of building a process able to unite diversified cultures of projects. The multiplication of projects generates a property pressure on the port wastelands as well as on the port areas that are still active. Another angle of analysis, focusing on territory representations, therefore appears to be necessary to help clarify the sources of tension that sometimes bring cities and ports into opposition.

> The project is an intention and not only an action [...] Since space is a social construct, mental space and material space are intertwined. The action on space is also motivated by the symbolic values attributed to space and the stakes linked to space are also stakes of representations (Rosemberg-Lasorne 1997, p. 2).

In addition, the first step is to understand the sense of the representations that go along with that action. Spatial representations provide the "prism" through which actors or public opinion perceive the construction issues; thanks to them, these stakeholders understand the reality of space, the challenges that it endures, and that threaten its future (Subra 2007). Particularly in the case of coveted spaces, such as port-city interfaces, the study of spatial representations enables a better understanding of the basis of the spatial challenges and social tensions by analyzing the strategies of individual and social stakeholders (Bailly 1995, p. 378).

Table 1 List of semi-directed interviews to study the representations of people in charge of port-city interface construction

Urban stakeholders	Brussels	Lille
Municipality employee (urban development)	2	2
Regional/Intercommunal employee (urban development)	3	3
Urban planning offices	3	1
Local development center	1	
Militant association for water management	1	
Port stakeholders or associated	**Brussels**	**Lille**
Port employee (strategic department)	3	2
Port Board member (or representative)	1	1
Administration, institute, association for the promotion of waterway transport	3	1
National or regional administration in charge of freight transport	1	1
Researcher in logistics	1	

Box 1: Build Territories: Methodologies used to Study the Representations of People in Charge of the Port-City Interface Construction

We studied representations that are spread through the interview method.

> When we study a social representation with the interview method, the analysis of the content of the interviews can help us to analyze the content of the social representation (Negura 2006, p. 5).

We led 30 semi-directed interviews with Lille and Brussels stakeholders who are involved in the management and construction of the port-city interfaces (Table 1). These interfaces can be linked to the urban sphere (local coordinators and creators working for municipalities, town planning agencies, design offices...) or to the port and industrial sphere (within the ports or their communities, from the Chamber of Commerce and Industry, river stakeholders, etc.). The purpose was to understand the representations from the port stakeholders about urban projects and, conversely, to understand the representations from urban stakeholders about port projects. For the development of the same space, how are the strategies and projects of the port stakeholders perceived by the urban stakeholders? And vice versa, how are the strategies and projects of the urban stakeholders perceived by port stakeholders? If these representations are both competing for the same space or are incompatible, they could, in fact, be the origin of preliminary obstacles to a peaceful cooperation. I use thematic analysis—explained by Negura (2006)—as a method of content analysis. The purpose is to find the semantic units that constitute the discursive universe of the utterance. In this case, this means reproducing a reformulation of the utterance content in a condensed and formal form. First, I outline the speech segments linked with the representations under study, i.e., the stakes and strategies of port and urban stakeholders and the factors that could block the development of these strategies. Second, these speech segments were put into thematic categories.

3.1.1 Replica of Compartmentalized Management in the Speech of the "Institutional" Stakeholders

The analysis of the representations shows that the fragmentation of the territory and the compartmentalized management between urban areas and port-industrial areas tend to be repeated in the speeches and representations of the stakeholders that organize these territories today. The definition of stakes and strategies for urban stakeholders can be obstacles to development for port stakeholders, and vice versa.

In this way, when we categorize the different records thematically, it appears that numerous representations of the urban stakeholders relate to the potential expansion of *the urban centrality by the mutation of industrial sites.*

> "The municipality clearly sees the development of a relatively dense and mixed area, even if the company is still there, but no longer with extremely heavy or dangerous activities;"

> "This site should become the center because we are no longer really on the outskirts. But everyone needs to find their place."

These urban stakeholders' representations are reflected in the expression of strategies on *opening, continuity with the city and nature in town.*

"We defend the idea of a watershed. We must come back to hydrographic logic, the bearer of landscape issues. We must reconnect with water, even poetically, regain awareness of the shape of the city;"

"the city wants to offer its banks to the promenade, which remains inaccessible."

For the urban stakeholders, the representation of the factors blocking the development of these strategies hinges on obstacles linked with *spatial isolation, action modes isolation, and difficulties in cooperating with the "extra-urban" port stakeholders.* In addition, the word "wasteland" is sometimes used by planners to designate the underused—but still active—port areas in a context of competition to develop these lands to give them urban functions.

"There is a way of working that remains very compartmentalized";

"They consider that they are not in town. Port stakeholders are not used to constraints and land optimization. They are used to spreading out over a field."

Reciprocally, the representations of the port stakeholders regarding the urban sphere can be perceived as a defensive attitude in the face of covetous designs on the urban redevelopment of port areas. These representations were mainly expressed by the *social acceptance of the port function, the will to give urban functions to the port areas and the loss of industrial and port properties becoming mixed areas.*

"The planning of economic port and industrial activities encounters reluctance on the part of elected officials and residents. Elected officials defend economic activities such as the tertiary sector and the creative industry, fashionable, trendy";

"We must constantly go against the enemy to make it clear that a port receives all types of products and products that are not necessarily polluting";

"The market conditions in this type of problem are decisive. In a logic of property development, the housing function is today profitable. The logic of diversity initially envisaged is abandoned (continuance of logistic activities and pedestrian continuity) in favor of this property development logic."

Based on a desire for reconversion, these representations are considered as obstacles to the implementation of their strategies, focusing mainly on the development of the freight transport function. *The optimization of the port plots and the massification of flows through multimodal uses and the multi-site management of a platform network are seen as critical issues.*

"We need to be able to reconcile urban planning concerns, but also to be able to fulfil our primary mission of carrying out transport, protecting storage, maintenance operations, etc.";

"our goal is to densify our spaces alongside the waterway and to find solutions to treat more tons per linear meter."

3.1.2 Competitive Strategies Stemming from a Division Between Residential and Productive Economies

Following this analysis, we suppose that the port and urban strategies on the future of the port-city interfaces can be understood as the spatial expression of a division between the residential and productive economies.

These areas—located between the port and the city—are the focus of bigger scale debates, such as the economic decisions regarding reindustrialization, the pursuit of the expansion of the service sector, and the rise of residential economy in European cities.

Indeed, one of the objectives pursued by the urban project is the revalorization of the territory. It is based on various communication tools that represent the space in the project. It involves a staging of urban values (centrality, mobility, and accessibility), lifestyle, and the aesthetic and landscape quality of public spaces and buildings. These elements, which are identified as new values, participate in the redefinition of the identity of the developed space (Bailleul 2008). These new values support presence and consumption as new factors of territorial economic development. As emphasized by Davezies (2008), "the productive economy, in which the logic of firms is to establish themselves in order to produce, weighs little locally compared to the residential economy, based on consumption, and on firms which establish themselves to sell."

This competition for land, with the backdrop of debates regarding the productive and residential economies, is reinforced by ground rent, which favors residential, commercial, and administrative uses. The difference in ground rent between urban (residential, commercial, tourist, administrative, etc.) and port-industrial functions, and the temptation to create added-value on this differential, can speed up the conversion of port territories to other functions (Mazy 2017, p. 6).

In Brussels, the situation has recently been nuanced with the creation of the "Canal Team" and the urban integration of productive activities, stimulated by the team of the Brussels Master Architect, through the promotion of the "productive city" and the organization of architectural competitions around subjects such as the redevelopment of a concrete plant or a company specializing in waste sorting (Mazy and Debrie 2018).

3.2 Regulation Areas: Between Desire and Feelings of Nature

Another part of the analysis focuses on the lived dimension of these port-city interfaces. Several questions remain. Are the tensions generated by the design phase found in the material dimension of the territory? How are these areas built as such, and seen by the people that live in them? Are these tensions to be found in the spatial practices? On the contrary, could these port-city interfaces paradoxically play a part in city regulations? Could they even function as an anti-gentrification strategy in a context of strong competition regarding the use of the ground (Rupprecht and Byrne 2018)?

Box 2: Lived Territories: Methodology for the Study of the Representations of Port-City Interface Users

To reveal the tendencies regarding the uses and representations related to the perception of space, the exploration support that was used was based on interview surveys.

> These surveys – which aim to understand a practical system (i.e. the practices and their links: ideologies, symbols,... – need modal and referential speeches (a speech that describes the way things are) obtained from interviews on the stakeholders' conceptions and description of the practices. (Blanchet and Gotman 2007, p. 30).

On site, 30 people (ten per site) "practicing these areas" between port and city were interviewed in a mid-season period, at different times of the day. The interviews were semi-directive, starting with easy and descriptive questions about their practices, questions regarding their knowledge of the on-site activities, and ending with questions to help determine their perceptions and opinions of the area and more particularly of the presence of port activities. The respondents were mainly 18–29 years old, and included workers, students, and employees that were representative of the neighborhood population.

As with the analysis of the "institutional" stakeholders' interviews, we used thematic analysis as a method of content analysis. First, we isolated fragments of the interview according to their theme: representations of the territory, uses, knowledge, and representations of the port activities. These parts were then grouped into thematic categories. For instance, regarding the spatial representations, four general themes stood out: quiet spaces, nature in town, areas with an interesting industrial style, and dirty areas. Some of the themes were then divided according to their negative or positive connotation. Finally, we created a form for each topic, recording the generally identified themes and the speech fragments related to them. The purpose was not to conduct a representative and quantitative analysis, but mainly to highlight the diversity of the main tendencies regarding the uses and representations of the people who live on these territories.

3.2.1 Many Differences Among Users' Representations of the Port-City Interface

The idea mainly associated with these port-city interfaces (Figs. 4, 5 and 6) is "quiet areas, giving a feeling of nature." According to the interviews, these areas primarily offer three spatial qualities (mentioned in 16 of 30 interviews):

- the possibility of withdrawing and finding solitude:

> "a quiet and peaceful place;" "You don't get bothered" and "You don't bother anyone";
>
> "A place where we can relax instead of being in the city center with people around talking a lot";

Fig. 4 People relaxing in front of Vergote Docks in Brussels, near sand deposits © K. Mazy

"I come here to gather my thoughts";

"It's nice to stay here when I want to be alone."

- the attraction of a "breathing" place that opens the senses, for example, when at sea:

"my motivation to come here is the water, the wind and the freshness";

"water clears your mind, as does the sea";

"There, you see water, boats and fish";

"When I see the barges, I start dreaming I am aboard and that I'm on my way to some other country";

"We come to stare at the scenery, the birds and the bikes that pass by";

"It makes me think of the sea, I feel the breeze."

- the representation of "nature" in town

"I like it. There is nature, there's everything. We meet each other. We have a little fun";

"I like to come here because it's natural, there are still animals. When we get closer to the center, there's nothing. There are herons and hedgehogs at night."

These feelings of retreat and the opening of senses, and a feeling of nature appear even more strongly since they are related to very dense environments that are isolated by transportation infrastructures, as discussed earlier. Accordingly, these areas can

Fig. 5 A fisherman, alongside Biestebroeck Docks, opposite a warehouse of construction materials © K. Mazy

Fig. 6 Panoramic view of the "Port de Lille" site © K. Mazy

play a regulatory role close to dense urban zones and act as a valve. These valves then function like a decompression chamber, a way of unwinding that allows people to channel what is not satisfied by everyday life (Lageiste 2008).

In contrast to this idea of positive tranquility, the second descriptive words that stand out are "a dirty place" (mentioned in 5 of 30 interviews):

> "It is quite dirty;" "it is gross;" "It looks like a dump";
>
> "The trucks bring a lot of dust";

> "It is dirty and chaotic";
>
> "It's annoying when we hear the horns of the boats and cars close by";
>
> "the smells and all that comes with them: pollution, boats, transport; people dump their trash."

Regarding the uses, the theme that stood out for the most part was "peaceful moments," mostly linked with sport and leisure activities (mentioned in 13 of 30 interviews):

> "We come here to fish and walk the dogs";
>
> "I play football here. Sometimes, I work out or I run in the morning";
>
> "We come to see the barges, for the activities, to ride our bikes or to see the boats. Sometimes, we climb on the sand dunes."

Another recurrent theme among uses that is linked to "peaceful moments" is "moments of retreat, contemplation, or even meditation" (mentioned in 5 of 30 interviews):

> "I think, it gives me time to think";
>
> "I come here to think, to empty my mind or to remember my childhood";
>
> "I come here to relax. I watch the landscape. I grew up here, it brings back memories";
>
> "I come here alone to think";
>
> "I come almost every day to see the animals. I smoke and I watch the view."

Finally, to heighten the given representations, a final question appealed to the interviewee's imagination: *If you could reinvent this place, what would you do with it?* A significant number of responses were in favor of preserving the existing situation for the peace it offers, but also because the space, as appropriate as it is, has become their space (mentioned in 10 of 30 interviews):

> "For the moment, I would leave it like this";
>
> "We would leave it this way. We are used to seeing it like this and it's good, this is our place";
>
> "I want to keep it like this. These buildings have always been here, and they are an integral part of the décor."

Between keeping the situation as it is and the wish for drastic change, some of the interviewees commented on potential improvements in the urban integration of the port (mentioned in 5 of 30 interviews):

> "We could give it a new lease of life; it is not really nice to look at";
>
> "The city has grown, but does the port have to disappear? We should perhaps work on its integration. Green areas are missing. We could try to combine port activities with the city by camouflaging them, for instance."

Finally, a considerable number of responses were in favor of drastic change of the port sites, with the ports being either relocated or replaced by a park (mentioned in 6 of 30 interviews):

> "The industrial activities on this riverbank could move to the other bank and a big park with playgrounds and activities can be built here. However, some neighbors do not agree as they think a park would attract thugs and smokers";
>
> "I don't know if it's possible to relocate these businesses. If it is possible, this would allow us to build a garden";
>
> "Let's demolish everything and make it a field. There is no need for a survey, since everyone thinks it's ugly."

According to some responses, a new urban project should replace the present site (mentioned in 4 interviews of 30):

> "Monuments, or something like that. I went to Paris and there I saw monuments on the riverbanks. Here, there is nothing";
>
> "It would be good to do something because there is nothing, it's empty. Social housing or new companies would be good here."

3.2.2 Should We Strengthen the Role of the Port-City Interfaces as Valves in a Dense Urban Environment?

There are three considerations that emerge from this analysis.

First, the representations of these sites, their uses, and their future are extremely diverse. They can be both very positive—quiet areas, breathing space, peaceful moments—and negative—dirty place—and are sometimes marked by a desire for drastic change. This shows that people can assess a landscape in many ways.

> A landscape is not only the portrait of a site or an act of nature, it is also made of conventions that are commonly shared by a society at a certain time (Paulhan, quoted by Paquot 2016, p. 60).

From this point of view, the diversity of representations of port-city interfaces echoes the variety of perceptions of urban wastelands, "*sometimes negatively valued or considered as abandoned spaces and sometimes positively valued and considered as natural spaces useful for multiple activities*" (Brun et al. 2018, p. 9).

Then comes the desire for "controlled" nature through the creation of parks, for example, to transform dirty, gray, polluted places. City-dwellers demand "clean" nature that is "under control" (Robert and Yengué 2018).

Finally, even though these post-industrial landscapes bear strong traces of human activity, the users' dominant representations paradoxically share some features of the "feelings of nature," as described by Paquot (2016). Throughout history, "natural" landscapes have always been associated with certain representations that can arouse the feeling of retreat, activating the imagination and stimulating the senses.

Since the Renaissance, natural landscapes have been experienced or thought of as privileged places for retreat, refuge, and introversion.

> "Natural" landscapes become the shelters of "spiritual" landscapes. They are more about personal journeys than panoramic views, an echo to the Socratic method of getting to know yourself (Paquot 2016, p. 49).

Yet the "empty" spaces of port-city interfaces are far from the common image of a nature show but are nevertheless mainly represented by their users as places of retreat where solitude can be found.

In the early twentieth century, often linked with the need for retreat and silence, places likely to stimulate the imagination were promoted by citizens who demanded "decent conditions for reverie and a retreat beneficial for introspection, for a feeling of well-being, in addition to silence" (Paquot 2016, p. 63). Moreover, "experience of the natural environment does not require grand and remote places, nor even a prolonged stay in a natural setting" (Kaplan et al. 1998). Despite their creation during the industrial period, and marked by an appeal to science and technology to the detriment of living, these empty spaces were still located at the interface of the city, with the port acting as a place of contemplation and meditation, far from the urban tumult.

Since the eighteenth century, the feeling of nature has been expressed as an emotion stimulated by the perceptions of our five senses (the smells, the salty taste of the sea air, the sun, the wind, dryness or moisture on the skin, etc.) (Paquot 2016, p. 54). The proximity of the water attracts residents to these places at the edge of the canal—called "breathing places"—that can activate the senses. Like a maritime area, the wind, breeze, and freshness which rush into this urban space, the canal, have an impact on the skin. The lapping of the water lulls our ears, while the wind blows a variety of odors from the waves to our nostrils. These pleasures for all the senses awaken an affective dimension in the relationship between human beings and things. The strong desire of a large proportion of users to preserve the situation shows a significant attachment to these hybrid landscapes.

4 Conclusion

The port-city relationship is characterized by a progressive disconnection. On the one hand, the fluvial ports have progressively detached themselves from the cities to be closer to efficient infrastructures (fluvial, rail, and, more recently, motorway networks) and land development opportunities. On the other hand, cities have repeatedly coveted port lands as areas to be reconverted to preserve and develop activities with higher added value. The port-city disconnection has led to the formation of wastelands. However, this stage is sometimes obliterated by speculative or anticipatory reconversion processes during this period of obsolescence.

Since the 1980s, cities with an inland port have accelerated the reconversion sequences of port spaces next to city centers for two main reasons. The first reason is based on a political will to bring back city dwellers to the city, in a context of territorial competition and internationalization strategies. Water is widely used as a natural and pleasant backdrop for contemporary recreation. The second reason is the spread of waterfront reconversion models developed in seaport cities from the 1950s. However, the geographic contexts of sea and inland ports differ. The narrow waterway does not allow the port to emerge over the water, while seaports are possible on the sea or coastal fringes (bluefield). In addition, port-city interfaces are integrated into tight urban networks, surrounded by transportation infrastructures. This configuration creates growing competition between urban and port uses for the development of these wastelands or underused spaces.

How are these places represented by their creators/stakeholders and users? The present analysis shows that the perspective of this port-city issue strongly depends on point of view, particularly those of creators and users. Some build them, others live in them. The main divergence concerns the role of conflict in the production of the territory. From the point of view of the conception of the port-city interface, the approach to these areas can be qualified as "geopolitical" where their contours move according to conflicts and consensus and mobilize many different groups of stakeholders.

The users' point of view of these interfaces located between cities and ports is somehow different from the contentious environment of the creation and decision spheres that relate to these spaces. On the one hand, some citizens beg for more "controlled" nature to transform an area that is seen as dirty, gray, or polluted. On the other hand, the representation of a peaceful, breathing space, capable of activating the senses, combined with moments of relaxation or retreat, dominates the comments of their users. This representation takes on more meaning in dense physical environments surrounded by transport infrastructure. The desire for temporary distance could be due to the need to unwind, to release daily stress. From this point of view, these spaces have the same relation to cities that attics have to house, i.e., "*places where we can idle far from the tumult, dream and put the world to rights*" (Bachelard, quoted by Dorso 2012, p. 51). By allowing these precious getaways, as mentioned by F. Dorso, these port-city interfaces offer a place for mechanisms of control of social life. In addition, despite not being considered as natural landscapes, the representations described here, however, share certain traits with feelings of nature, such as a sense of retreat, to dream and activate the senses. At a time when waterway spaces are considered as *cultural infrastructures* (Farinella 2005) or *blue urban ecosystem services* (Haase 2015), it seems important to integrate their quality of emptiness, in very tight urban network, creating a quiet environment for retreat and solitude, activating the imagination and stimulating the senses. Future investigations by planners could, therefore, focus on exploiting the role of these interstitial breathing spaces within urbanized spaces in dense urban environments as a secondary function for ports.

References

Bailleul H (2008) Les nouvelles formes de la communication autour des projets urbains: modalités, impacts, enjeux pour un débat participatif [New forms of communication around urban projects: modalities, impacts, stakes for a participative debate]. Métropoles 3. https://metropoles.revues.org/2202

Bailly A (1995) Les représentations en géographie [Representations in geography]. In: Bailly A, Ferras R, Pumain D (eds) Encyclopédie de la géographie. Economica, pp 369–381

Baudouin T, Collin M (1996) L'après friches portuaires [The post-harbour wastelands]. Urbanisme 291:24–32

Bétin C, Cottet-Dumoulin L (1999) Mémoire fluviale, Planification stratégique, Le cas d'Oullins dans l'agglomération lyonnaise [Fluvial memory, strategic planning, the case of Oullins in the Lyon metropolitan area]. Annales De La Recherche Urbaine 82:118–129

Bird J (1963) The major seaports of the United Kingdom. Hutchinson

Blanchet A, Gotman A (2007) L'enquête et ses méthodes. L'entretien [Survey and its methods. The interview]. Armand Collin

Boubacha E, Davoult D, Gueguen E (1997) Ville et port mutation et recomposition. Note de synthèse et bibliographie [City and port mutation and recomposition. Summary note and bibliography]. Éditions Villes et territoires. https://catalogue.bnf.fr/ark:/12148/cb370009731

Brun M, Di Pietro F, Bonthoux S (2018) Residents' perceptions and valuations of urban wastelands are influenced by vegetation structure. Urban For Urban Green 29:393–403. https://doi.org/10.1016/j.ufug.2017.01.005

Chaline C (1988) La reconversion des espaces fluvio-portuaires dans les grandes métropoles [The reconversion of fluvio-port spaces in large metropolises]. Annales De Géographie 544:695–715

Davezies L (2008) La République et ses territoires [The Republic and its territories]. Editions du Seuil

Dorso F (2012) Le pas de côté. Logiques de l'écart et régulations informelles dans l'espace urbain [The side step. Logics of the gap and informal regulations in urban space]. In: Boucher M, Malochet, V (eds) Regards croisés sur la régulation des désordres [Crossed views on the regulation of disorders]. L'Harmattan, pp 165–182

Farinella R (2005) I fiumi come infrstrutture culturali, Rivers as cultural infrastructures. Editrice Compositori

Fremont A (2011) Fleuve et métropole: la grande déconnexion [River and metropolis: the great disconnection]. Research program FLUIDE, Agence Nationale de la Recherche [National Research Agency], Paris-Est University, Unité Systèmes Productifs, Logistique, Organisation des Transports, Travail [Productive Systems, Logistics, Transport Organization, Labor Unit]

Haase D (2015) Reflections about blue ecosystem services in cities. Sustain Water Quality Ecol 5:77–83

Hall PV (2012) Connecting, disconnecting and reconnecting: port logistics and Vancouver's Fraser river. L'espace Géographique 3(41):223–235

Hoyle BS (1989) The port city interface—trends, problems and examples. Geoforum 20(4):429–435

Kaplan R, Kaplan S, Ryan R (1998) With people in mind: design and management of everyday nature. Bibliovault OAI Repository, The University of Chicago Press

Lageiste J (2008) La plage, un objet géographique de désir [The beach as a geographical object of desire]. Géographie Et Cultures 67:7–26. https://doi.org/10.4000/gc.1002

Le Sueur B (1997) La voie d'eau, machine hydraulique, outil polyvalent d'aménagement du territoire [The waterway, hydraulic machine, multi-purpose tool for land use planning]. Annales De Géographie 106(593–594):195–204

Mazy K (2014) Villes et ports fluviaux: le projet comme dispositif de reconnexion ? Regards croisés sur Bruxelles et Lille [River cities and ports: the project as a re-connection device? Crossed views on Brussels and Lille]. PhD thesis, Free University of Bruxelles—University of Lille. https://www.theses.fr/2014LIL10049

Mazy K (2017) Rethinking the ties between Brussels and its port: a development issue for the canal area. Brussels Stud. https://doi.org/10.4000/brussels.1514

Mazy K, Debrie J (2018) La ville productive: un changement de paradigme pour le projet urbain ? L'exemple des interfaces ville-port [The productive city: a paradigm shift for the urban project? The example of city-port interfaces]. 20e Rencontres Internationales en Urbanisme "Que reste-t-il du projet?" [20th international meeting on urban planning "what remains of the project?"], Lille

Negura L (2006) L'analyse de contenu dans l'étude des représentations sociales [The content analysis in the study of social representations]. SociologieS, Théories et recherches

Paquot T (2016) Paysages et sentiments de la nature [Landscapes and feelings of nature]. In: Paquot T Le paysage [Landscape]. La Découverte, pp 45–68. https://www.cairn.info/Le-paysage--9782707166982.htm

Prelorenzo C (1999) Les édifices et espaces portuaires ou l'invitation à la grande dimension [Port buildings and spaces or the invitation to the big dimension]. Annales De La Recherche Urbaine 82:101–108

Robert A, Yengué JL (2018) Les citadins, un désir de nature «sous contrôle», «fleurie et propre» [City dwellers, a desire for a nature "under control," "full of flowers and clean"]. Métropoles 22. https://doi.org/10.4000/metropoles.5619

Rosemberg-Lasorne M (1997) Marketing urbain et projet de ville: parole et représentations géographiques des acteurs [Urban marketing and city project: speech and geographical representations of the actors]. Cybergeo 32. https://cybergeo.revues.org/1977

Rupprecht C, Byrne J (2018) Informal urban green space as anti-gentrification strategy? In: Curran W, Hamilton T (eds) Just green enough, urban development and environmental gentrification. Taylor and Francis Group, pp 209–226

Subra P (2007) Géopolitique de l'aménagement du territoire [Geopolitics of land-use planning] (FRBNF41135077000008). Armand Colin

Kristel Mazy Architect and urban planner by training, **Kristel Mazy** is Associate Professor at the Faculty of Architecture and Urban Planning of the University of Mons (Belgium). Her research themes focus on the evolution of strategies, urban planning tools, and stakeholder systems in coveted contexts such as city-port interfaces or urban wastelands. She contributes to the steering committee of the network "Inter-friches" (Inter-wastelands).

Part III
Planning Nature With Urban Wastelands

The Regeneration of Urban Riverbanks: A Dilemma Between Environmental and Social Issues

Lessons from French and Brazilian Case Studies

Jean-Paul Carrière and Stuart Farthing

Abstract In this research, we investigated the environmental as well as social impacts of the regeneration of urban riverbanks and waterfronts, localities previously occupied by port-industrial activities established during the twentieth century or earlier, but more recently abandoned and occupied by informal housing and wasteland or brownfield land. The urban plans and strategies of metropolises as well as of medium-sized cities have become increasingly underpinned by the "paradigm of attractiveness," in a context of globalization and competition between cities at an international scale. In order to become more attractive for investment, mainly foreign direct investment, cities need to improve their image and their international ranking. The recycling of the abandoned wasteland in these areas has become a policy priority. Under the banner of sustainable development, strategic plans in many cities throughout the world have thus promoted flagship projects for the regeneration of these waterfront areas, incorporating new environmental amenities.

However, the implementation of such planning strategies reveals significant conflicts within local planning policies: such projects seem to be difficult to reconcile with "the right to the city," following Henri Lefebvre. Indeed, these projects have not avoided the problem of gentrification, but have reinforced urban socio-spatial segregation. We illustrate these conflicts and the lack of social sustainability in such projects where significant environmental issues are concerned, using three case studies, two in France (Nantes and Bordeaux) and one in Brazil (Recife).

Keywords Wasteland and brownfield land · Regeneration of waterfronts and riverside landscapes · Planning strategy and urban marketing · Attractiveness · "Right to the city" · France · Brazil

---

J.-P. Carrière (✉)
UMR CITERES-CNRS, Université de Tours, Tours, France
e-mail: jean-paul.carriere@univ-tours.fr

S. Farthing
Urban Planning, University of the West of England, Bristol, UK
e-mail: farthingstuart@gmail.com

F. Di Pietro and A. Robert (eds.), *Urban Wastelands*, Cities and Nature,
https://doi.org/10.1007/978-3-030-74882-1_11

1 Introduction

This chapter presents the main conclusions of research (Carrière and De la Mora 2018) conducted in the framework of the French–Portuguese Dialogs in Urban Planning (program Attilio).[1] In many cities throughout the world, the banks of rivers were previously occupied by either low-income people living in slums or by old and declining port-industrial activities. Industry and port facilities, as well as formal and informal housing for poor people and workers, were a common feature along rivers in many cities after the industrial revolution. After the Second World War, however, these areas rapidly became obsolete, being transformed into wastelands or brownfields.[2]

In the last decades of the twentieth century, we saw a significant movement in favor of the re-use of these areas and neighborhoods. Cities are "rediscovering" their rivers, and planners and developers are conceiving new projects on the riverbanks, aiming to respond to a growing "desire for nature" (Bourdeau-Lepage 2005), the river being looked at as an "interface" between nature and the city.

However, the environmental, as well as social impacts of the regeneration of urban riverside areas, cannot be understood without considering the general context of globalization, which exacerbates competition between cities at a world scale. Indeed, urban planning strategies are increasingly underpinned by the "paradigm of attractiveness," promoting flagship projects such as the regeneration of riverbanks and the redevelopment of waterfronts, in order to restore or to improve the international image of cities and to attract investment, mainly direct foreign investments (Carrière and Demazière 2002a and b).

Riverside areas offer high-value environmental and landscape potentialities for construction companies and real estate agencies. But they also constitute good opportunities to renew the city's image with a sustainable dimension. As a result, many large French cities have given priority in their urban plans to the regeneration and the 'renaturation' of their waterfronts, areas previously occupied by obsolete industrial activities and harbors, as in Bordeaux and Nantes, located, respectively, on the rivers Garonne and Loire. Both cases provide clear illustrations of this metropolitan planning strategy. In this chapter, we use the findings of both case studies to illustrate our analyses. However, we could have found many other examples of French metropolises, like Lyon, or of smaller cities, such as Orléans, Tours, Nancy, Rouen… which have copied the same kind of strategy, seeing riverbanks as assets for enhancing their attractiveness.

---

[1] This program brings together Portuguese-speaking and French-speaking researchers and practitioners in town planning. It is supported by French-speaking and Brazilian associations bringing together training and research institutes in town planning. This program gives rise to bi-annual meetings in Europe and Brazil. Our chapter supplements and updates the work that was presented during the Salvador de Bahia meetings in 2016, published in the proceedings of this meeting (Carrière and De la Mora 2018).

[2] Wasteland, located along the rivers, as defined here, can include brownfield land, but also areas that have not been developed. Both types exist in our three cases.

In Brazil, despite important differences from France in its social and urban context, we have observed the same convergence between market forces and local planning strategies having as an aim the regeneration of the waterfront to make the cities more attractive. The case of the estuaries and seafronts of the metropolitan area of Recife is a useful representative Brazilian case, that we discuss in this chapter. However, one of our main conclusions is that the implementation of this strategy conflicts with another key objective: "the right to the city" (after Henri Lefebvre)[3] a right which is enshrined in Brazilian law. The "Estatuto da Cidade" (2011) specifically recognises the social dimension to city life and the right of each person to stay in their neighbourhood after renewal or rehabilitation, but this does not happen in practice particularly on river banks, and in low-lying areas where there are concentrations of favelas and poor housing.

In France, no such legal reference to the right to city exists, though all planning documents refer to the "sustainable city," which includes a notion of social—not just environmental—sustainability. But, de facto, our case studies show that the redevelopments of riverside neighborhoods provoke the same kind of dilemma as in Brazil: how to reconcile a strategy of increasing the city's attractiveness with the need not to increase socio-spatial fragmentation. In both cases, the redevelopment of the riverbanks, often managed by private investors, produce perverse consequences: the preservation of the environment and of the "natural" character of the river does not favor a more sustainable city, because it provokes processes of social exclusion and increased gentrification. The consequence is that the socio-spatial fragmentation of the city has been reinforced, which is not consistent with the aim of increasing attractiveness! That is one reason why the claim to the "right to the city" has become more widely supported.

As indicated above, in the first part of the chapter, we draw together some of the main conclusions from the three case studies of Bordeaux, Nantes, and Recife to show how the riverbank regeneration projects are a key part of the paradigm of attractiveness. Next, we analyze the impacts of these projects and strategies in terms of socio-spatial fragmentation of the cities and their conflict with the "right to the city."

2 The regeneration of urban riversides: three case studies in France and Brazil

During the twentieth century, estuaries and rivers lost their main function as communication axes and became less attractive for new developments. The banks of rivers were then occupied by the former harbor and industrial areas becoming brownfield and wasteland, as in Bordeaux or Nantes, or as in Recife occupied by favelas and inadequately constructed areas for a deprived and very poor population (Figs. 1, 2, 3).

Fig. 1 The "Île de Nantes" an abandoned industrial area on a large island of the Loire River, in process of regeneration (in 2005) © SAMOA 2005

Fig. 2 The quays and docks, in Bordeaux, in the 80s before regeneration. *Source* F. Ducasse, B. Broustet: 1970–1995 Bordeaux—Ombres et lumières. © François Ducasse)

Despite the differences and the contrasts between these three cases, there is a similar consensus among the stakeholders involved in the design and the implementation of urban development and planning projects, that these riverbanks offer

Fig. 3 Recife, a riverbank in the center, occupied by *palafitas* (makeshift houses on stilts in flooded areas) © K. Hochart (2014)

important advantages and assets for urban renewal, in terms both of a reserve of land and as places with a distinctive identity and great landscape and environment quality.

In other words, the regeneration of the riverbanks allows the creation of something distinctively different from the "banal" building projects of suburban areas, but also provides an opportunity to promote a new sustainable city image, the river giving the projects a "sustainable feel" (Demazière 2015)!

This phenomenon has been noted worldwide, for some time. The regeneration of waterfronts and riverbanks originated in the United States, in the 1960s, in cities like Boston and Baltimore, but such a strategy focusing new urban development on riverside areas has spread through the world, mainly in Europe: Bilbao, Lisbon, and Barcelona are good examples of new urbanism focused on the capitalization of the assets of waterfronts, but in the big cities of other continents, like Cape Town, Sydney, and Buenos Aires, these exist too. Moreover, similar urban projects imitating these are in progress in small or medium-sized cities.

To understand the importance of this evolution in the relationship between waterfront areas and the city, supported by planning visions and documents as well as by the projects of private urban developers, we need to take into account changes in the global socioeconomic system. The way of looking at the river in urban development projects has not only changed because there is a new "desire for nature" in the post-modern city, or because of the rediscovery of the pleasure of living on the waterfront. Our judgement written 18 years ago (Carrière 2002, pp. 9–10. Transl. by the authors) still remains relevant: "*The new conception of the "urban project" which has accompanied the globalization of the economy [leads to] a strengthening of the image and the competitiveness of the city by a re-evaluation of run-down areas, which offer great urban and environmental opportunities [...]. The challenge is to participate in international competition and to attract a flow of investment mainly tertiary and internationally mobile. [...] These operations combining "symbolic features" and urban renewal become the underpinnings of global urban management led increasingly by the initiative of private–public partnerships.*"

From this perspective, the banks of rivers should be regarded as particularly propitious areas for implementing a new resilient urban marketing strategy. They offer distinctive features and have become "...elements of a new urban landscape" (Lechner 2006, p. 87). Neoliberal urbanism, particularly influenced by the British model of private–public urban development corporations, has given a strong impulse to "rediscover" the rivers and their banks, as bases for a "new city."

Beyond this, what findings can we report from studying these three cases?

2.1 Three Cases with Anthropized Landscapes but Quite Different Natural Features

How can we describe the landscapes of the rivers in our three cases and the characteristics of nature we find there? Anthropization is a significant and common feature with a landscape that has resulted from long-run urbanization processes in which rivers themselves have been the primary means of transport and economic exchange as well as acting as major energy sources. However, we should note that the ecosystems and the natural features of these areas are quite different.

In both French Atlantic cities, there are no real traces of wild nature, with the exception of the water itself. Waterfronts have been occupied for centuries by continuous, high-density port or industrial activities. Consequently, the nature that remains is largely "artificial" and "tamed." In both cities, it has been 'planned' and endowed with parks, gardens and paths. As in most European cities, "flood plain nature" has thus given way to an ordered nature with little biodiversity in terms of either flora or fauna.

The tropical city has different characteristics with a much more heterogeneous and less managed nature. The floodplain on which Recife has been expanding since the sixteenth century is crossed by 5 rivers, the Capibaribe being the most important, but also by many streams and by 66 canals, crossed by 110 bridges. Hence the terms "*Brazilian Venice,*" "*amphibious city*" or "*aquacentric urban space*" are often used in the literature (see the website www.inciti.org[3]). Over the centuries, the relationship between the rivers and the city has changed significantly. With the expansion of the city, away from the earliest settlements, and with the advent of new modes of transportation, the close relationship found in the past has changed

[3] Consultation: 2020/05/25.

to one where the city, it is said, "turns its back on the banks" (Carvalho Filho et al. 2015). The banks themselves have become zones of mixed and fragmented land uses, forming many varied landscape units and natural habitats (Sá Carneiro et al. 2009). The Capibaribe's delta was historically the only route for the export of sugar. As a consequence, along the 35 km of the river's sinuous course through the city and along other watercourses, some vestiges of sugar cane mills, called "*engenhos,*" and sugar factories remain. Here too, the earliest housing developments were located, as well as more recent tall skyscrapers producing in these areas a mix of informal housing, including *favelas* (slums) and *palafitas* (makeshift houses on stilts in flooded areas) and luxury housing. From the beginning of the twentieth century, as a result of the deficiencies in the sanitation system, there was, therefore, a rapid decline in the environmental and biological quality of the waterways with attendant high levels of pollution. In this process, many areas with high ecological and landscape value have become fragile, residual, and invisible spaces, undermining the idea of rivers acting as green and ecological corridors, and thereby, also restricting public access to them. So, various sections of the riverbanks can be regarded as sites of "abandoned and camouflaged nature" (Sá Carneiro and Menezes 2011). Finally, we can say that rivers and waterways are ubiquitous in the city, and that their banks are characterized by a patchwork of natural and built environments, with high-density built-up areas, historic places with some parks and gardens (such as those designed by the well-known Brazilian landscape architect, Burle Marx), and brownfield land, but also with areas of Atlantic ancient forest, sandbanks, mangrove, and coastal tropical vegetation. Despite their ecological degradation, Recife's rivers are likely to have more abundant, and more diverse, wildlife, than the Loire or the Garonne. As a result of this variety of "natural places" in Recife, the main issue from a "green" point of view is to inter-relate all the landscape units in an urban landscape plan, allowing the creation of a system of public open spaces. Accordingly, the Capibaribe Park project (Diniz et al. 2016, Carvalho Filho et al. 2015, Dos Santos Calvacanti et al. 2015) aims to link old built-up areas, wasteland, and areas of vegetation in a continuous green corridor (see below).

2.2 *Urban Wastelands on Riverbanks: Long Neglected Development Opportunities, but with High Potential*

When they are located in the urban core, riverbanks offer land reserves and opportunities for urban renewal or rehabilitation, with the added benefit of avoiding excessive urban sprawl. Indeed, the riverbanks are often undeveloped or occupied by wasteland which can be "recycled," land abandoned after the closure of industrial activities, or occupied by land liable to flooding, though some land may be protected.[4] So,

[4] However, this does not preclude a high concentration of people in informal dwellings, as is the case of the favelas of Recife; a majority of them being located in high-risk areas (slopes, flooded areas…) along the rivers.

reclaiming this land for development is a way of rebuilding the urban core, as we note in our 3 case studies, even in Recife, where the costs of works for the prevention of flooding is very high.

In Nantes, the Île de Nantes, surrounded by the arms of the river and in the heart of the city, provided a great opportunity for a considerable expansion of the metropolitan center. On its western side, the island is occupied by an old neighborhood including much brownfield land, particularly the immense area of the so-called "les Chantiers de l'Atlantique" shipyards, disused in 1987, and on the eastern side, by a mixed area of offices and housing. By 2012, there were 18,000 residents living on the island and some 22,000 jobs, when the first phase of the project was completed. After the closure of the shipyards, a plan was adopted to rehabilitate the site, and implementation began in 2001. A "*Société d'Economie Mixte*,"[5] the SAMOA, was created, in order to manage the project. The main partners involved in the creation of this company were The Metropolis of Nantes, the Municipality of Nantes, the main communes surrounding Nantes, the Chamber of commerce and industry, the *Caisse des Dépots et Consignations*,[6] and six private or public banks (Barthel 2009; Devisme 2007).

For the first stage of the project (2003–2012), a new urban area was designed, including public zones, and real estate programs, giving priority to the rehabilitation of existing buildings. New buildings were constructed for housing, service, and commercial activities, with new university buildings, a School of Architecture, and also some large facilities such as a courthouse.

The second stage, which should be completed in 2030, is more ambitious, with the aim of making the island a "metacenter," an attractive focus of the metropole, by the construction of 1,500,000 m^2 of real estate, including 10,000 new dwellings and 450,000 m^2 dedicated to industrial buildings or offices, and 350,000 m^2 of other facilities. Moreover, 1,600,000 m^2 of public and leisure areas are planned, as well as new forms of mobility (cycle paths, 3 tramway and bus lines, etc.).

Being a spectacular redevelopment of port-industrial brownfield land on the banks of the main French river, this project is one of the most important urban planning initiatives in France, outside of Paris. Pascal Galinier, in Le Monde (2019/12/17), considers, the island renamed: "Quartier de la creation" (i.e., Creation Neighborhood) as an "urban laboratory" and as one of the most noteworthy French examples of an innovative smart city, mainly thanks to the number of start-ups and new university facilities, located in the old rehabilitated industrial buildings and dedicated to digital disciplines and research.

Bordeaux, located on the Garonne estuary, heir to a rich port history,[7] was the subject of an ambitious operation to regenerate its left bank, starting in the year 2000. Next, the project involved the right bank. The first step was to work on a large port facility in the city center, which had been abandoned in 1987. More than 3 km of

[5] Or, in English, a "mixed economy company," meaning a private law company whose capital is mainly of public local origin.

[6] A public financial institution, responsible for collecting the savings of the French population.

[7] In 1789, at the time of the French Revolution, Bordeaux was the main French port.

quays 80 m wide, 55,000 m^2 of dock sheds, and 200,000 m^2 of quays were regenerated (Allaman 2003), with the aim of allowing the city to recover its international attractiveness, undermined by the obsolescence of port facilities installed in the nineteenth or early twentieth century. We could describe this as a flagship project aiming to restore the former luster to a prestigious space: the royal square, with its opulent façades of classic style built in the eighteenth century on the shore to impress foreign sailors. The main objective of the project was to restructure the interface between nature, the river, and the city by means of a large-scale development project in the center of the city, based on the improvement of the "natural" as well as built landscape. From the 1980s, this plan has provoked a fierce debate within the city (Lechner 2006).

For the second stage, a total of 5 km of port wasteland is being regenerated on the left bank, which is part of a more general urban plan at the metropolitan scale (Bonnin 2007), including a new bridge on the river, new tramway lines, the rehabilitation of old run-down neighborhoods along the river and of former industrial or railway zones. The Mayor Alain Juppé wanted to provide Bordeaux with a central area in line with its metropolitan ambitions. This was to be achieved by rehabilitating most of the dock buildings as well as the World Heritage listed waterfront, a project launched in 1995, by the destruction of 4.5 km of ironwork security fencing which isolated the port and separated the river from the city center. So, an important aim of the plan was to restore green areas, to highlight the built heritage, to reduce traffic, and to create places for leisure. The riverbank thus became one of the most vibrant spaces in Bordeaux, and consequently a very attractive area in the Central Business District of a European metropolis. The "mirror of water" installed in place of an old run-down port warehouse, in front of the most beautiful classical façade of the quays, is the key symbol of this renovation (Fig. 4).

In Recife, the regeneration of the riverbanks has been focused on well-located but insanitary areas liable to flooding, very close to the city center. It provides land and building plots for luxury real estate operations, in highly strategic locations given their accessibility to the core of the city. It has provoked fierce conflict between private developers and representatives of social movements, while municipal urban policy, like Janus, faces two ways and remains schizophrenic: on the one hand, the Municipality approves private highly profitable real estate projects, as in the project Novo Recife (i.e., New Recife), in the Bacia da Pina (the central bay of Recife), and on the other hand formally keeps the PREZEIS,[8] a social program intended to allow land regularization and development of currently insecure informal areas of housing. In theory, this program is planned to guarantee the maintenance of the poor population on the spot. But, the PREZEIS is unable to guarantee the right to the city for the poorer people in these very attractive areas, because market forces are strong in these areas of the city along the fronts of 5 rivers and the bay.

[8] The PREZEIS, or "Regularization of Special Zones with Social Interest Program," is designed to rehabilitate and redevelop favelas and informal precarious areas, allowing their inhabitants to continue living in their homes.

Fig. 4 The quays of Bordeaux after regeneration: in the foreground, the "mirror of water" in place of the old dock warehouses, and in the background, the classic facades of the twentieth century rehabilitated © authors, 2020

The three case studies show how in the context of worldwide urban competition, planners and local politicians have implemented strategies of urban marketing based on their distinctive local character, by "rediscovering" their riverbanks. To regenerate them and their brownfield or wasteland is both a way of preserving local memory and built heritage, and to recreate a new image and a new urban identity. It is a way of profiting from the symbolic capital of the river. Moreover, regenerating the riverfront is not limited to real estate operations alone, but integrates the planning of new land uses: touristic, cultural, and leisure activities. So, in Recife too, urban planning becomes an essential component of a global strategy of "urban marketing."

The landscape and environmental aspects of riverside development from the perspective of urban sustainable development transform river corridors into "*trames vertes et bleues*," i.e., green and blue networks, which are meant to preserve water quality and biodiversity. This argument is used to justify regeneration projects, in France, as well as in Brazil! However, it is also clear that the creation of new landscape and environmental amenities also creates positive externalities, much appreciated by private developers and stakeholders who can take advantage of the benefit of these new locations. Indeed, the riverbanks and waterfronts provide opportunities for developers "to play" with renovated buildings, rehabilitated heritage (including industrial structures), landscape, and water in order to increase their supply on the market, in ways not possible in other kinds of areas. These opportunities, we observe in our case studies, are maximized when the "natural" amenities of the river can be combined with symbols of local identity and/or historic memories of the site. They can be regarded as "mixes" of sustainability and profitability.

However, it is obvious that, despite the reasons given in local plans, which cite purely environmental aims, there are mainly socioeconomic logics and financial objectives for adopting new landscape and environmental norms for regenerating urban river frontages (Gérardot 2004), as the 3 cases show.

In Nantes, following the architects A. Chemetoff and J-L. Berthomeu, authors of the Plan guide (i.e., guide plan) published in 1999, "*It is a question of finding a way, a living material, which can cultivate the memory of a history of past activities that have marked the relations between the river and the city, and whilst ensuring that the entire urban area develops the landscape of a city which is open to the river in its geographical center.*" (Chemetoff and Berthomeu 1999, transl. by the authors). This is the reason for having listed and preserved symbols of shipbuilding activities on the site, for example, a monumental crane, while the sheds were transformed into offices or leisure and culture facilities, and planting and landscaping were planned to create a landscape framework to "*restore the relation between city and water*" (Chemetoff and Berthomeu 1999, transl. by the authors). Moreover, the whole project is designed to make the island a new attractive central space for a European metropolis. The symbolic scope of the restructuring of the island is reinforced by the construction of a sustainable neighborhood (*écoquartier*, in French), aiming to get the marketing label of ecocity, as the brand image of a city providing quality of life and optimal environmental conditions. However, we should not forget that more than sustainability the entire project aims to renew the image of Nantes, involving private and public stakeholders in a profitable megaproject.

In Bordeaux, whereas in the eighteenth century, the development of the quays had been designed with very high aesthetic concerns, during the industrial revolution the installation of dock warehouses and traffic routes created a visual barrier, separating the city from its river. This is why the primary purpose of the redevelopment project of the port-industrial area, in the city center, was to rehabilitate the historic façade of the left bank and restore landscape quality. Then, for the right bank, a general plan, called the *Plan Garonne*, approved in 2000, has involved all riverside communes of the agglomeration. Priority was given on the one hand to reducing car use,[9] promoting pedestrian and cycle paths as well as public transport (tramways), and on the other hand, to the aesthetic restoration of public space. As in Nantes, the development was thought of as a landscape project, once again aiming to reinforce the attractiveness of the area, in a manner worthy of a European metropolis. The planning competition, won by landscape architect Michel Desvignes, was explicitly asked "*to exploit the scenic and landscape dimension*"[10] of the river (Lechner 2006, p. 49). The "natural" landscape, as well as the built heritage of the banks, was subject to strategic orientations, in the framework of a larger project for the whole metropolis: so, on the right bank, the Garonne plan aimed to build on brownfield and wasteland bought from the autonomous port, a new neighborhood, a university campus, offices and public or private facilities (a cinema multiplex, for example), a three-hectare

[9] The proposed objective is to halve the traffic on the shore, which in 2000 reached 80,000 vehicles per day.

[10] *«Exploiter la dimension scénique et paysagère»*.

Fig. 5 The twin towers, on the edge of the historic center, on the left bank of the Capibaribe River, in Recife © Mila Montezuma

park, a botanical garden, cycle paths, while a new bridge and a new tramway line link both rivers. So, the whole regeneration project for both rivers seeks to attract higher level metropolitan jobs and functions with international significance, and to create a "development arc" with the *TGV*[11] station located in the south of the city.

In Recife, the high potential of the riverbanks has begun to be exploited, in recent decades. The enhancement of the landscape of the river and the development of wasteland on the banks is one of the key elements of the planning strategy and of the public as well as private development projects. Private real estate companies have even influenced the urban policy guidelines, through strategic and flagship projects, such as, for example, the building of the twin towers on the waterfront in an export industrial area, on the edge of the historic and cultural city center (Fig. 5).

Other similar skyscraper projects have transformed the riverside landscape in the center, and simultaneously accelerated the gentrification of it, in accordance with the strategy of boosting attractiveness (Carrière and De la Mora 2012). Most of these luxury buildings are located in environmental protection zones on the riverside, areas previously occupied informally by communities of poor people, who are obliged to move to distant peripheral neighborhoods. This is the case with the *Novo Recife* (New Recife) mentioned above, which includes 14 skyscrapers of 40 floors.

[11] TGV is the French acronym for high-speed train.

Another megaproject built on wasteland with high environmental quality is the shopping center *Rio Mar* built on the front of the *Pina bay*, which offers an unrivaled view over the edge of the estuary and gives access to *Manguezais* Park, the largest natural aquatic park in Brazil. This mall frequented by wealthy consumers has had a big impact in terms of gentrification of the surrounding neighborhoods, despite or rather as a result of the redevelopment of the banks. We could quote many other examples where environmental improvements have been combined with real estate developments of riverbanks in Recife! However, we should emphasize that in all these operations, the main problem has been above all social. One of the most famous cases is the ZEIS (Special Zone with Social Interest) of the favela *Brasilia Teimosa*, a symbol of urban segregation in Recife, located on a peninsula with high development potential, close to the most prestigious beach, the *Praia de Boa Viagem*. Developers tried to oust the poor of the favela in order to build a 5-star and 40-floor hotel. The residents demonstrated to pressure the municipality into blocking this project which violates the law of PREZEIS, which in theory should stop poor populations facing the risk of eviction.

In addition to these private projects, the creation in 2012 of the Linear Park of the Capiraribe River, initiated by the Municipality, in partnership with the Federal University of Pernambuco, illustrates the close links between public intervention and real estate developers in upgrading the banks of the river. The aim of this project is to allow the population to rediscover the environment of the river, by means of a mix of improvements (green spaces, places for cultural activities, cycle paths, pedestrian paths, navigable routes) over a length of 35 km, along the two banks of the main river crossing the city, and finally to make the city more attractive by improving its "natural" amenities.

2.3 *Riverbanks: Places with Strong Governance Challenges*

All our case study cities are set within broader functional urban areas. In this type of situation, the redevelopment of the river banks should not be seen as falling within the political and administrative responsibility of a single center, but should require the establishment of inter-municipal governance. A river constitutes a particularly useful mechanism of "support" for establishing such new inter-municipal cooperation based on a shared project for the rehabilitation of heritage, both "natural" and built landscapes. From this perspective, the French experience seems more advanced than that in Brazil, even if this country played a pioneering role with the creation of its "Metropolitan Regions" in 1974 (Carrière 2012).

In Nantes, the "reconquest" of the banks of the River Loire started a process of wider reflection and cooperation, in the 13 municipalities belonging to the former Urban Community[12] far beyond the perimeter of the island of Nantes. A joint atlas

[12] The Urban Community of Nantes is the old structure of intermunicipal cooperation that existed before the creation of the Metropolis in 2015, including 24 communes, with 646 522 inhabitants

of the banks of the Loire has been drawn up to identify possible redevelopment opportunities and derelict sites over a length of 47 km and a width of 1 km on each side. The river has thus become the instrument of a cooperative strategy and the vehicle for metropolitan restructuring bringing together the cities of Nantes and St Nazaire, 70 km apart, and other intermediate municipalities (Masboungi 2003).

The urban area of Bordeaux, with a long tradition of inter-municipal cooperation through the Urban Community of Bordeaux (CUB),[13] was able to initiate its redevelopment project for the quays of Garonne at an inter-municipal scale, allowing it to coordinate public stakeholders on the left bank. But the old autonomous port, downstream of the city, and the land on the right bank also part of the new developments belonged to separate municipalities. This has resulted in permanent talks and interactions between actors with often divergent interests; all the more so as some refurbishment of platforms or dock buildings has been entrusted to the private sector. This is why in the whole project many activities were carried out by public–private partnerships, while the development of the Bastide district, on the right bank, was entrusted, as on the island of Nantes, to a mixed economy company.

In the case of Recife, inter-municipal cooperation has remained limited to transport issues, development projects being considered within municipalities. No regeneration operations on the banks of the five rivers which cross the Metropolitan Region of Recife have been carried out within an inter-municipal framework, with the exception of the Pro-Metropole project, carried out by the cities of Recife and Olinda, respectively on the right and the left banks of the River Beriberibe. In Brazil, municipalities have great difficulty in conducting coordinated operations, especially since these are often carried out on a case-by-case basis, by large private firms such as the construction company Odebrecht.

In talking, in general, about the regeneration process for the banks of rivers, the question of governance is not just a question of inter-municipal cooperation, because this type of redevelopment involves rethinking the multiplicity of uses of all the areas concerned, and requires the mobilization of a large number of other institutional or economic stakeholders with often divergent interests, such as consular chambers, the state, port authorities, firms… In addition, there is the question of the participation of civil society and its representatives, who are often excluded from participating in the design of projects.

Drawing together the lessons from the French and Brazilian examples, it seems that the findings are similar, whatever the context.

In all our cases, the projects have a strong symbolic dimension. As a result of their great heritage, environmental, and landscape value, the areas around rivers are particularly well suited for the implementation of development strategies aimed at restoring the attractiveness of the city and renewing its image. De facto, cities in

(2017), while the whole urban area includes 108 municipalities and exceeds 960,000 inhabitants (8th largest in France).

[13] The CUB, transformed in 2015 into a metropolis, includes 28 municipalities in the Bordeaux agglomeration with just over 750,000 inhabitants, while the whole urban area includes 255 municipalities and exceeds 1,232,000 inhabitants, in 2016 (5th largest in France).

emerging countries like Brazil have not adopted fundamentally different strategies from those of more advanced countries for their integration into global competition.

The attention of urban planners and local authorities has focused on sites and landscapes in crisis, whether because of the decline of a past industrial activity producing strong negative externalities or an uncontrolled informal urbanization. Consequently, the problem of the development of riversides is not limited to the question of their embellishment or their renaturation!

Riverfront urban development meets the criteria of "*urban entrepreneurship*", as defined by Harvey (1989), that is to say involving a large intervention from the private sector, in particular, in the form of public–private partnership.

Beyond these findings, the question arises of the compatibility of the environmental and social aims of urban riverbank regeneration plans.

3 Regenerating the Riverside Areas Versus Guaranteeing the "Right to the City:" A Permanent Contradiction of Local Urban Policies?

The right to the city cannot be restricted simply to the capacity of individuals to access decent housing, and to urban resources or amenities, as Harvey (2009) points out. It is also a collective right to transform the city. It is about enabling the population to create a city where everyone is guaranteed access to city facilities and urban life in all its dimensions, to places for meeting and exchange, etc. The right to the city is, therefore, opposed to the fragmented vision of the city, where land uses are physically separated by the planning of functionalist inspiration, and also where there is social segregation (Costes 2010). The right to the city is achieved through the re-composition of urban space and a socio-spatial mix.

From this point of view, both in France and in Brazil, riverbank development projects do not meet the requirements of the right to the city, despite environmental improvements, which are the main justification advanced by decision makers. This is explained by three main factors.

1. The main project decision makers are dependent on market forces and are less concerned about the need to reduce socio-spatial fragmentation. In Bordeaux, the projects aim to "showcase" the river, which leads to the reservation of the immediate surroundings of the water for public spaces (gardens, promenades, leisure parks) or for prestigious buildings. This makes it possible to enhance architectural homogeneity, a direct reflection of the social homogeneity of the rehabilitated spaces, by producing the effect of a continuous façade, thanks to the rehabilitation of eighteenth century buildings; this is in contrast to the dilapidated and run-down districts still remaining further back behind the waterfront. In Recife, the construction of luxury buildings with direct access to rivers for the wealthiest households produces an even more spectacular physical and social contrast between the waterfront and the backlands of these areas. In both cases, there is an increase of socio-spatial division

within the riverside neighborhoods, a phenomenon that is also to be found on the Île de Nantes.

The increased social segregation produced by riverbank development projects is all the stronger where social inequalities are markedly high, as in Brazil; this does not mean that where there is an appearance of more general prosperity, as in France, strong urban segregation processes are not also being strengthened.

In the case of Recife, the spatial organization of the city does not conform to the classic model of center-periphery contrast (Carrière and Hochart 2016). The physical distance between condominiums for "wealthy households" or the upper middle class and islets occupied by the poor are very small. In the center of the city, the poor and rich live in close geographic proximity, though at the local scale walls enforce segregation. This is the consequence of an informal housing production system which has led the poor to occupy public or private land in districts of the city core. Informal housing (64.7% of the total housing stock of the Metropolitan Region in 2010, according to the database of the Metropolitan Observatory), very often built in clandestine self-production in favelas, occupied by the poor, are found in all the districts, close to the most prestigious buildings. For historical reasons (Carrière and De la Mora 2014) Recife is distinguished by the importance of its makeshift housing stock: in 2010, 852,000 inhabitants of the Metropolitan Region lived in such housing estates, or 23.2% of the total population (In Rio, the same figure is 14.2% and in São Paulo, 11%). Within these enclaves of poverty scattered throughout the city, and more particularly in the center, lives a socially excluded population, 80% of whom are concentrated in Special Areas with Social Interest, protected in principle from the threats of real estate agents, as seen in Sect. 2.2.[14] These areas, located for the most part on environmentally damaged riverside areas, which are often flooded, are in fact highly coveted by large national and international real estate groups, due to their strategic location.

All in all, it is obvious that the process of socio-spatial fragmentation, far from being reduced by urban renewal and the construction of gated communities, is increasing along the riverfronts of Recife. In the cases of Nantes and Bordeaux, the gentrification of the riverside redeveloped areas has less impressive physical effects, but takes the form of social homogenization in favor of those with high incomes, the less wealthy having to retreat to the peri-urban outskirts.

2. The regeneration of riverfronts has had a significant impact on property values, increasing prices rapidly in and around the areas concerned.

In Nantes, property prices increased on average by 20.3% from 2008 to 2017, while in Bordeaux there was a 43.8% increase over the same period, the price per m^2 today exceeding 5000 Euros along the quays (source Notaries and INSEE), this being the main factor behind urban segregation. In both cases, unsurprisingly, we noted that property values increased rapidly after the start of regeneration activities. As a result, obsolete areas abandoned by the port industries and the shipyards of the Île de Nantes have reached prices today close to those in the city center: over 3000

[14] For more details (and a map) on the spatial distribution of precarious housing in Recife, see Carrière (2018).

Euros per m^2. In Bordeaux, while prices on the left bank quays have reached record levels, values on the re-urbanized right bank often exceed those of central districts. As a result, the poorest and even middle-class households can no longer afford such sums and their numbers have decreased in these districts.

In Recife, the same kind of process is at work on the re-urbanized banks, with a similar logic as in France. Only, households of class A or B, i.e., with the highest incomes, according to the classification of the Brazilian Institute of Geography and Statistics, have access to comfortable and luxurious condominiums.

3. The third reason for the gap between the claim to the right to the city by the inhabitants of these areas and actual development projects is the priority given to environmental objectives in order to renovate the image of the city. Even if this could be justified, it is difficult to combine with that of citizen participation in the "fabric of the city"! We have seen this both in Brazil and in France. The collective re-appropriation of urban space, desired by H. Lefebvre, has been interpreted in different ways. In Brazil, the "*city statute*" was passed by Parliament in 2001 and participatory budgets have been implemented, notably in Recife. In France, there have also been some experiments with participatory democracy, which allows Costes to talk of an "*instrumentalization of Henri Lefebvre's thought*" (Costes 2010, transl. by the authors). But, in both cases, the vision of the right to the city has faded faced with the logic of profitability that has guided brownfield or wasteland redevelopment operations on the waterside, while local planning policies have remained dominated by the thinking behind the attractiveness paradigm, whatever the party political orientation of the municipalities or inter-communal structures.

According to the Urban Planning Agency of Nantes, the Île de Nantes project has been developed in consultation with all stakeholders, including residents and their associations, with a consultation mechanism implemented by the municipality. But we can legitimately speculate about the real weight of citizens' voices in relation to other heavyweight participants, such as central government, regional and departmental organizations, the autonomous port, the national railway company, and all private investors. Citizen dialogue was initiated with the creation of a Neighborhood Council in 2010 to give opinions on the development of the project up to 2030. In parallel, a workshop was organized with a panel of residents supposed to represent the population of Nantes, to formulate proposals in terms of social life, mobility, environment, housing, commerce, and employment. This desire to involve residents is part of the eco-neighborhood project, which has a "*strong image at a European level and (is) a showcase for sustainable development*," and has been undertaken to encourage private businesses to commit to good practice development activities, according to the terms used by the Vice-President of the Urban Community of Nantes, cited by Barthel (2009). But according to the same author, this showcase is gradually "*being transformed into the major tourist focus of the urban area*," whilst moving away from "*a more social planning approach aimed at working with the users of the island on future mechanisms to promote residents' initiatives or those of formal associations even in the management of development*" (Barthel 2009, transl. by the authors). This means that the right to the city, as a principle by which residents take control of their living space, is not, in fact, a top priority.

In Bordeaux, despite the referendum organized by the Sud-Ouest newspaper in the early 1980s to endorse the idea of building a modern façade on the brownfield land on the right bank, opposite the classical architecture of the quays, consultation with the public was mainly used to communicate and publicize the objectives of the project to recover the river and was much less about citizen mobilization from the perspective of a participatory project, where the population could choose between different options. So, it is difficult to claim that the decision-making processes have taken account of the wishes of the inhabitants.

In Recife, the participation of the inhabitants in the options for regeneration of the riverfronts took the form of meetings organized by the Commissions of Urbanization and Legalization of the Special Areas with Social Interest and the provision of a budget for participation, as in the case of *Brasilia Teimosa*. However, this did not really allow the inhabitants to have an influence on the decision-making processes relating to the major real estate projects which would have a big impact on the reshaping of the landscape of the river and the use of space, as we saw in part I. As a result, the feelings of the inhabitants have been manifested through social movements of resistance to projects, as is the case with *Brasilia Teimosa*; this happens even when this resistance stops any possibility of upgrading the favelas, because the inhabitants prioritize the defense of their rights to remain in their properties, even if they are unhealthy. This outcome is a result of the dual mode of governance of local urban policy: strategic planning decisions about large planning projects are taken outside regulatory and democratic mechanisms, the latter being only really respected for small-scale decisions, which do not challenge the neoliberal system of land allocation.

So, as shown by Wolch et al. (2014), urban greening strategies alongside rivers and on wasteland there, as in other places, can have unintended effects. Regeneration aimed at addressing problems of environmental justice, even with the participation of inhabitants and NGOs (as in Recife), can make neighborhoods more pleasant, more attractive, and healthier, but can also lead to gentrification, increasing housing costs, and require the displacement of marginalized and vulnerable people.

4 Conclusion

In France, as well as in Brazil, wastelands on waterfront river sites offer excellent opportunities in the cores of cities and metropolises for new urban development projects. Various stakeholders are directly interested in their implementation.

Local and central public authorities see it as a chance to create a new central urban focus and a way of enhancing urban attractiveness, by renewing the image of the city emphasizing "nature" and the river.

Private real estate operators find particularly profitable investment opportunities there because they meet the demands of affluent social groups or of tertiary activities in search of locations allowing them to benefit from high-quality amenities and strong positive externalities.

As a consequence, the development of riverfronts can be seen as revealing the intensity of the dilemma faced between the objectives of creating the attractive city and the satisfaction of the inhabitants and their right to the city.

What lessons emerge from the France—Brazil comparison about how this dilemma can (not) be resolved?

Our case studies show that, whatever the national context, rivers and their waterfronts had lost their historic urban functions as simple trade axes and as a location for polluting industries. As a result, these places have been characterised by neglected zones of declining and abandoned port-industrial activity and the location of low cost housing. With the rise of the pursuit of attractiveness in urban policy such areas have now become strategic development locations, because of their perceived landscape and environmental assets. They have become arenas for the display of high level urban functions and/or of developments which enhance the image of the city at an international scale. In these new circumstances, the pursuit of attractiveness, with its associated profitable real estate investment, seems to have downgraded the importance attached to another key urban policy objective: the right to the city. This has happened despite attempts to involve the local population in planning decisions, attempts that seem to amount to mere communication strategies.

Our observation is incontrovertible: in France as in Brazil, the poorest people are consigned to locations outside the regenerated areas. This can be the result of the increase in land values which causes social segregation, as we see in Nantes or Bordeaux, or a consequence of the dismantling of informal housing areas and the decision not to grant rights of occupation to poor squatters without title deeds living on public land, as in the case of Recife. But the phenomenon of gentrification brought about by the regeneration of riverfront sites, which conflicts with the principles of the right to the city, is also to be seen in many cities, including small and medium ones. Our examples are not exceptional cases in France or in Brazil, or probably in other countries.

References

Allaman M (2003) L'honneur retrouvé des quais de Bordeaux [The regained honor of the quays of Bordeaux]. Rev Diagon 163:29–34

Barthel P-A (2009) Faire la preuve de l'urbanisme durable: les enjeux de la régénération de l'île de Nantes [Demonstrating sustainable urban planning: the challenges of regenerating the Ile de Nantes]. Vertigo 9:2

Bonnin S (2007) Fleuves en ville: enjeux écologiques et projets urbains [Rivers in city. Ecological issues at stake and urban projects] Rev Strat (online)

Bourdeau-Lepage L, Huriot J-M (2005) La métropolisation: thème et variations [Metropolization: theme and variations], in Buisson M-A. Mignot D (eds), Concentration économique et ségrégation spatiale [Economic concentration and spatial segregation], Ed De Boeck, p. 39–64

Carrière J-P (2002) Villes et projets urbains en Méditerranée [Cities and urban projects in the Mediterranean], Ed. MSH Villes et Territoires, Université de Tours

Carrière J-P (2012) Regards croisés sur les dynamiques métropolitaines au Brésil et en Europe [Exchange of views on metropolitan dynamics in Brazil and Europe]. In: Tellier L-N (ed)

Métropoles des Amériques en mutation [American Mutating Metropolises], Presses Universitaires du Québec, pp 31–46

Carrière J-P (2018) O Direito à cidade, [The right to the city]. In: Fernandes A. et Chagas M (eds) O direito à cidade na França e no Brasil: uma nova agenda urbana [The right to the city in France and Brazil: a new agenda], Ed. UFBA, Salvador de Bahia, pp 763–769

Carrière J-P, De la Mora L (2012) Regards croisés France-Brésil sur les processus et enjeux de la métropolisation au Brésil [Exchange of views on processes and issues at stake of metropolization in Brazil], Symposium Dialogues France – Brésil APERAU – ANPUR – São Paulo, 27–30 March 2012

Carrière J-P, De la Mora L (2014) Précarité urbaine et fragmentation socio-spatiale au sein des métropoles brésiliennes: le cas de Recife [Urban precariousness and socio-spatial fragmentation within Brazilian metropolises: the case of Recife]. Revue Géographie, Economie, Société 4:373–397

Carrière J-P, De la Mora L (2018) O dilema atratividade metropolitana—redução da fragmentação socioespacial: reflexões sobre a requalificação das zonas ribeirinhas nas áreas metropolitanas na França e no Brasil [The dilemma of metropolitan attractiveness versus reduction of socio-spatial fragmentation: reflections about the rehabilitation of riverside zones in metropolitan areas of France and Brazil]. In: Fernandes A et Chagas M (eds) O direito à cidade na França e no Brasil: uma nova agenda urbana [The right to the city in France and Brazil: a new agenda], Ed. UFBA, Salvador de Bahia, pp 385–415

Carrière J-P, Demazière C (2002a) Projets urbains et grands projets emblématiques : réflexions à partir de l'exemple d'Expo 98 à Lisbonne [Urban planning and flagship development projects: lessons from Expo 98 in Lisbon]. In: Carrière J-P (ed) Villes et projets urbains en Méditerranée [Cities and urban projects in the Mediterranean], Presses Universitaires François Rabelais, pp 31–51

Carrière J-P, Demazière C (2002b) Urban planning and flagship development projects: lessons from Expo 98. Lisbon, Plan, Pract Res 17(1):69–79

Carrière J-P, Hochart K. (2016) Un développement urbain durable et socialement inclusif est-il possible dans le contexte de fragmentation socio-spatiale des métropoles des pays émergents? Réflexions à partir des cas de Recife (Brésil) et de Chennai (Inde) [Is sustainable and socially inclusive urban development possible in the context of socio-spatial fragmentation of metropolises in emerging countries? Reflections from the cases of Recife (Brazil) and Chennai (India)], Geografia e Ordenamento do Territorio, vol 10

Carvalho Filho L, Ferraz de Sá W, Puttini C, Monteiro C (2015) The capibaribe park: restructuring the urban fabric of the city of Recife by articulating public spaces. In: Proceedings of the 10th international space syntax symposium, pp 1–17. https://www.sss10.bartlett.ucl.ac.uk/wp-content/uploads/2015/07/SSS10_Proceedings_063.pdf

Chemetoff A, Berthomieu J-L (1999) L'île de Nantes, le plan guide en projet [The Île de Nantes, the plan guide], Éditions Memo

Costes L (2010) Le Droit à la ville de Henri Lefebvre : quel héritage politique et scientifique? [Henri Lefebvre's Right to the City: what political and scientific heritage?], Espaces et sociétés n, pp 140–141, pp 177–191, www.cairn.info/revue-espaces-et-societes-2010-1-page-177.htm, https://doi.org/10.3917/esp.140.0177

Demazière C (2015) Les enjeux de la planification spatiale en Angleterre et en France. Regards croisés [The issues at stake in spatial planning in England and France. Exchange of views], Espaces et sociétés 1:160–161, 67–84

Devisme L (2007) Île de Nantes: une ville se construit sous nos yeux. Histoire du projet [Île de nantes: a city that is being built before our eyes. History of the project], Revue Place publique 4

Diniz FR, Ferraz de Sá W, Alencar AK (2016) Na fonte das cidades, as águas e as pessoas: a experiência do projeto do Parque Capibaribe, Recife, Brasil [At the origin of cities: waters and people. The experience of the Capibaribe Park project, Recife, Brazil]. In: Atlas da 5ª Conferência da Rede Lusófona de Morfologia Urbana, Escola de Arquitectura da Universidade do Minho, Guimarães, Portugal, pp 727–736

Dos Santos CR, De Oliveira Melo LC, Gama Monteiro CM (2015) Como resgastar a relação da cidade com os ambientes naturais: projeto Parque Capibaribe, Recife–PE [How to save the city's relationship with its natural environment: Capibaribe Park project, Recife–PE]. Cidades Verdes 03(08):33–48
Gérardot C (2004) Les élus lyonnais et leurs fleuves: une reconquête en question [Lyon's elected officials and their rivers: a reconquest in question]. Géocarrefour 79(1):75–83
Harvey D (1989) From managerialism to entrepreneurialism: the transformation in urban governance in late capitalism. Geografiska Annaler 71 B:7–19
Harvey D (2009) Social justice and the city, Revised edn. University of Georgia Press
Hochart K (2014) Lecture croisée de dynamiques urbaines en Inde et au Brésil. Habitat informel et fragmentation socio-spatiale à Chennai et Recife [Cross-reading of urban dynamics in India and Brazil. Informal housing and socio-spatial fragmentation in Chennai and Recife], Rapport de fin d'études, Département Aménagement, Polytech'Tours, 92 p
Lechner G (2006) Le fleuve dans la ville: la valorisation des berges en milieu urbain. Note de synthèse. Centre de documentation urbaine. DGUHC. 123 p
Lefebvre H (1968) Le droit à la ville [The right to the city]. Éditions Anthropos, Paris, p 164
Masbougi A (2003) Nantes, la Loire dessine le projet [Nantes, the Loire River draws the project]. Éditions de la Villette, Paris
Sá Carneiro AR, Menezes P (2011) O sistema de espaços livres e a conservação da paísagem do Recife – PE [The system of open spaces and the landscape conservation in Recife–PE]. In: Soares Macedo et al (eds) Quadro dos sistemas de espaços livres nas cidades brasileiras [Framework for open spaces into Brazilian cities], USP, pp 159–169
Sá Carneiro AR, Duarte M, Marques EA (2009) A conservação da paisagem na perspectiva de um sistema de espaços livres públicos do Recife [Landscape conservation in the perspective of an open system of public spaces in Recife], Paisagem ambiente ensaios 26:127–141, S. Paulo
Wolch JR, Byrne J, Newell JP (2014) Urban green space, public health and environmental justice: the challenge of making cities 'just green enough.' Landsc Urban Plan 125:234–244

Jean-Paul Carrière, PhD in Economics, is Professor Emeritus in Regional and Urban Planning at the University of Tours (France), research team CITERES. In addition, he chairs the *Section Prospective du CESER Centre-Val de Loire* (Forward Studies Section of the Economic, Social, and Environmental Council of the Centre-Val de Loire Region). His main research interests are sustainable territorial development and comparative international analysis, chiefly focusing on Brazilian and Romanian cases. He has published many papers and books, several of which deal with the issue of urban wastelands.

Stuart Farthing has a first degree and a PhD in Geography, but taught and researched for many years in the Planning School at the University of West England, Bristol (United Kingdom) on issues connected with urban development and planning. He has conducted research, funded by UK Government bodies, on the factors affecting the re-use of derelict urban sites and buildings and on the barriers to the development of vacant land in urban areas for social housing.

Renaturation and Ecosystem Services of Contaminated Urban Wastelands in France

Yves Petit-Berghem, Elisabeth Rémy, and Marine Canavese

Abstract In urban areas, many wastelands deriving from former industrial activities may contain degraded and polluted soils. When reconversion of these wastelands is included in a renaturation project, it opens the way to more extensive approaches in favor of biodiversity and ecosystems. The designers of a future brownfield redevelopment projects must therefore think upstream about the clean-up strategy that is least harmful to biodiversity. The stated objective is to breathe new life into degraded soils while designing a new landscape based on the dynamics and resilience of ecological systems. Choices are not easy to make insofar as projects do not have the same time horizon as the dynamics of ecological systems. While ecosystem services can structure projects, we show that the renaturation of polluted wastelands is a very complex subject because it also comes up against the complexity of urban territories and the diversity of ways of thinking, which causes tensions and sometimes incomprehension about the future of the environment that is to be built. It is therefore necessary to gather feedback from in situ experiments carried out in projects for the phytomanagement of formerly polluted wastelands.

Keywords Polluted soil · Industrial wasteland · Urban nature · France · Ecosystem services · Stakeholders

Y. Petit-Berghem (✉)
Landscape Project Research Laboratory (Larep), National School Superior of Landscape, Versailles, France
e-mail: y.petitberghem@ecole-paysage.fr

E. Rémy
INRAE-AgroParisTech, UMR SAD-APT, Team "Proximity", Paris, France
e-mail: elisabeth.remy@agroparistech.fr

M. Canavese
Max Weber Center, UMR 5283, University Lyon 2, Lyon, France
e-mail: m.canavese@yahoo.fr

F. Di Pietro and A. Robert (eds.), *Urban Wastelands*, Cities and Nature,
https://doi.org/10.1007/978-3-030-74882-1_12

1 Introduction

Urban wastelands are scattered throughout the urban environment and are subject to pressures linked to the city structure like fragmentation, and also present forms of alteration linked in particular to unregulated uses (rubbish dumps, unregulated housing, soil pollution, etc.). The requalification of wastelands has become an important issue, calling on politicians concerned about reusing degraded land, as well as citizens concerned about the quality of their living environment (France and Ouellet 2002). It is integrated in urban renewal operations and is part of global thinking on energy and ecological transitions in a perspective of sustainable territorial development and the control of climate change, particularly with regard to the creation of green and blue networks (Bailly 2013; Berthelin 2018).

In urban areas, many wastelands deriving from former industrial activities may contain soils of poor quality. These impoverished, degraded and polluted soils are a barrier to the total expression of potential ecosystem services (Bünemann et al. 2018; Erkossa et al. 2007). When requalification is included in a renaturation project, it opens the way to more extensive approaches in favor of biodiversity and ecosystems. However, taking biodiversity into account requires a good knowledge of the characteristics of the wasteland and raises the question of legal obligations connected in particular with the procedures applied to polluted sites and soils (Ademe 2014).

The aim of a renaturation project is to restore a suitable place for the fauna and flora by acting on the soil quality which must be improved. How can this quality be assessed? In general, environmental recommendations are accompanied by a decontamination strategy adapted to the local context and by health risk assessment in order to ensure that the environment is perfectly compatible with the new uses. What are the difficulties found in these soil decontamination operations and the ecological questions posed by the project leaders?

By integrating ecological issues into the project, the stated objective is to breathe new life into degraded soils while designing a new landscape based on the dynamics and resilience of ecological systems. Renaturation, however, raises questions (ecological, social) for the city players, both the public operators who define the upstream orders and the planners involved in carrying out the projects. However, like any environmental problem with high political stakes, the renaturation of wastelands does not escape the logic of a political ecology that brings biodiversity into the realm of the market economy (Lévêque 2017; Maris 2014).

This renaturation therefore tests contaminated brownfield ecosystems that experts qualify and assess to highlight the benefits of a rich biodiversity by providing essential services for the public welfare.

The ecosystem services approach to biodiversity thus shifts the protection of nature from a logic of preservation against the people to a logic of conservation for people (Kareiva and Marvier 2007). However, ecosystem services also reflect the power relationships of project participants, the values they hold, and their interest in defending a practice or action. Indeed, the same function provided by a brownfield ecosystem can be a service for some and a disservice for others.[1] On the other hand, by implicitly endorsing a certain notion of the good, services miss out on the idea of the existence of a per se value of nature, which can neither be quantitatively assessed nor exchanged, and which stems from the diversity of our links with it. How can this notion of ecosystem services be used in brownfield projects? How can it guide projects? What are its strengths and limitations?

While biodiversity as a provider of services is guiding the debates, other questions arise because renatured wastelands are not only a reservoir of nature, they are also thought of by considering the city as an interaction of environments and a favorable field for proposing new relationships between humans and nature (Masboungi 2018). These relationships are complex, especially when it comes to projecting the future of a territory in a global urban project perspective that takes into account the social needs and individual/collective aspirations of the local populations (residents, associations, human groups defending interests). Brownfields can become grounds for experimentation where questions between nature and society also arise: What is the place of the citizen in relation to the way in which projects are presented in the public space? Considering that a project is also a collective work bringing together experts and laymen, what is the orientation to be given to the projects as well as their ecological and social relevance? Widely used to help communities decide between different land uses or to prevent the degradation of ecosystems, can the notion of ecosystem service help prepare reliable projects to reintroduce biodiversity in cities and fight against land artificialization?

Facing these problems, a development in three parts is proposed. It is initially necessary to recontextualize the pollution in the wastelands that have developed in towns or in peri-urban areas. The need for decontamination is raised, together with the difficulties posed for developers. Secondly, the question for renaturing polluted wastelands is introduced, by discussing the controversies surrounding a vegetalized nature whose orientations in the projects are not always debated. The notion of ecosystem service is used both to understand why biodiversity is favored in projects and also why it is associated with the soil whose functionality must be restored. Finally, having demonstrated that the concept of ecosystem services removes the complexity of the relationship between humans and nature, it will be necessary to look to the future, in particular by using past lessons and feedback from experiments on contaminated sites. The need to question the challenges of greening by going beyond the notion of ecosystem services, which is too restrictive to initiate a public debate on the positioning to be given to projects, will be discussed.

On a methodological level, the reflections carried out are part of research that we position between the biotechnical sciences and the social sciences. This research is based on a review of the literature on the renaturation projects of polluted urban wastelands, and on the discourses and practices they convey in the context of actual

[1] For example, the function of maintaining an uncultivated wastelands ecosystem can be seen as an advantage for an ecological engineer (maintenance of natural habitats), whereas it will be considered uninteresting for some form of agronomy (no production system)—note that agroecology attempts to link the two disciplines, see below.

or potential contamination of urban or peri-urban soils. The projects are also fed by interdisciplinary collectives and working groups that bring together researchers and civil society actors.[2] Since the stakes related to the depollution and renaturation of wastelands are not always well understood and, above all, are not shared with all the actors concerned, we rely in the last part on feedback from a field survey (in situ observations, semi-directive interviews) which is useful here to appreciate these stakes and difficulties.

2 Industrial Heritage and the Question of Soil Pollution

A large number of urban wastelands derive from industrial or service activities that have left their mark on landscapes and soils. After giving some details on the environmental impact of these activities, soil quality is questioned in order to guide their use in the context of a future project. Taking into account the new services provided by the soil (support for biodiversity, food, recreation) requires detecting the origin of pollutants and then assessing the health risks before taking action to decontaminate them.

2.1 *Emergence of Brownfield Sites and the Environmental Consequences Thereof*

In cities, many sites that have been used in the past for industrial or service activities have been abandoned; these sites are potentially renaturable derelict sites but also provide land opportunities for urban renewal. Where future use is not yet planned, derelict sites constitute waiting areas available for projects that address the challenges of energy and ecological transition and local urban planning policies. In France, brownfields constitute areas of varying size of at least 2,000 m^2 according to INSEE.[3] When they are inherited from abandoned industrial sites, they are inventoried in public databases that will help public authorities to carry out preventive or curative actions. BASOL is a national database on polluted or potentially polluted sites and soils, which lists nearly 6,800 polluted sites in 2018. BASIAS concerns former industrial sites and service activities, and this database lists 300,000 sites (active or abandoned) likely to generate pollution. These two databases are not exhaustive but are a means of warning about possible soil pollution and can thus guide the studies to be carried out in anticipation of changes in use.

Before they are reassigned to other uses, wastelands must be subject to expert assessments aimed at identifying pollution likely to generate a risk for humans, fauna, flora, groundwater, landscape, etc. (Limasset et al. 2015). The sources of pollution can

[2] For example, the French working group "Risks related to collective and private urban gardens."

[3] National Institute of Statistics and Economic Studies.

be extremely varied and generate toxicities depending on their concentrations in the soil, their chemical form and their bioavailability.[4] They include inorganic pollutants, especially metallic trace elements (MTEs), some of which are essential to biological processes at low doses (e.g. copper, chromium, zinc), while others play no useful role for living organisms (e.g. cadmium, lead, mercury). Carbon-based pollutants such as hydrocarbons, solvents or pesticides, whose degradation in the environment is slow, with the particularity for some of them that they infiltrate the food chain, are directly related to human activities (Douay et al. 2008). By accumulating in soils, they can persist for a long time and also be transported by runoff water or in groundwater (El Khalil et al. 2013). In the same time, poor quality fill or topsoil (in agronomic or chemical terms) can constitute other sources of pollution, leading urban project stakeholders to limit topsoil inputs today.

The polluted nature of a soil has become a criterion to be considered in the development of an urban project; since 2014, the law for access to housing and renovated urban planning in France has been calling on planners to take greater account of soil as a three-dimensional system (Blanchart et al. 2017). The pollution criterion is considered as early as the upstream phase of a project, and it is the research consultancies specializing in polluted sites and soils that carry out the diagnoses. The level of soil pollution at a project site has an impact on the feasibility of the project, on the strategic choices and on the total budget of the operation—all of which can be subject to debate and controversy. If it proves incompatible with the considered uses, the level of pollution requires major management work (containment, excavation and treatment on site or sending to a landfill center).

Wasteland is therefore, at first sight, a land whose environmental quality is depreciated. This loss of value is linked in particular to the cost of clean-up work, a cost that is all the more significant as the redevelopment of the wasteland will involve opening it up to the public and leisure and/or gardening activities in the city. Green spaces and vegetable gardens can indeed contribute through their uses (playgrounds, relaxation, picnics, and consumption of vegetables) to the exposure of populations to metallic and organic pollutants. When environmental quality is not reduced to the chemical quality of the soil, wastelands are nonetheless, in theory, interesting for biodiversity, the arrival of a new nature in the city that provides the inhabitants with a new contact with the earth and the soil. Faced with this legitimate demand, which is also expressed by citizens in participatory project approaches, it is important to carry out a factual assessment of the state of the environment before envisaging scenarios for the requalification and/or renaturation of polluted wastelands. This assessment involves an evaluation of soil quality, a very complex operation in such urbanized environments.

[4] This parameter defines the fraction of a contaminant that will actually be absorbed by a living organism.

2.2 Assessing Soil Quality

When urban brownfields develop, it must be noted that the quality of the soil is clearly influenced by the urban context and its history. The problem of too poor or contaminated soil is often encountered in urban areas. Urban wastelands are indeed home to degraded soils that have lost some of their original properties due to the artificialization of the environment. These urban soils are generally chemically poor with nitrogen, phosphorus and potassium deficits and low organic matter contents. Some soils have become compacted and impermeable and no longer allow biophysical exchanges between the underground and aerial layers. Their loss of fertility as a result of disturbed and artificial environments does not allow them to perform their nourishing functions, which handicaps the potential regeneration of vegetation, especially that which needs a supply of fertilizers to develop. Improving soil quality means, if possible, restoring its fertility and improving its biogeochemical functioning (Joimel et al. 2016; Walter et al. 2015).

In an urban context, projects must take into account French planning documents such as the ScoT,[5] PLU,[6] and PLUi,[7] which advocate thrifty use of land while mentioning the need to preserve its quality. However, soil quality is difficult to assess because it is built on the basis of temporal dynamics (Beck 2020) and the soil's capacity to remember former uses or activities that may have affected its functionality, particularly through contamination. It is the history of urban soils that must be questioned as some urban soils have been built from materials of varied and sometimes dubious origins (embankments) whose traceability is not known. In the Ile-de-France region, it has been shown that a lack of knowledge of the history of the development of Parisian green spaces minimizes the potential risks of contamination related to input land (Gitton et al. 2018). The sanitation of the various Parisian districts was carried out in the nineteenth century in particular by evacuating untreated urban "sludge," potentially loaded with pollutants including heavy metals (Barles 1999) in proportions that are difficult to quantify, into peripheral market gardening communities. These supplied the Parisian neighborhoods from which the exported sludge originally came with vegetal soil.

When a project plans to transform a wasteland whose historical trajectory is not perfectly identified, soil pollution and its potential danger to public health are therefore not easy to assess (Chalmandrier et al. 2017). Laboratory soil analyses are useful for detecting the presence of pollutants, but the high cost of these analyses limits their number, which is problematic when the site is large and/or not easily accessible. Moreover, in the absence of regulatory threshold values for assessing the level of contamination of soil intended for cultivation, suspicion may arise regarding the assessment of the risk of contamination.

---

[5] Territorial coherence scheme.

[6] Local urban plan (municipality scale).

[7] Local intermunicipal urban plan.

Faced with these difficulties, depollution treatments will be chosen according to the risks to public health and the environment, the proposed uses and the costs and time required for the decision and implementation of projects by public operators.

2.3 Clearing up a Site: Problems Raised

Excavation techniques that remove soil and then dump it elsewhere are the preferred route. These techniques substantially eliminate pollution, but they are still very expensive and can be problematic when large volumes of material have to be processed.

The decontamination of the soil of allotment gardens (today family gardens) or shared gardens is a complex operation. The task is not simple because the level of pollution can change over time and the pollutants are more or less transmissible depending on the plant species that are cultivated. The problem is that each cultivable plot of land reflects its history: unique with its own cocktail of pollutants. This is why many experts advocate a case-by-case management.

In general, soil remediation solutions can be time-consuming and expensive. Phytotechnologies (the use of living plants) have the ability to stabilize inorganic pollutants such as metals. However, these technologies are complex and not applicable everywhere, because pollutants may have a timeframe that is not compatible with the timing of public and private actors, associations and citizens looking for concrete feeder city solutions (Damas and Coulon 2016), thus raising questions about the relevance of the projects and their temporality.

Furthermore, these techniques raise the problem of pollutants that accumulate in biomass that must also be treated, for example by combustion in boilers equipped with efficient filtration systems. When burned, this biomass from polluted soils still produces by-products (ash) that cannot be spread in the environment if they still contain heavy metal residues. But doing nothing with these metal-laden plants is tantamount to turning them into new waste. Leaving depolluting plants to die on the spot is not a viable solution because the metal pollution is likely to return to the soil in the short term (Ademe 2014). Rehabilitating a site by phytoextraction takes decades. Public funds are lacking and these operations often require the involvement of chemical industrialists who partly finance the recycling operations of the metals stored in the plants.

For many elected officials, aware of the lobbying of developers and the insistence of the State to build housing on former industrial sites, there is no such thing as complete clean-up.

In Clichy-Batignolles (Fig. 1), a former brownfield (railway right-of-way) has been transformed into an eco-neighborhood with a 10-hectare park created in 2007. Upstream of the project, very close attention was paid to soil quality through quantitative health risk assessments (Mailliard 2016). The potential impact on future uses was assessed, but complete soil remediation is difficult due to the presence of former storage areas near the site (coal and other fuels).

Fig. 1 A former brownfield transformed into an ecological district. Clichy Batignolles: Martin Luther King park, Paris 17th arrondissement © Petit-Berghem, 2019

In Limeil-Brevannes (department of Val-de-Marne), clean-up work presiding over the construction of the "Les Temps Durables" eco-neighborhood in 2016 has not completely removed the 18 m of buried waste, including heavy metals (Peuportier 2016). A geotextile tablecloth and a new supply of clean soil over a few dozen centimeters were not enough to remove the pollution that led the Regional Health Agency (ARS) to prohibit residents from putting edible fruits and vegetables in their gardens. In Lys-lez-Lannoy (Nord department), developers are adapting to polluted soils by proposing a new architecture without underground parking or ground floor housing to reduce the contact between residents and pollutants from former industrial landfills (Decocq 2010).

The problem is not always related to the construction of new neighborhoods. Former industrial wastelands are sometimes illegally occupied by families who have not been able to obtain social housing. This is the case, for example, in Champigny-sur-Marne (department of Val-de-Marne) where dozens of families (Portuguese emigrants) occupied wood and metal sheet huts until the late 1960s before being evicted following a motorway project. The abandonment of this project, however, led to the return of former residents to these lead- and mercury-polluted lands to set up illegally fenced and privatized vegetable gardens (ETP Paris-Est-Marne and Bois 2016). This raises the question of the information given to the various publics concerned by these health risks.

3 Renaturation of Polluted Wastelands: An Issue Under Debate

Faced with the difficulties of depollution, the renaturation of wastelands can serve as an alternative to development projects that have not seen the light of day because of pollution problems or related costs. Renaturation makes it possible to meet land recycling requirements and helps to reintroduce nature into the city, the stakes and modalities of which need to be discussed.

3.1 Brownfield Vegetation: From Diagnosis to the Restoration of Biodiversity

The issue of revegetation is not new for wasteland, which by definition is naturally called upon to be vegetated. This reservoir of nature is of interest to naturalists, leading them to identify plant and animal species; the ecological engineer will make an in-depth ecological diagnosis with a functional approach and will study the dynamics of the existing ecosystem, and its interaction with other possible ecosystems.

Supported by precise mapping of habitats and soils, this work makes it possible to precisely define the ecological potential of the site and to specify the upstream issues of a construction site, from the very start of the design phase. It is thus possible to highlight several stages of evolution in the wasteland, to identify ecological corridors, but also to project the consequences of the project on the natural environment and the present biodiversity.

Diagnoses are most effective and usable by developers when they are applied at the scale of the territory, infrastructure or neighborhood because it is possible to see ecosystems in their entirety and the connections between several habitats at this scale. Conversely, at the scale of the plot or sub-plot, the diagnosis is limited to the inventory of species and does not make it possible to identify landscapes occupied or crossed by communities of living organisms (Larramendy et al. 2014).

Taking into account the present or future ecology and biodiversity studies of polluted wastelands therefore opens the door to a large amount of information on the site, a set of contextual data that does not only refer to social or landscape issues, but above all concerns living organisms linked by dynamic interactions. Despite the wide variety of situations encountered, it must be noted that the ecological dynamics of wastelands leads to the arrival of pioneer and nitrophilic species, often accompanied by invasive exotic species (e.g. *Erigeron canadensis, Senecio inaequidens, Buddleia* ssp.). This trend leads to a depletion of species diversity, leading some ecological engineers to refer to a degradation of ecosystem functioning with a net loss in terms of the provision of ecosystem services.

While the ecological dynamics of wasteland are of great scientific and epistemological interest for the conservation sciences, they are not generally discussed in urban or landscape architecture projects because they would imply the implementation of a project leading to the return of an ecosystem to its historical trajectory. Synonymous with ecological restoration in the strict sense defined by the *Society for Ecological Restoration International* (SERI 2004), this process of assisting the self-regeneration of ecosystems that have been degraded, damaged or destroyed has no place in cities, where nature is simply an element in the functioning of the urbanized system. At best, it is the rehabilitation of environments that is evoked with an emphasis on the services that are provided rather than the restoration itself (Aronson et al. 1993; Cristofoli and Mahy 2010). Actions may, for example, consist of reconstructing soils or not waterproofing them in order to restore their ecological functions.

When it comes to requalifying a neighborhood or carrying out a voluntary landscaping operation, it is renaturation in the sense of the design of landscaped public spaces that takes place with a view to restoring the ecological and landscape state of sites that are deemed to be degraded by human activities. Nature thus participates in the functioning of the sustainable city and the challenge is to build landscapes where biodiversity can develop by providing the expected ecosystem services (Carron et al. 2009).

In cities, in the context of contaminated sites, it is obvious that the search for a natural initial state is not feasible and the priority is rather to make urban ecosystems sustainable by limiting the effects of anthropogenic disturbances. For this reason, renaturation is mainly considered from the perspective of human-assisted revegetation and soil rehabilitation, often in the form of de-artificialization. In France, this trend is now in compliance with regulatory requirements concerning the "restoration" of biodiversity and the rehabilitation of polluted or degraded sites and soils (RECORD 2018).

3.2 *What Place for Biodiversity and Soils?*

The law for the recovery of biodiversity, nature and landscapes, promulgated in August 2016, enshrines in French law a renewed vision of biodiversity. This vision is now part of the nation's common heritage, whose protection, enhancement, restoration, rehabilitation and management are considered of general interest and contribute to the objective of sustainable development. Article L110-1 of the Environment Code states that one of the objectives of sustainable development is precisely the preservation of biodiversity, environments, resources and the safeguarding of related services and uses.

Nevertheless, despite a favorable context, biodiversity is not always protected as such, particularly when it is hosted in soil for which there is no regulatory definition in France. Consequently, when it comes to integrating this compartment, which is a carrier of biodiversity and associated services, the national methodology for the management of polluted sites and soils (SSP in French) comes into force, which is very much oriented towards issues related to human health and health risk assessment.

The designers of a future brownfield redevelopment project must therefore anticipate the clean-up strategy that is the least harmful for biodiversity.

For this reason, biodiversity is rarely considered independently of soils. In many projects, the search for high biodiversity is associated with the agronomic improvement of soils, either through total or partial reconstitution of the soil, or through its treatment followed by an amendment (Seybold et al. 1999; Benayas et al. 2009). In the case of reconstitution, it is nowadays frequent to use soils built from organic and mineral by-products coming in particular from neglected or poorly valorized materials. They can be composed of compost, urban or industrial sludge, incineration ashes, or construction materials. The materials are mixed together and the objective is to ensure a high level of functionality to enhance biodiversity (vegetation support, biological habitat, food reservoir).

Meeting these criteria, Technosols[8] are an alternative to the use of large quantities of topsoil from agricultural or forest land often used to create urban green spaces. But these constructed soils must have adequate agronomic properties and drainage capacity to ensure sufficient fertility; they must also be compatible with the land use and comply with environmental restrictions to prevent the release of pollutants into the environment. It is therefore important to evaluate the agronomic properties of these soils under field conditions and to verify the safety of the mixtures for the environment and human health. These objectives are stated in research programs funded by ADEME, such as the SITERRE program (2011–2016) devoted to the construction of soils with urban waste for the greening of cities (Vidal-Beaudet 2018). Can Technosol become both a growing medium, a vector of biodiversity (soil microorganisms and habitats), and also a source of balance for human physical and mental health? Is it possible to look for soils that are capable of optimally fulfilling the main land uses in terms of fertility and at the same time capable of performing other functions such as providing social bonding, providing a source of emotion and pleasure, and guaranteeing ecological responsibility in terms of public health? This is the challenge which faces the reclamation of brownfield soils, which is all the more difficult to meet when Technosol takes place on formerly polluted sites. Restoring life to degraded soils or building new ones that provide associated ecosystem services is therefore becoming a challenge for wastelands transformed into experimental sites dedicated to biodiversity and sustainable development, where certain stakeholders (researchers, experts) meet to observe, study, create and promote a more comprehensive ecological approach that also includes the design of a new landscape. The aim of this approach is to reconcile human activities and biodiversity, in particular by taking ecosystem services into account.

Nevertheless, however legitimate it may be in the context of the functioning and maintenance of urban wasteland ecosystems that stakeholders consider to be of high scientific value (but with remaining uncertainties), is it tenable when it comes to

[8] Technosols are a new category of soils, first proposed in 2006 to the World Reference Base for Soil Resources. They can be formed by natural or anthropogenic materials in variable proportions and, above all, they are dominated or highly influenced by human activity. They can be considered as artificial anthroposols (Monsérié 2009).

carrying out a project for the reconversion of polluted wasteland? This project should also contribute to urban renewal by seeking a new use for a polluted site and by creating a local dynamic from the initial stages of reflection, and then at each stage, as part of a participatory planning process. In this sense, do ecosystem services enable participants to take ownership of the project, imagine it, and project themselves into a new living space?

3.3 Limitations of an Ecosystem Services Approach

Widely used to quantify the "benefits" provided by an ecosystem, ecosystem services are an economic tool to help communities evaluate different ways of managing land and decide between different uses.

Services provided to society by wasteland ecosystems have become multiple over the years. These ecosystems not only participate in the regulation of natural processes but also in economic activity (Costanza et al. 1997). Benefits may be material through provisioning services (food and plant production) or regulating services (climate, floods, diseases) or immaterial through spiritual enrichment, recreation, social relations or aesthetic values. Human beings, as part of ecosystems, benefit from these goods and services produced by their functioning. Ecosystem services thus define a relational concept that also involves different epistemic cultures (Maris 2014). The various scientists concerned use the concept to qualify the functioning of such an ecosystem as representing a benefit for such a beneficiary or beneficiaries. By applying it to their subject of study, these scientists assess the usefulness of a service for the purpose of developing (or reviewing) public policy or promoting action to manage an ecosystem—conflicts of interest can arise. A Technosol specialist, for example, will be involved in a logic of functional soil restoration combining physical and biological engineering; he or she will steer the debate by considering that this new soil is a solution to requalify a polluted wasteland and that it offers an opportunity to propose alternative fertile soils adapted to the city of the future without necessarily first holding a public debate on the merits of this project.

An ecosystem service therefore only makes sense in relation to a beneficiary who will defend his or her interests and the benefits that he or she may derive from the provision of that service. It is therefore important to identify the services provided and their beneficiaries in order to better understand the needs of the various stakeholders and to ease the tensions that may arise when deciding to give priority to a particular service.

It is often difficult or even impossible to provide services that are capable of satisfying all the stakeholders in a project and the future users of a redeveloped site for economic, ethical or societal reasons. The public operator makes choices between different services. A willingness to improve one of the services affects the ecosystem and usually results in changes to other ecosystem services. For example, measures taken to increase food production often result in greater use of water and water becomes of poorer quality or less available for other uses. Similarly, the quest for greater ecosystem biodiversity may have negative trade-offs on cultural services, for example if it is decided to prohibit access to a protected site or limit its use (Bally 2017). But synergies are also possible, and measures to ensure the sustainability of an ecosystem can also produce benefits for many services or stakeholders. For example, the creation of a nature area meets spiritual and recreational needs while generating other amenities such as providing natural habitats for wild flora and fauna, biodiversity conservation and carbon sequestration.

The main challenge when conducting an ecosystem services approach is therefore to make choices by integrating synergies or antagonisms between services. The issue is complex when it comes to arbitrating between supply and regulation/support services, which are often opposed in the short term but associated in the medium and long term (the sustainability of the former depends on the state of the latter). The task is not simple and the same service can be valued differently according to the utility that each one grants to it.

Moreover, this notion omits negative effects or disservices and, above all, it overlooks the fact that the same function can be a service for some and a disservice for others. The quality of a partially remediated soil, for example, could be perceived positively by an expert (environmental engineer, agronomist) if the soil regains acceptable biogeochemical functioning (biomass, return of earthworms)—this could lead to confusion if it is suggested that a "renatured" soil is a healthy environment (Chalmandrier et al. 2019)—whereas it will be judged mediocre for an association wishing to set up alternative forms of agriculture such as agroecology or vegetable gardens.

The difficulty of making a choice sometimes leads planners to propose a package of ecosystem services upstream and to measure the value of these services or the cost associated with their degradation. This economic approach can be useful to justify the financial costs of these projects and promote their acceptability. However, in order to convince people of the merits of the projects, it often relies on indicators whose implementation and interpretation remain complex and reserved to researchers. Moreover, some indicators do not identify beneficiaries. Others are designed to conduct large-scale evaluations and are not adapted to the more local level of conversion projects (Le Clec'h et al. 2016). Finally, others adapt to the context of ecological restoration and in this case neglect cultural services based on non-use values, which they consider as natural heritage (Maes et al. 2016). The use of indicators is all the more questionable as cultural services are much less covered than regulating services that are more easily correlated with ecosystem functions.

It must be noted that indicators are associated with exclusively quantitative approaches; measurement becomes a reference for evaluating performance and monitoring progress towards the expected results. If a soil that has been cleaned and restored from a wasteland can become the basis for new crops, in particular through the creation of shared gardens, an indicator will, for example, quantify the volume of food produced each year thanks to the gardens set up on the site (subject to checking their edibility/absence of chemical contamination) and the number of members of these gardens. Another indicator will quantify the contribution of the natural environment to the production of honey and other products of the hive; this indicator requires the prior identification of beekeepers and the local fruit and vegetable producing population interested in these products, provided that they do not bring additional costs in the case of short circuit sales.

In a cost–benefit approach, the beneficiaries and the sector providing the service must be known in advance, as the profitability of operations is not always guaranteed. For example, some plants such as miscanthus (*Miscanthus* ssp.) used in phytotechnologies produce biomass that can be used for different purposes (production of heat or electricity, animal bedding, mulch, fibers for eco-construction), but the difficulty is to find agricultural users integrated into a chain and ready to modify their practices for a species whose price is not attractive enough because the plant produces a lot of volume for a relatively low weight,[9] which forces it to be used very locally.

In another register, a previously polluted and treated site can (again) become attractive to local people as well as to tourists who wish to walk there or to keen naturalists, provided that exposure to the pollutants has been properly treated. But how can the role played by the site's landscapes, fauna or flora in the development of recreational activities be quantified? This requires the setting up of complex and relatively time-consuming surveys targeting the socio-demographic categories that will need to be selected beforehand. If qualitative protocols are not used, it is easier in this case to use a parameter that is more easily measured, e.g. the number of schoolchildren or adults who have been given a guided tour per year and/or questionnaires with closed questions that are easier to process to obtain clear-cut opinions on the appreciation of landscapes. The results from these approaches are generally simplified and misleading because they balance immeasurable values resulting from complex interactions between humans and nature. Going beyond this approach opens up a new space for reflection and discussion that invites us to take into account the diversity of the relationships that people have with non-human living things (Tassin 2020), i.e. the diversity of the ways in which nature is represented and valued.

4 Polluted Wasteland and New Renaturation Projects

4.1 Taking into Account the Stakeholders (Experts vs. Users) in the Orientations of the Project

A contaminated site project cannot be thought of independently of the actors who participate in its co-construction in a territorial logic where land pressure, transport problems and urban sprawl are major issues for the present development of cities (Mehdi et al. 2012). When a brownfield redevelopment project is orientated towards renaturation, several options are possible and can be submitted to public debate. Discussions are oriented towards the reconquest of biodiversity and the place of nature to be preserved. Tensions may arise over the meaning given to the word

[9] One m^3 of micanthus is equivalent to about 130 kg.

nature, the balance of nature and the future to be built, taking into account the stakes of biodiversity in political, economic and societal choices. Shouldn't the choice of uses also coincide with the quality of urban, peri-urban and rural soils, considering their reciprocal articulation? Should the wasteland be conserved without developing it under the pretext that it intrinsically contains a high level of biodiversity, or should it be requalified by considering that it could give rise to an urban project where new biodiversity responds to new uses? In all cases, the ecosystem is a spatialized socio-ecosystem at the scale of a territory that needs to be taken into account by considering the complexity of existing interactions between humans and nature. Environmental engineers defending functional aspects mention dynamic ecological systems, which constantly evolve in time and space (Lévêque 2017).

According to these experts, renaturation does not aim to return to the situation that existed before the system was disrupted by contemporary human activities, but it must ensure that it favors heterogeneous wasteland landscapes composed of different evolutionary stages. The latter do indeed produce a greater number of different species than wastelands that do not offer the same landscape heterogeneity. Studies have shown that this heterogeneity plays a major role in the ecological services of pollination and natural pest control by predators. Under the current conditions of climate change and worrying damage to biodiversity, wastelands can thus (again) become regeneration areas for fauna and flora—if the latter are not impacted by pollutants present in the soil.

Other experts will not have the same way of considering renaturation because their scientific culture is not the same as their vision of (peri-) urban nature. Some agronomists consider that it is possible to reconcile agricultural production and environmental preservation, particularly in the context of agroecological projects. When the soil is cleared, the wasteland is seen as a cultivated plot with a biological community where biological regulations (such as pest control or soil fertility) are exercised and can be improved by increasing biological diversity (Huyghe 2020).

At the scale of a territory, these two categories of experts can be brought together by considering that the increase in functional diversity and the constitution of new landscape mosaics appear to be a major determinant in promoting biodiversity in landscapes. When a project is in the design phase, these experts can dialogue with practitioners (landscape designers, town planners, etc.) who will set negotiated objectives that take into account both the ecological and social issues of biodiversity management. In order to grasp the complexity of these issues, it is often necessary to listen to what users have to say and to provide time for mediation and collective discussions (participatory workshops). For the designer, ecology is still very often considered from the point of view of a nature to be (re)built and developed; a nature between biophysical materiality and subjectivity that must be made functional by uses and not by ecosystems that should be preserved. The project is a way of rebuilding a relationship between humans (inhabitants-users) and their surrounding environment. If species take part in the renaturation of a site, they can also be chosen according to human preferences and the aesthetic or social value that some stakeholders (planners, organizations involved in land management) will attribute.

Biodiverse nature can be built with the population living in the neighborhood as it constitutes its living environment. Designers pay attention to scholarly knowledge, but also know how to detach themselves from it by integrating the inhabiting practices at the basis of a certain mode of knowledge production (epistemic value). Every human group or tribe (naturalists, gardeners, and hikers) will be able to defend its vision of nature and its "project" for the future. Local authorities can integrate as best as they can the sharing of ideas that do not necessarily meet with consensus, but they do not escape a form of political instrumentality because there is a strong pressure to rehabilitate an industrial site, especially in cities where wasteland is above all a land reserve and a neglected area to be developed.

Thus, biodiversity is very frequently instrumentalized in projects aimed primarily at revitalizing economically urban centers that are running on empty and/or recreating social ties. Biodiversity will not be thought of as such, but will be used to provide solutions that respond to the engineering and political injunction to recreate green, to use local materials, to avoid having plant barriers that are too opaque, soils that are too impermeable, without always measuring the impact of these measures, which are certainly commendable, but which sometimes mobilize polluted soils. The specifications often contain the same recommendations, for example concerning local species that optimize pollination efficiency, or soils that need to be reclaimed by improving their functionality, a project that is sometimes unrealistic given the poor quality of wasteland soils.

At this stage, our objective is to draw the first lessons, based on a case study (Chalmandrier et al. 2019) of the approaches mobilized by scientists and territorial engineering networks when they requalify an urban neglected area that has turned out to be contaminated.

4.2 *The Example of Mazingarbe (France): Orientations and Controversies*

The former mining basin of the Nord-Pas-de-Calais region bears the scars of many of the carbochemical industries that accompanied coal mining from the end of the nineteenth century onwards. The many industrial wastelands still present in this region bear witness to these past activities. Today, their requalification is a challenge in terms of managing polluted sites and soils in order to prevent environmental and health hazards.

Among these petrochemical plants, the vast Mazingarbe complex, which specializes in the recovery of coke oven gas and the production of nitrogen fertilizers and plastic materials, bears witness to this heavy legacy of industrial activities.

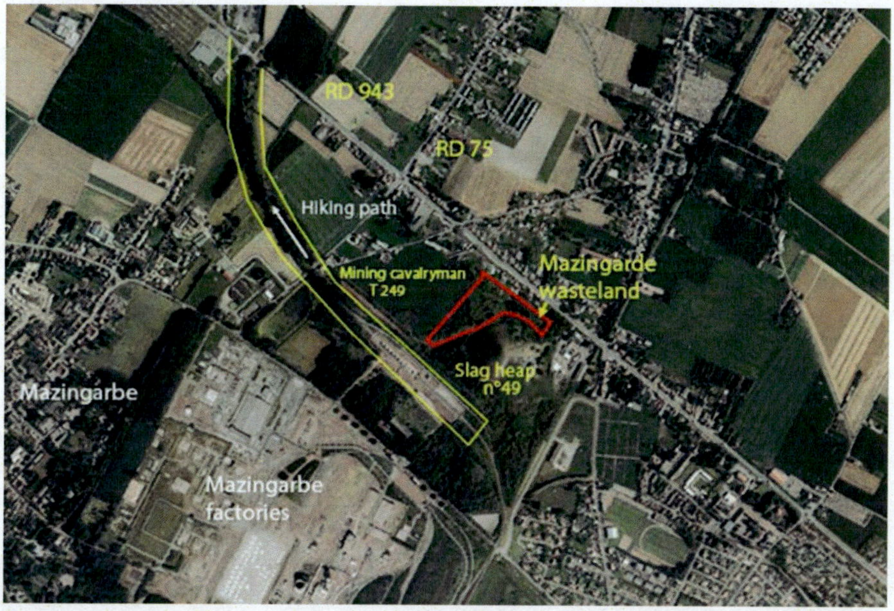

Fig. 2 Location of landscape features around the Mazingarbe wasteland (*Source* Modified after Chalmandrier et al. 2019)

To the north of this industrial complex, at the foot of slag heap 49, lies a wasteland partly owned by the EPF Nord-Pas-de-Calais.[10] By 1960, multi-contaminated sludge from the neighboring industrial complex was stored in this wasteland. Accessible by a former mine track converted into a rural road, the wasteland is bordered by a row of houses along the former national road linking Lens to Béthune (Fig. 2).

In 2017, the site was chosen to be the subject of a multidisciplinary research program, called MisChar,[11] supported by ADEME[12] and involving academics, institutions, design offices and private operators. The aim of the experiment was to assess the value of phytotechnologies in managing a multi-contaminated industrial wasteland, i.e. improving soil functionality, reducing environmental and health hazards while playing a role in ecological connectivity and the constitution of the green grid.

[10] In France, an EPF (Établissement Public Foncier) is a public agency that acquires land for development by a third party responsible for the construction of housing, new neighborhoods or public facilities (https://www.cohesion-territoires.gouv.fr/les-etablissements-publics-fonciers-epf). The acquisition by EPF Nord-Pas-de-Calais of the 2,200 ha of slag heaps and their rights-of-way is part of the development of the regional green network and in particular that of the former mining basin, which has a large deficit of green spaces.

[11] MisChar Project—Reconditioning of multi-contaminated soils using miscanthus biochar: ecological viability and socioeconomic value of management methods in urban and agricultural environments. https://mischar-43.webself.net/.

[12] ADEME (Agence De l'Environnement et de la Maîtrise de l'Energie) is a French public Agency for the Environment and Energy Management.

This interdisciplinary research focused on monitoring the experiment by putting it into perspective with the perceptions of local residents who will be indirectly affected by the project. Starting from the principle that the approaches mobilized by scientists and the territory's engineering networks (EPF, associations, etc.) when they redefine contaminated wastelands shape both a space and a representation of the public, the multidisciplinary team asked itself the following questions: who is the "actor" of the wasteland? Is it a place worthy of interest and for whom? What answers can be given to the difficult question of how to manage highly contaminated soils in peri-urban areas?

When examining the data on the practices of local residents and stakeholders (Chalmandrier et al. 2019), a discrepancy emerges between the status of urban neglect, where waste is disposed of, and the potential for a green grid to which territorial engineering associates the site. The problem in terms of a socially acceptable approach is the lack of upstream discussion on the appropriateness of the project and the absence of debate on its implementation. Applied to the case study, the survey quickly showed the discrepancy between the status granted to the wasteland locally and the orientations chosen by the project leaders.

The particularity of our upstream approach is that it gives a voice to the inhabitants: local residents can express their perception of the wasteland by raising other questions. This was the case here with a question about the current impact of air pollution associated with industries still operating in the surrounding environment. It is true that soil pollution has the particularity of being historical and invisible, i.e. not directly perceptible unlike some atmospheric pollution.

The local residents do not easily make the link with the soil, which is to be restored with the aim of renaturing the site afterwards. Surveys have shown that these inhabitants consider this site to be a non-place. While one scientist after another has occupied this place for their experiments, the representations of the local residents have not even evolved as the project has progressed, even though it has made the land more accessible by eliminating certain invasive and toxic plants (e.g. *Heracleum mantegazzianum*).

On issues as technical as soil contamination, the stakes involved in soil remediation and renaturation are not always well understood and shared by all the stakeholders. It is true that soil contamination is a very complex subject, which is not widely disseminated—and which can be "hidden" for various reasons. An entry through representations and social acceptability generally underlies a unilateral questioning of project leaders towards local recipients who are considered by definition as passive or even irrational. For this reason, it is easier to ignore or dismiss them without involving them in the project's reflections. The risk is then the generation of a feeling of mistrust or indifference to the actions that are undertaken. At a time when we are talking about restoring degraded and polluted environments, shouldn't we start by restoring confidence in the local populations by first asking their opinion on the future of the site after having informed them of the state of its contamination?

5 Conclusion

The renaturation of polluted wastelands is a very complex subject that is not easy to deal with, especially when it involves issues of biodiversity recovery and the reconstruction of landscapes co-produced by nature and society. Choices are not easy to make insofar as projects do not have the same time horizon as the dynamics of ecological systems. However, in an urban or peri-urban context, ecological systems are in fact anthropized systems, i.e. the result of a significant amount of work in shaping landscapes to meet particular uses. There is no state of nature that would be a kind of historical reference to which we should refer in terms of ecological restoration. The project leaders understood the need to restore nature as a provider of services that are theoretically indispensable for the sustainability of our well-being, as well as for future economic and social development. While ecosystem services can structure projects, it must be noted that they also come up against the complexity of urban territories and the diversity of ways of thinking, which causes tensions and sometimes incomprehension about the future of our environment to be built.

When discussion is initiated on a renaturation project that aims to increase recognition of the contribution of ecosystems to human well-being and economic activity, the discussion remains confined to reduced circles of experts. The latter guide the debates and select their interlocutors by mobilizing the notion of service, which loses its relevance once confronted with social reality (Bally 2017; Maris 2014). Indeed, ecosystem services are not part of a common language, making it difficult to take into account citizens' initiatives that are not expressed as services but according to their own knowledge, perceptions and values that they ascribe to nature, in the same way that few elected officials mobilize the notion to designate their actions and arrangements. The difficulty is even greater when it comes to evoking the cultural services that refer in particular to the aesthetics of landscapes and the benefits of nature, which are not limited to useful functions or those deemed useful by humans. In this sense, if a project of renaturation of polluted wasteland can lead to the creation of an ecosystem that generates services, it must also address the entire population and establish common rules in order to preserve and perpetuate this ecosystem while providing everyone with the possibility and the right to use it.

Initial feedback on the implementation of an experiment on a polluted peri-urban wasteland shows the differences in perceptions and uses of a peri-urban wasteland between what some people (residents) do—or rather do not do—and what other stakeholders (developers, experts, scientists) plan to do. It will be necessary to develop and deepen this type of feedback from in situ experiments carried out on phytomanagement and natural area management projects.

References

ADEME (2014) Biodiversité et reconversion des friches urbaines polluées [Biodiversity and reconversion of polluted urban wastelands]. Connaître pour agir, Angers

Aronson J, Floret C, Floc'h E, Ovalle C, Pontanier R (1993) Restoration and rehabilitation of degraded ecosystems in arid and semi-arid lands. I. A view from the south. Restor Ecol 1:8–17. https://doi.org/10.1111/j.1526-100X.1993.tb00004.x

Bailly E (2013) From public spaces to the landscaped spaces of the sustainable city [From public spaces to the landscaped spaces of the sustainable city. Articulo J Urban Res. https://doi.org/10.4000/articulo.2233

Bally F (2017) Quels services écosystémiques culturels sont produits par des citoyens et quelles valeurs y associent-ils? [What cultural ecosystem services are produced by citizens and what values do they associate with them?]. Environ Urbain Spec Iss 4. https://journals.openedition.org/eue/1590

Barles S (1999) La ville délétère. Médecins et ingénieurs dans l'espace urbain [The city is deleterious. Doctors and engineers in the urban space]. XVIIIe-XIXe siècle. Champ Vallon, Seyssel

Beck C (2020) Les sols aussi ont une histoire [Soils also have a history]. In: Mougin C, Douay F, Canavese M, Lebeau T, Rémy E (ed) Les sols urbains sont-ils cultivables? Quæ, Versailles (in press)

Benayas JMR, Newton AC, Diaz A, Bullock JM (2009) Enhancement of biodiversity and ecosystem services by ecological restoration: a meta-analysis. Science 325:1121–1124. https://doi.org/10.1126/science.1172460

Berthelin J (2018) La requalification des friches industrielles et urbaines pour préserver les sols agricoles [The requalification of industrial and urban brownfields to preserve agricultural soils]. Revue De L'académie D'agriculture 16:29–56

Blanchart A, Sere G, Cherel J, Warot G, Stas M, Consalès JN, Schwartz C (2017) Contribution des sols à la production de services écosystémiques en milieu urbain – une revue [Contribution of soils to the production of ecosystem services in urban areas—a review]. Environ Urbain. https://journals.openedition.org/eue/1809

Bünemann EK et al (2018) Soil quality—a critical review. Soil Biol Biochem 120:105–125. https://doi.org/10.1016/j.soilbio.2018.01.030

Carron C, Salomon Cavin J, Ruegg J (2009) La préservation du sol dans un espace de plus en plus urbanisé : vers des stratégies intégratives ? Le cas des défenseurs de la nature [Soil conservation in an increasingly urbanized space: towards integrative strategies? The case of conservationists]. Research report, IPTEH, Lausanne

Chalmandrier M, Canavese M, Petit-Berghem Y, Rémy E (2017) 'L'agriculture urbaine' entre concept scientifique et modèle d'action. Une notion mise à l'épreuve par le jardinage et le sol urbains [Urban agriculture' between scientific concept and action model. A concept put to the test by gardening and the ur-bath soil]. Géographie Et Cultures 101:119–138. https://doi.org/10.4000/gc.5052

Chalmandrier M, Douay F, Louvel B, Petit-Berghem Y, Rémy E (2019) Retour d'expérience sur la mise en place d'une expérimentation sur une friche périurbaine contaminée : dépasser la notion « d'acceptabilité sociale » pour faire du sol un bien commun [Feedback on the implementation of an experiment on a contaminated peri-urban wasteland: going beyond the notion of "social acceptability" to make the soil a common good]. Topia. https://cloud.topia.fr/index.php/s/C5mC0tzm3MEQqrp#pdfviewer

Costanza R, d'Arge R, De Groot R, Farber S, Grasso M, Hannon B, Limburg K, Naeem S, O'Neill RV, Paruelo J (1997) The value of the world's ecosystem services and natural capital. Nature 387:253–260. https://doi.org/10.1038/387253a0

Cristofoli S, Mahy G (2010) Restauration écologique: contexte, contraintes et indicateurs de suivi [Ecological restoration: context, constraints and monitoring indicators]. Biotechnol Agron Soc Environ 14(1):203–211

Damas O, Coulon A (2016) Créer des sols fertiles: du déchet à la végétalisation urbaine [Creating fertile soils: from waste to urban greening]. Le Moniteur, Paris

Decocq C (2010) Friches industrielles et pollutions historiques, Mission d'information et d'évaluation [Industrial wastelands and historical pollution, Fact-finding and assessment mission]. Review report, Lille Métropole Communauté Urbaine, Lille

Douay F, Roussel H, Pruvot C, Waterlot C (2008) Impact of a smelter closedown on metal contents of wheat cultivated in the neighbourhood. Environ Sci Pollut Res 15:162–169. https://doi.org/10.1065/espr2006.12.366

El Khalil H, Schwartz C, El Hamiani O, Kubiniok J, Morel JL, Boularbah A (2013) Distribution of major elements and trace metals as indicators of technosolisation of urban and suburban soils. J Soils Sediments 13(3):519–530. https://doi.org/10.1007/s11368-012-0594-x

Erkossa T, Itanna F, Stahr K (2007) Indexing soil quality: a new paradigm in soil science research. Soil Res 45(2):129–137. https://doi.org/10.1071/SR06064

Etablissement Public Territorial (ETP) Paris-Est-Marne & Bois (2016) Plan Local d'Urbanisme [Local Urbanism Plan]. Review report, ETP, Champigny-sur-Marne

France D, Ouellet C (2002) La réhabilitation des friches industrielles: un pas vers la ville viable? [Brownfield remediation: a step towards the sustainable city?]. VertigO 3(2). https://doi.org/10.4000/vertigo.3812

Gitton C, Verger Y, Brondeau F, Charvet R, Nold F, Branchu P, Douay F, Lamy I, Mougin C, Petit C, Rémy E (2018) L'économie circulaire: cercle vertueux ou cercle vicieux ? Le cas de l'utilisation de terres maraîchères pour aménager des espaces verts urbains [The circular economy: virtuous circle or vicious circle? The case of the use of market gardening land to develop urban green spaces]. Géocarrefour 92(2). https://doi.org/10.4000/geocarrefour.11950

Huyghe C (2020) Production agricole et préservation de l'environnement: est-ce possible ? [Agricultural production and preservation of the environment: is it possible?]. Science Et Pseudo-Sciences 331:46–51

Joimel S et al (2016) Physico-chemical characteristics of topsoil for contrasted forest, agricultural, urban and industrial land uses in France. Sci Total Environ 555:40–47

Kareiva P, Marvier M (2007) Conservation for people. Sci Am 297:50–57

Larramendy S, Huet S, Micand A, Provendier D (2014) Conception écologique d'un espace public paysager – Guide méthodologique de conduite de projet [Ecological design of a landscaped public space—methodological guide for project management]. Plante & Cité, Angers

Le Clec'h S, Oszwald J, Decaens T, Desjardins T, Dufour S, Grimaldi M, Jegou N, Lavelle P (2016) Mapping multiple ecosystem services indicators: toward an objective-oriented approach. Ecol Ind 69:508–521

Lévêque C (2017) La biodiversité, avec ou sans l'homme? [Biodiversity, with or without man?]. Quæ, Versailles

Limasset E, Fourny S, Collet JL, Michel P, Alary C, Laboudigue A (2015) Approche REFRINDD pour accompagner les acteurs de la requalification des friches industrielles potentiellement polluées dans une démarche durable: guide méthodologique et prototype d'outil d'accompagnement [REFRINDD approach to accompany the actors of the requalification of potentially polluted industrial wastelands in a sustainable approach: methodological guide and prototype support tool]. ADEME, Angers

Maes J et al (2016) An indicator framework for assessing ecosystem services in support of the EU Biodiversity Strategy to 2020. Ecosyst Serv 17:14–23. https://doi.org/10.1016/j.ecoser.2015.10.023

Mailliard T (2016) SYCTOM-Centre de tri Paris 17ème – Phase APD – Dossier de demande d'autorisation d'exploiter – Etude d'impact [Phase APD—application for authorization to operate—environmental assessment report]. NEODYNE, Paris

Maris V (2014) Nature à vendre. Les limites des services écosystémiques [Nature for sale. The limits of ecosystem services]. Quæ, Versailles

Masboungi A (2018) L'urbanisme des milieux vivants [Urban planning of living environments]. Parenthèses, Paris

Mehdi L, Weber C, Di Pietro F, Selmi W (2012) Evolution de la place du végétal dans la ville, de l'espace vert à la trame verte [Evolution of the role of vegetation in the city, from green space to green frame]. VertigO 12(2). https://doi.org/10.4000/vertigo.12670

Monsérié MF (2009) Processus d'agrégation dans un technosol : contribution des constituants anthropiques (oxydes de fer, laitiers, HAP, ETM) à la formation des associations organo-minérales [Aggregation process in a technosol: contribution of anthropogenic constituents (iron oxides, slag, PAH, MTE) to the formation of organo-mineral associations]. PhD thesis, Sciences agricoles. Institut National Polytechnique de Lorraine

Peuportier B (2016) Eco-design for buildings and neighbourhoods. CRC Press, Taylor and Francis Group, London

RECORD (2018) Mesure de la biodiversité et évaluation des services écosystémiques des milieux restaurés. Méthodes et retours d'expériences [Measurement of biodiversity and assessment of ecosystem services in restored environments. Methods and feedback]. Final report, 17-1021/1A

Seybold CA, Herrick JE, Brejda JJ (1999) Soil resilience: a fundamental componant of soil quality. Soil Sci 164(4):224–234. https://doi.org/10.1097/00010694-199904000-00002

Society for Ecological Restoration International Science & Policy Working Group (2004) The SER International Primer on Ecological Restoration (SERI). www.ser.org & Tucson: Society for Ecological Restoration International

Tassin J (2020) Pour une écologie du sensible [For an ecology of the sensitive]. O. Jacob, Paris

Vidal-Beaudet L (2018) Du déchet au Technosol fertile: l'approche circulaire du programme français de recherche SITERRE [From waste to fertile Technosol: the circular approach of the French research program SITERRE]. VertigO 31. https://doi.org/10.4000/vertigo.21887

Walter C et al (2015) Les services écosystémiques des sols: du concept à sa valorisation ? [Soil ecosystem services: from concept to valuation?]. Cahier Demeter 15:51–68

Yves Petit-Berghem is a geographer at the National School of Landscape in Versailles (France). By mobilizing an interdisciplinary approach between ecology, geography (interface between environments and societies), and the landscape project, his research activities question the socio-landscape dynamics as well as the processes that underpin these dynamics. His recent work focuses on the evolution of the practices of landscape professionals, particularly those faced with the challenge of reconverting polluted urban wastelands.

Elisabeth Rémy is a Sociologist at INRAE (Institute for Agronomy and Environmental Research), Paris (France). Since 2014, she has been working on controversies and risks related to (peri-)urban soils. She leads the interdisciplinary and interregional Working Group "Risks related to collective and private urban gardens" bringing together researchers in biotechnical sciences and the human and social sciences. She participates in the MisChar project on the requalification of a peri-urban wasteland. Her current research focuses on the quality of composts produced by amateur gardeners of (peri-)urban collective gardens (POTAGERS project).

Marine Canavese is a doctoral student in sociology at the University Lumière Lyon 2, Max Weber Center (France). She is writing a PhD dissertation retracing the public socio-history of polluted and potentially polluted soils in France, with an interest in urban soils and the trajectories of urban wastelands such as, for example, the past of collective garden sites. Her fieldwork focuses mainly on the Lyon metropolitan area, but also on sites in Paris, Nantes, and Lille.

Approaches to Developing Urban Wastelands as Elements of Green Infrastructure

Juliane Mathey and Stefanie Rößler

Abstract For cities, it is essential to have richly structured and multifunctional green systems which satisfy ecological needs and at the same time are attractive, usable, low cost or even profitable. Indeed, it is doubtful that sufficient green infrastructure can be realized solely by pursuing traditional models of green space development. At a time of changing demands, constrained public finances and limited urban space, new and unusual types of green spaces can supplement traditional elements of urban green infrastructure. Urban wastelands with their various stages of vegetation can provide a number of ecosystem services to tackle challenges such as stopping the loss of biodiversity, adapting to climate change and creating recreational and healthy urban environments. The conservation of spontaneous biotopes, on the one hand, and the active greening of urban wastelands, on the other, offer the potential to develop urban green systems that provide a range of essential ecological, social and aesthetic services. This chapter outlines the potential of urban wastelands to supplement urban green infrastructure. We consider how different "designs" of urban wastelands are perceived and used by residents. Based on findings regarding biodiversity, ecosystem services and the perception/acceptance of vegetation-covered wastelands, various planning and development approaches are presented.

Keywords Urban ecosystem services · Urban biodiversity · Landscape planning · Urban green spaces · Urban wastelands

1 Introduction

Sustainable urban development faces manifold challenges such as adapting to climate change, stopping the loss of biodiversity, creating healthy environments for urban populations as well as ensuring fair and efficient land use. In order to meet these

J. Mathey (✉) · S. Rößler
Leibniz Institute of Ecological Urban and Regional Development (IOER), Dresden, Germany
e-mail: j.mathey@ioer.de

S. Rößler
e-mail: s.roessler@ioer.de

F. Di Pietro and A. Robert (eds.), *Urban Wastelands*, Cities and Nature,
https://doi.org/10.1007/978-3-030-74882-1_13

challenges, it is essential that cities have richly structured and multifunctional green systems which satisfy ecological needs and at the same time are attractive, usable, low cost or even profitable.

Urban green infrastructure is assigned a crucial role in providing such benefits and thus influencing the quality of life of the urban population (Tzoulas et al. 2007; Niemelä et al. 2010; Gómez et al. 2011; Naumann et al. 2011; Elmqvist et al. 2013; Mathey and Rink 2020). Green infrastructure can be understood as "a strategically planned network of natural and semi-natural areas with other environmental features designed and managed to deliver a wide range of ecosystem services" (COM 2013). In urban areas, such infrastructure encompasses a large variety of green features ranging from urban parks and natural areas to street trees and green roofs (DG Environment 2012).

The scope for developing this green infrastructure within existing, dense settlement structures is becoming ever more limited, particularly in the face of global urbanization (Tian et al. 2012; Zhong et al. 2020). Indeed, it is doubtful that sufficient green infrastructure can be realized solely by pursuing traditional models of green space development. At a time of changing demands, constrained public finances and limited urban space, new and unusual types of green space can supplement traditional elements of urban green infrastructure such as parks and playgrounds.

Urban wastelands, particularly if they are covered by vegetation, constitute vital components of a city's green infrastructure. Undisturbed sites are often biologically diverse, and their unregulated status can encourage innovative spatial activities by a wide range of users (Unt and Bell 2014). Additionally, they frequently constitute the only available sites for new green spaces. Green space on wasteland can provide many social and environmental benefits, such as opportunities to walk and cycle, increased flora and wildlife as well as reduced noise and air pollution. It can also contribute to economic regeneration (Doick and Hutchings 2007; Wirth et al. 2018), functioning as a soft location factor to attract investors as well as highly qualified workers.

In this way, we see that urban wastelands with spontaneous vegetation can help tackle current challenges such as preserving biodiversity (see, e.g., Konopka and Wüstendörfer 1995; Dettmar 2005; Sukopp 2005; Wittig 2010; Kowarik 2011, 2013; Bonthoux et al. 2014), adapting to climate change (see Gill et al. 2007; Bowler et al. 2010; Mathey et al. 2014) and fostering recreational and healthy urban environments (see Schemel et al. 2005; Völksen 2005; Hannig 2006; Arndt and Rink 2013; Unt and Bell 2014; Rupprecht et al. 2015; Rink and Arndt 2016) in dense urban environments (Brun et al. 2017; De Valck et al. 2019).

Consequently, many cities around the world have started to look closely for solutions for preserving or developing urban wastelands as green spaces (Zhong et al. 2020). Urban wastelands, defined as "areas whose previous, usually structural use has been abandoned and which are not actively put to a defined subsequent use for a certain period of time [or permanently]" (Rößler and Mathey 2018), are no longer merely understood as an environmental burden but rather as a valuable and non-renewable land resource (Banzhaf et al. 2018). Although they are often used for parking, makeshift athletic fields or play areas and even junkyards, they are occasionally incorporated into the urban fabric via community gardening or habitat restoration projects (Kremer et al. 2013; Németh and Langhorst 2014; both cited in Anderson and Minor 2017).

In view of the multifunctional and shifting demands for green spaces, the question arises: Which planning approaches are most appropriate to increase the ecological value of urban wastelands and at the same time offer recreational functions?

The conservation of spontaneous biotopes, on the one hand, and the active greening of urban wastelands, on the other, offer the potential to develop urban green systems that provide a range of essential ecological, social and aesthetic services. Of course, if the approaches of ecological succession or urban wilderness are considered as development options, it is important to know how wastelands are perceived by the potential users, namely local inhabitants, in particular, whether spontaneously developed wilderness on wastelands is accepted, exactly how (and to which degree) urban wastelands are used and which preferences for use are expressed.

This chapter outlines the potential of urban wastelands to supplement urban green infrastructure. Based on findings regarding biodiversity, ecosystem services and the perception/acceptance of vegetation-covered wastelands, various planning and development approaches are presented. The findings on biodiversity and ecosystem services were obtained from Central European studies. Our own studies on the perception/acceptance and use of wastelands were conducted in Dresden/Germany and are comparable with results for Leipzig/Germany (Mathey et al. 2018).

2 The Biodiversity of Urban Vegetation-Covered Wastelands

As a rule, urban wastelands display a strong anthropogenic influence and can be understood as a typical “Novel Ecosystem”, namely one that has been permanently altered by human impact in both its abiotic and biotic components (Hobbs et al. 2006; Kowarik 2011). The loss of native species and the immigration of new species are an essential aspect of Novel Ecosystems. In particular, they host unique compositions of species that have no historical reference and cannot be reversed (Francis 2014 as cited in Werner 2019).

If left relatively undisturbed, unsealed wastelands generally exhibit dynamic forms of natural development, i.e., they undergo a process of ecological succession that produces a particular kind of urban vegetation that differs from the man-made and artificially maintained green spaces normally found in cities. Over a period of years, such sites can develop habitat structures unique to urban areas (Hohn et al. 2007; Kowarik 2013), providing valuable retreats as well as substitute or stepping-stone habitats for animal and plant species.

Often a mosaic of diverse habitats appears depending on the particular soil substrate types, nutrient levels, the local climatic conditions as well as the diverse stages of plant development. This mosaic is also caused by human activity (the type/intensity of former and current land use, period of abandonment), the site size, the degree of soil sealing, neighborhood effects as well as the high pace of ecological change (Schadek et al. 2009; Hansen et al. 2012; Rebele 2013). Depending on the length of the fallow period and the intensity of current use, under Central European conditions we can distinguish roughly four types of urban wastelands that accord with the various stages of ecological succession (Dettmar 1995; Rebele and Dettmar 1996) (Table 1). Each stage harbors a particular mix of plant and animal species, allowing different forms of nature experience (Bonthoux et al. 2014). This dynamic spatio-temporal mosaic results in a high level of plant and animal species diversity (see, e.g., Wittig and Zucchi 1993; Rebele and Dettmar 1996; Arlt et al. 2003; Muratet et al. 2007; Buglife 2009; Wittig 2010; Kattwinkel et al. 2011).

The ecological potential of wastelands is determined not only by the suitability and importance of their vegetative features, but also by the position within the network of green infrastructure (Hobbs et al. 2014 cited in Riley et al. 2018). Certain sites that are located in or close to key patches or corridors are especially significant for improving the ecological structure and functions of green infrastructure.

3 Ecosystem Services of Vegetation-Covered Urban Wastelands

Vegetation-covered urban wastelands can provide a number of ecosystem services, i.e., direct and indirect benefits that residents derive from the functioning of ecosystems within the urban environment (Costanza et al. 1997; Gómez-Baggethun et al. 2013).

Differences in the quantity and quality of ecosystem services will depend on the specific inventory of vegetation and structural parameters of vegetation-covered wastelands. Since the structural parameters of wastelands reflect the stages of ecological succession, the supply of individual ecosystem services (which depends on a specific vegetation structure) varies for each stage (Rebele and Dettmar 1996; Schadek et al. 2009; Mathey et al. 2015) (Table 1, Fig. 1).

Of course, urban wastelands can also constitute a danger to health due to the physical legacy of previous activities (such as tunnels, house ruins, holes, polluted soils, and poisonous substances) or provoke anxiety from undesired current forms of use (by homeless people, vagabonds, thugs, crooks, drug addicts, etc.). Spontaneous vegetation can pose ecological and/or health challenges when invasive plant species (e.g., *Reynoutria japonica, Ambrosia artemisiifolia*) flourish or support undesirable wildlife (e.g., disease vectors) (Dickie et al. 2014; Goldstein 2001; both cited in Riley et al. 2018). Therefore, this type of novel urban nature can be perceived negatively by neighborhood residents (Lafortezza et al. 2008; Brun et al. 2017) so that, in particular, cultural ecosystem services remain unused, whether nature experiences, environmental education, outdoor meeting places, spiritual attachments or aesthetic functions.

Table 1 Typology of vegetation-covered urban wastelands. Lines 2–3: stages of ecological succession on urban wastelands based on the fallow period and data on types of habitats; lines 4–5: cooling effects of different succession stages (temperature differences compared to a surrounding asphalt area on a hot summer day); and lines 6–9: acceptance of different stages of ecological succession by residents (survey in Dresden/Germany: n = 305; for more details, see below in Chap. 4). Stages of urban wastelands: derived from Rebele and Dettmar 1996 among others; modified after Mathey and Rink 2010; Mathey et al 2018; photomontages: Tittel, Wahl; background Volz

Urban wastelands: stages of succession and habitat types			
 Young wasteland with pioneer vegetation	 *Older wasteland with wild meadow*	 Old wasteland with tall herbaceous vegetation	 Wasteland with wild woodland
Initial 3-year fallow period: open, fragmentary, ruderal pioneer populations with short-lived, annual species	Fallow period of 3–10 years: closing extensive vegetation cover, increasing proportion of persistent ruderal vegetation, single bushes and groves higher than 5 m	Fallow period of 10–50 years: mainly persistent species, ruderal tall forbs, bushes, single groves higher than 10 m	Fallow period of more than 50 years: dense groves, highly growing herb layer (where canopy is patchy), characteristic woodland
Cooling effects of different stages of ecological succession			
1.4 K	1.5–1.6 K		1.7 K
Acceptance of different stages of ecological succession in Dresden			
Approval: 9.3%	Approval: 26.2%	Approval: 34.1%	Approval: 26.6%
Rejection: 61.5%	Rejection: 42.6%	Rejection: 36.9%	Rejection: 47.9%
No comment: 29.1%	No comment: 30.3%	No comment: 29.1%	No comment: 25.7%

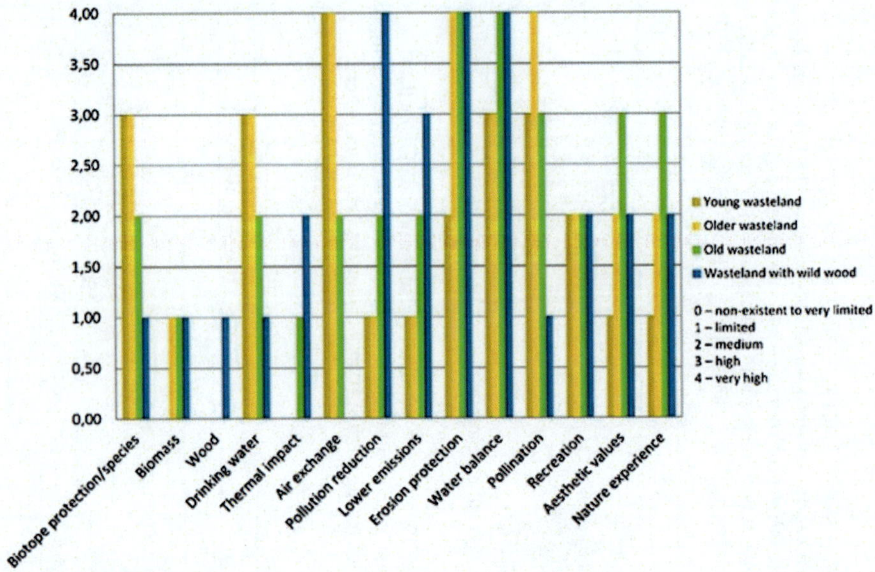

Fig. 1 Levels of ecosystem services provided by four successional stages of urban wasteland (general qualitative assessment based on literature review by Plappert; *Source* Modified after Plappert 2013)

4 Perception, Acceptance and Use of Urban Wastelands

If the ecosystem services provided by spontaneous vegetation are to be exploited in a coordinated planning process, it is certainly helpful to have information on how urban wastelands and their different stages of succession are perceived, accepted and used by residents (Mathey et al. 2018; Brun et al. 2017; Rupprecht et al. 2015 cited in Riley et al. 2018).

In the following, we present some findings of a study conducted in Dresden/Germany on the perception, acceptance and use of urban wastelands (for the description of methods, see Box 1 and Box 2).

Box 1: "Survey on the perception, acceptance and use of urban wastelands".

The residents' survey was carried out in 2009 in the City of Dresden/Germany (population: 519,055 as of December 31, 2009, LH DD 2019). In 2008, 3,131 wastelands were registered, encompassing approximately 4,352 individual plots, with a total area of approximately 1,603 ha (LH DD 2008). Standardized written questionnaires were completed. Respondents in the respective districts (one area of detached housing and five districts of multiple-family housing) were chosen as random samples based on the number of available flats. 305 evaluable questionnaires were returned.

4.1 *Perception and Acceptance*

The findings of the residents' survey conducted in Dresden/Germany reveal a mixed view of urban wastelands: although negative attitudes dominate, some respondents made positive statements. Thus, 16% described wastelands as "interesting sites" and 13% claimed that they provide "an unusual view". The largest share of respondents, namely 38%, expressed a clearly negative opinion, describing these areas as "disturbing the cityscape" (38%). In contrast, less than 10% of respondents viewed urban wastelands as an "asset to the cityscape".

More specifically, 22% of respondents said that the perception of the wasteland depends on the type of site. While wastelands with spontaneous wild vegetation are perceived as "real" (natural) places, they may also be viewed as "rubbish-strewn" areas (Mathey et al. 2015, 2018). Regarding their local area, the majority of respondents were of the opinion that urban wastelands with spontaneous wild vegetation tend to reduce rather than increase the value of the residential area. The survey also asked whether spontaneous wild vegetation on urban wastelands is worth preserving. Here, 53% of respondents answered "no" versus 32% who answered "yes". Breaking down the group of respondents by age, we find some differences in their evaluation (Fig. 2). Thus, younger people (<35 years) viewed spontaneous, wild vegetation on urban wastelands more positively than older people (35–70 years). The opinions of

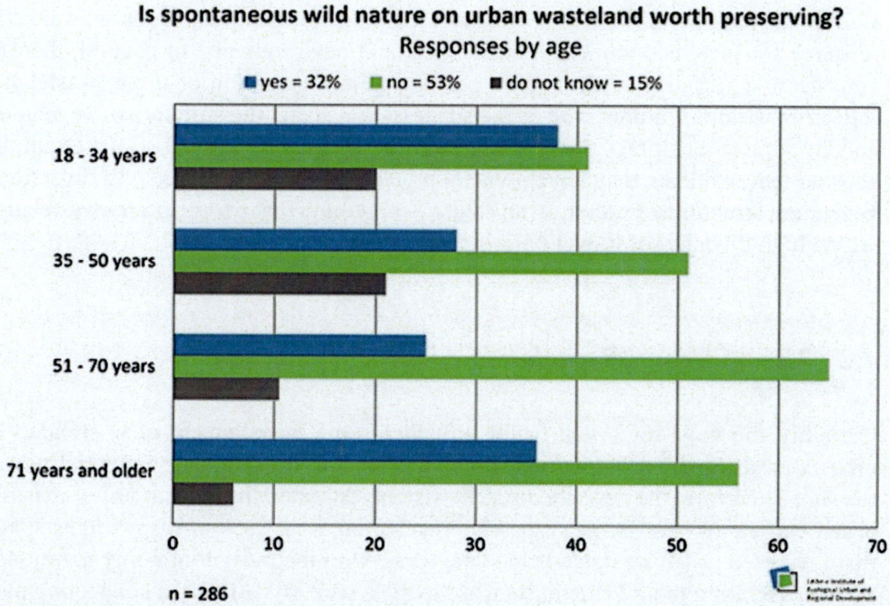

Fig. 2 Answers to the question "Is spontaneous wild nature on urban wastelands worth preserving?" broken down into four age classes of respondents (n = 286 respondents in Dresden/Germany) (*Source* Modified after Banse and Mathey 2013)

those aged over 70 were also somewhat more positive. In Dresden/Germany, the opinion of the aesthetic value of wastelands as well as its perception and acceptance is very diverse, with negative attitudes dominating. The valuation depends on the type of site/successional stages with different vegetation structures. These findings match the results of Konopka and Wüstendörfer (1995) for Nuremberg/Germany and of Brun et al. (2017) for two French cities (Tours and Blois). Even if the studies were conducted at different times (1995, 2009, and 2017), and despite assumed cultural differences, the results are largely consistent.

Regarding the provision of ecosystem services, it is interesting to consider the respondents' reactions to vegetative development in each of the four stages of ecological succession (Table 1). Urban wastelands characterized by a high degree of surface sealing and low vegetative cover (pioneer vegetation) were viewed unfavorably by those questioned. Interest in using the site increased with advanced greening of wastelands by persistent ruderal vegetation and tall herbaceous plants. The level of interest then dropped for later successional stages featuring more or less dense and wild wood-like vegetation structures. These results confirm that the density and structure of vegetation on urban wastelands significantly influence their acceptance (Mathey et al. 2015).

Some authors have found a more negative perception of urban wastelands due to residents' emphasis of various factors such as their fear of physical threat as well as the unattractiveness or neglected appearance of sites (Herbst and Herbst 2006; Rink and Arndt 2016; Riley et al. 2018). In contrast, Anderson and Minor (2017) cite many studies that have identified a positive understanding of urban wastelands as a valuable resource for local economies, communities, and environments. In general, there is considerable variation in the public perception and valuation of urban wasteland, reflecting various features such as the successional stage, the structure of vegetation and the attractive colors of flowers. The study by Hofmann et al. (2012), for example, showed that residents may have a visual preference for wastelands with diversified vegetation structures. Further, Brun et al. (2017) found that people prefer wastelands with a high diversity of flower coloring.

4.2 Use of Urban Wastelands

Certainly, the most important factor influencing the management of wastelands in cities as well as the attitude/interests of residents is the actual or potential use of the site. Therefore, the described survey also investigated the current use of existing urban wasteland sites in the respective residential area. 28% of respondents stated that they used urban wastelands in some form, with the most commonly mentioned activities relating to movement, such as having a walk (68%) or as a connecting path (46%) (Fig. 3). In contrast, social activities were much more rarely mentioned; these included meeting with friends or family (meeting places), barbeques and other social events or playing with children (place for children) (Mathey et al. 2018).

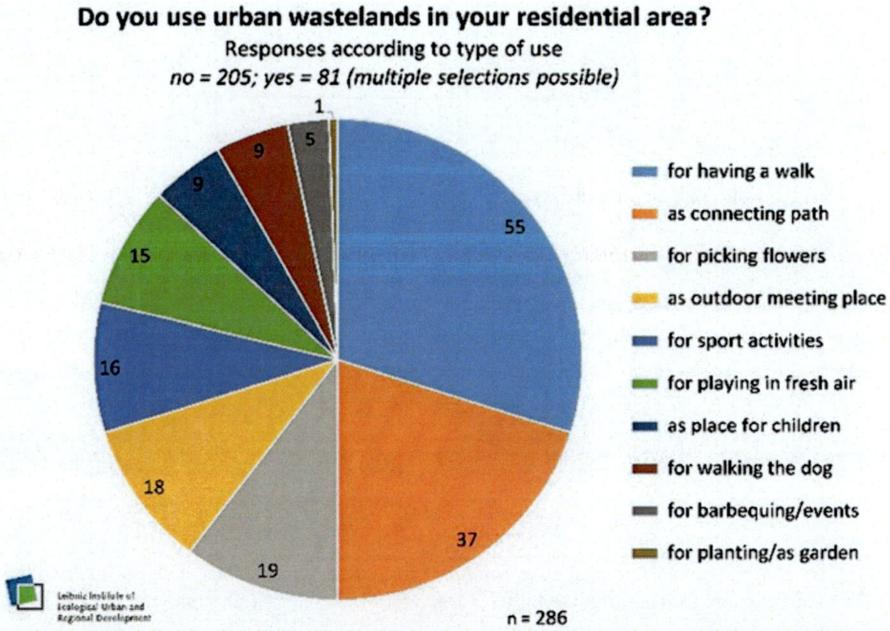

Fig. 3 Types of use of vegetation-covered urban wastelands by residents in Dresden/Germany (modified after Mathey et al. 2015 and after Banse and Mathey 2013)

These results are in line with other studies showing that urban wastelands are hosts for various informal activities (see, e.g., Konopka and Wüstendörfer 1995; Keil 2005; Unt and Bell 2014). Clearly, such informal green spaces provide great opportunities for residents, especially children and teenagers, to keep in contact with nature and to enjoy a feeling of adventure. Brun et al. (2017) found a correlation between the positive evaluation of wastelands and residents' uses: residents who used a wasteland for recreational activities assigned it a clearly positive value.

4.3 Preferred Uses and Design of Urban Wastelands

With respect to the management of urban wastelands, it is vital to determine the wishes of residents regarding the future use of sites. In the discussed survey, the residents were asked to assess which design and utilization options they could imagine for the wastelands in their local area. The main focus was on those types of use that could serve both to improve the quality of life in the city and to protect urban nature. A range of 17 design and use variants were presented (such as green spaces, parks, forests and gardens), to be classified as "desirable", "acceptable" or "undesirable" (Fig. 4, Box 2).

Fig. 4 Suggested design and use options for urban wastelands during interviews in Dresden/Germany (*Source* Modified after Banse and Mathey 2013; photomontages: Tittel, Wahl; background Volz)

Box 2: "Photo method—operating with visual methods".

Photomontages were developed for the survey to depict different wasteland situations against a standard inner-city backdrop. These images were then linked to different questions in the survey. An effort was made to ensure similar lighting, backgrounds as well as perspective and color of the montages so that these extraneous factors would not influence the respondents' perception and evaluation.

48% of the respondents wanted the wasteland sites in the residential area to be made usable again. Only 13% wished to leave them unchanged while about 24% thought that it depended on the type of wasteland. The majority of those questioned desired a planned and designed after-use such as laid-out, public parks as well as green spaces with trees. Sites for nature conservation were also frequently mentioned as variations for design or use. Further possible after-uses were playgrounds, flower meadows or lawns (Fig. 5). Rebuilding or surface sealing in the form of car parks or garages were, in contrast, rarely desired, as were utilizations as flower beds/allotments or extensive types of cultivation, e.g., to grow renewable resources such as poplar or rapeseed (Mathey et al. 2018).

There are a few previous studies on local residents' ideas for the use of urban wastelands. One notable exception is the study by Washbourne et al. (2020), which quantitatively surveyed the demand of all major stakeholder groups for ecosystem services from wastelands. The results showed a strong trade-off between demand for services related to property development (e.g., low flood risk) and all other services,

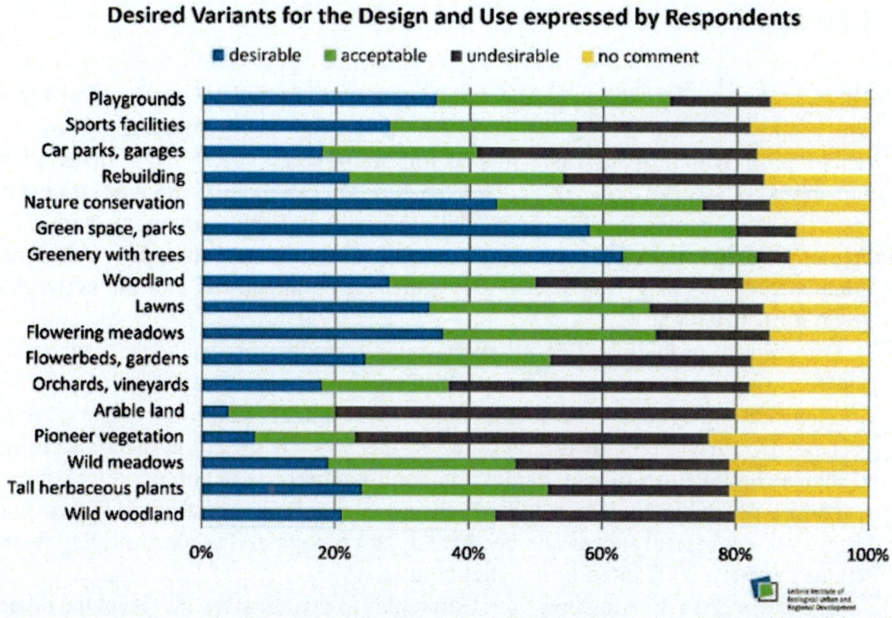

Fig. 5 Desired variants for the design and use of urban wasteland sites as expressed by respondents in Dresden/Germany (multiple answers possible) (*Source* Modified after Banse and Mathey 2013)

which were linked to vegetation-covered sites. Regarding such vegetation-covered wastelands, we can distinguish between demand for the services of more "natural" vegetated sites (e.g., habitat services) and those linked to aesthetics and recreation. The study findings indicate that most stakeholders desired the creation of multifunctional green spaces from wastelands (Washbourne et al. 2020).

5 Exploiting Urban Wastelands for the Development of Green Infrastructure

In the previous sections, we have shown how vegetation-covered urban wastelands offer a variety of ecosystem services and can contribute to biodiversity. Additionally, they are accepted and valued by at least some parts of the urban population. If these potentials are to be preserved and exploited to aid sustainable urban development, they must be anchored into urban and landscape planning processes and decisions. In the following, we present some planning approaches at the city and site levels that can support the protection and development of vegetation-covered urban wastelands as elements of urban green infrastructure.

5.1 City Level

There are two major options for the redevelopment of wasteland sites: revitalization, which aims to reuse these wastelands for residential or commercial purposes, or renaturation, which treats wastelands as important green spaces for humans, plants and animals (Banzhaf et al. 2018). This renaturation can include strategies that either leave the wasteland within the ecological succession process or intervene with minor or major design measures in order to provide different functions. If we designate urban wastelands as green spaces, they can then be integrated into the network of green infrastructure.

One current problem is that urban wastelands undergoing ecological succession and featuring spontaneous vegetation in Germany are not protected by conservation legislation because these factors are not recognized within the urban planning process. Urban wastelands are recorded in the municipal land registers according to their original form of use, namely as commercial or residential sites. Although awareness of the beneficial services of urban wastelands exists at the local urban planning level, ecological succession is currently not recognized as a proactive greening strategy (Ahern 2007 cited in Banzhaf et al. 2018).

The first step in drawing up general strategies to exploit at the city level the diverse ecosystem services of urban wasteland as well as their potentials for biodiversity is to survey their quantity and quality (Mathey et al. 2015). This should also be done with a view to deciding whether a wasteland is more suitable for revitalization or renaturation. In view of the general lack of appropriate guidelines (Banzhaf et al. 2018), an approach shall be proposed here to aid the decision-making process.

In order to consider the demand to foster a compact city as well as to maintain and develop its green infrastructure, the city authorities in Dresden/Germany placed its landscape plan under the "Leitbild" (overall concept) of "Dresden—the compact city in an ecological network", conceived as a guiding principle for the entire city development (LH DD 2018). The idea here is to achieve a spatial concentration of urban functions by means of a compact, cell-like structure. The resulting ecological network structure reflects natural and infrastructural features as well as taking account of vital biotope networks and watercourses, interlinking existing and new elements of the green infrastructure, which also includes vegetation-covered wastelands (Artmann et al. 2017; Mathey et al. 2017). Such overarching strategies support decision-making on whether a wasteland should be designated for revitalization or renaturation.

In Germany, municipal biodiversity strategies are drawn up for the effective and efficient protection and development of local biodiversity. Through these strategies, it is possible to systematically record, describe and negotiate objectives as well as activities to promote urban nature with respect to legal, economic, planning and ecological factors. Here also, urban wasteland with spontaneous vegetation can be specifically addressed (UrbanNBS-Team 2020).

In order to improve the quality of life and the environment in deprived areas, comprehensive urban regeneration concepts should consider the potential of wastelands for green space development. Integrated urban development concepts can provide an appropriate framework to guide urban regeneration in densely built-up areas (Rößler et al. 2019a, b). Landscape planning creates city-wide concepts and seems to be very helpful in conceptualizing such green infrastructure strategies while integrating urban wastelands (Wende et al. 2012).

The approaches introduced here should be underpinned by an appropriate knowledge base of wastelands in each municipality. One source of data is, of course, wasteland registers or land-use databases maintained by local authorities. Survey parameters should be selected with regard to biodiversity potentials and the ecosystem services that are of interest. Parameters such as degree of soil sealing, vegetative structures as well as species inventories derived from the corresponding data sources can help to evaluate the capacity to provide the envisaged ecosystem services. If suitable, the residents' wishes for the planned and designed after-use of urban wastelands, such as public parks or green spaces with trees, should be complied with. But even if not so popular in the survey in Dresden/Germany, it may make sense to develop urban wastelands into other types of green spaces such as flower beds, allotments, more natural-vegetated sites and extensive types of cultivation (e.g., to grow renewable resources). As it has been shown that the level of provision of services varies for each stage of ecological succession, this must be considered when evaluating the inventory of green urban brownfields in a city. These aspects should be integrated into analytical and evaluation algorithms, which can then be drawn upon to devise strategies for a city's treatment of wastelands, and which can also provide a solid basis for decisions regarding individual planning measures (Mathey et al. 2015). A detailed evaluation can be made in the framework of an environmental assessment.

5.2 Site Level

The objective of developing urban wastelands can be to provide different ecosystem services. To strengthen the link to aesthetics or recreation needs an active designing of the site. Figure 6 shows an example of active design. It is the area of a demolished prefabricated high-rise building, which was transformed into a lawn with trees and wildflower strips planted by volunteers from the residential area. To focus on habitat services, more "natural" vegetated sites with ecological succession are appropriate. Figure 7 shows a succession park on a former military wasteland, where ecological succession is combined with a nature-like path and some benches. Therefore, it is necessary to determine which planning approaches will serve to increase the ecological and social value of urban wastelands. While various options for the exploitation of wastelands to complement and expand traditional notions of urban open space are currently being discussed (Rößler and Mathey 2014), there is no universally applicable solution to the reuse of urban wastelands as forms of green space (Pediaditi et al. 2010). Yet all envisaged approaches should take into account

Fig. 6 Left: urban wasteland in Dresden/Germany shortly after the demolition of a high-rise building in 2006. Right: the same area after transformation into a lawn with wildflower strips and trees in 2013 © R. Bendner

Fig. 7 Succession park on a former military brownfield in Dresden/Germany. Left: in 2010. Right: the same area in 2020 © J. Mathey

the existing vegetative structure, the location of wastelands in the city, the issue of accessibility and the specific requirements regarding design and use (Unt and Bell 2014) as discussed in Chap. 4. New forms of green space usage enjoying increasing popularity in cities are urban agriculture, urban woodland, gardens, sport/leisure pursuits, venues for cultural projects/public events, low-intervention parks, nature experience areas and urban wilderness. These forms of use also apply to urban wastelands in order to develop green spaces. They offer various potentials to foster urban biodiversity as well as ecosystem services such as micro-climatic regulation services and recreational services (Mathey et al. 2015) (Table 2).

Based on a survey of the demand for ecosystem services, Washbourne et al. (2020) found that the creation of multifunctional green spaces from wasteland sites would be desirable to most stakeholders. This multi-functionality can be achieved at a single wasteland site that offers several ecosystem services.

Table 2 Options for reusing urban wastelands as green space and the literature-based qualitative evaluation of their habitat services (after, e.g., Dettmar 1995; Kattwinkel et al. 2011; Muratet et al. 2007; Rebele and Dettmar 1996; Wittig and Zucchi 1993; Wittig 2010, and own knowledge), micro-climatic regulation services (after Lehmann et al. 2014) and recreational services as well as their acceptance by residents in Dresden/Germany

Options for green spaces on urban wastelands	Habitat services	Micro-climate regulation services	Recreational services	Acceptance by residents
Urban agriculture	+/−	+/++	+/−	+/−
Urban woodland	+/++	++	+/++	+/++
Gardens	+/−	++	++	++
Sports/Leisure pursuits	+/−	+	+ +	+
Cultural projects/Public events	−	+	+/++	+/−
Low-intervention parks	++	++	+/++	+/−
Nature experience areas	++	++	++	+/−
Urban wilderness	++	+/++	+/−	+/−

Source Modified after Mathey et al. (2015, 2018). ++ "well suited", + "suited", − "unsuited", +/− "detailed investigation of individual site necessary", +/++ depending on specific characteristics of the site

Table 3 Comparison of requirements for high biodiversity versus high acceptance by residents concerning green spaces in cities

Parameter	Requirements for high biodiversity	Requirements for high acceptance
Size	Habitat mosaics, bigger areas	Bigger usable areas
Shape	"Wilderness"	Designed areas, visible aim
Intensity of use	Undisturbed/disturbed areas	Undisturbed, usable and safe areas
Connectivity	Connected areas	Connected areas
Structure	Variety of structures	Aesthetic composition (structures)
Plant diversity	Variety of typical/rare plant species	Variety of attractive plants
Animal diversity	Variety of typical/rare animal species	Interesting animals (no foxes/raccoons)

Source Modified after Mathey and Rink (2010)

But how can the design options for urban biodiversity meet the acceptance of residents? Table 3 compares the requirements for green space design to achieve high biodiversity with those to ensure high acceptance by residents.

While many of these requirements are conflicting, others are compatible or offer a pathway toward compromise. Urban wastelands undergoing ecological succession and featuring spontaneous vegetation generally offer many possibilities. For example, semi-natural grasslands offer a particular set of ecosystem services depending on whether they are naturally vegetated (supporting biodiversity) or areas of mown lawn (providing recreational services).

In accordance with the idea of "Novel Ecosystems", Kowarik (2011) suggested new models of desirable development for urban wastelands, also encompassing natural dynamics such as ecological succession. If, however, certain fallow stages are to be preserved for reasons of nature conservation or to maintain specific ecosystem services, then only some nature-compatible uses can be permitted, adapted to the respective stages of succession. To a certain extent, this will replace active maintenance activities while still ensuring the continuation of the desired vegetation structure (Fig. 8).

Some other approaches can be named here, such as a conceptual framework combining an assessment of ecosystem services, economic cost–benefit analysis and analysis of spatial patterns to enable the ex-ante evaluation of the ecosystem services of green space alternatives (Zhong et al. 2020); an ex-ante approach to elaborate urban design concepts for wastelands and to plan dense urban structures in the case that a wasteland is designated for rebuilding (Koch et al. 2019); or indeed the prioritization of wasteland regeneration interventions (Cortinovis and Geneletti 2018). This latter assessment focuses on two ecosystem services of critical importance to the city, namely micro-climate regulation and nature-based recreation.

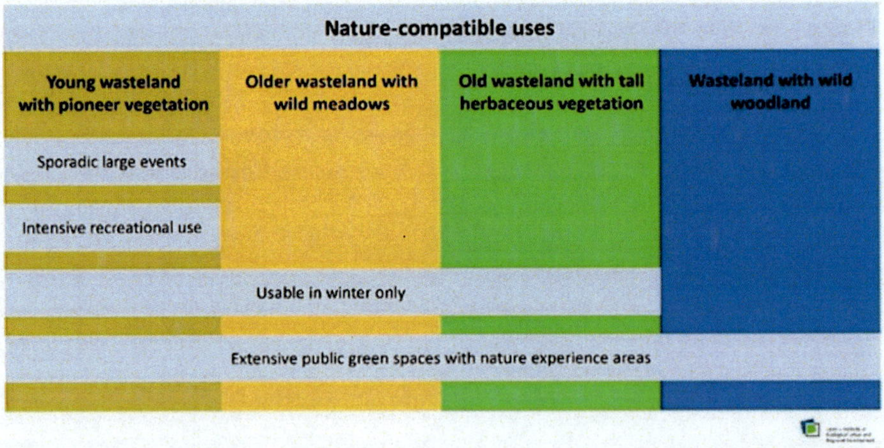

Fig. 8 Nature-compatible uses to preserve successional stages (*Source* Modified after Mathey et al. 2003)

6 Conclusion

The sheer variety of vegetation-covered urban wastelands brings manifold ecological benefits for urban environments. By providing habitats for diverse plant and animal species, they play a crucial role in supporting urban biodiversity. Additionally, vegetation-covered wastelands supply a number of ecosystem services, not least recreational services and an experience of nature.

Urban development strategies and measures should thus exploit urban wastelands to meet current challenges such as climate change or to ensure environmental justice. But we need ways of preserving these sites or defining them as official green spaces so as to incorporate such areas into existing networks of urban green infrastructure.

Strategic approaches and overall planning concepts must be devised to address the potentials and options of urban wastelands for urban green infrastructure development. At the city level, this means carefully analyzing the current situation and drawing up general planning guidelines.

Urban green spaces are desired by local residents, who often prefer their retention to the redevelopment of wasteland sites. However, such green spaces must be well managed in order to answer the needs of the residents for orderliness and security. If unmanaged wastelands are to be included in a development concept, an unusual degree of public communication is required along with a participative process. A vital precondition here is to include citizens in the planning and preparation of the design process and motivate them to get involved (see Doick 2010; Rall and Haase 2011). In this way, areas of succession could be better accepted and more closely integrated into future plans for urban wastelands in residential areas.

References

Anderson EC, Minor ES (2017) Vacant lots: an underexplored resource for ecological and social benefits in cities. Urban For Urban Green 21:146–152

Arlt G, Kowarik I, Mathey J, Rebele F (eds) (2003) Urbane Innentwicklung in Ökologie und Planung [City-interior development in ecology and planning]. IÖR-Schriften 39, Dresden

Arndt T, Rink D (2013) Urbaner Wald als innovative Freiraumstrategie für schrumpfende Städte. Zur Akzeptanz und Wahrnehmung von Aufforstungen urbaner Brachen in Leipzig [Urban forest as an innovative open space strategy for shrinking cities. On the acceptance and perception of reforestation of urban wastelands in Leipzig]. CONTUREC 5:23–37

Artmann M, Bastian O, Grunewald K (2017) Using the concepts of green infrastructure and ecosystem services to specify Leitbilder for compact and green cities—the example of the landscape plan of Dresden (Germany). Sustainability 9(2) 198:1–26

Banse J, Mathey J (2013) Wahrnehmung, Akzeptanz und Nutzung von Stadtbrachen. Ergebnisse einer Befragung in ausgewählten Stadtgebieten von Dresden [Perception, acceptance and use of urban wastelands. Results of a survey in selected urban areas of Dresden]. CONTUREC 5:39–56

Banzhaf E, Arndt T, Ladiges J (2018) Potentials of urban brownfields for improving the quality of urban space. In: Kabisch S, Koch F, Gawel E, Haase A, Knapp S, Krellenberg K, Nivala J, Zehnsdorf A (eds) Urban transformations, sustainable urban development through resource efficiency, quality of life and resilience. Future City 10. Springer

Bonthoux S, Brun M, Di Pietro F, Greulich S, Bouché-Pillon S (2014) How can wastelands promote biodiversity in cities? A review. Landsc Urban Plan 132:79–88

Bowler DE, Buyung-Ali L, Knight TM, Pullin AS (2010) Urban greening to cool towns and cities: a systematic review of the empirical evidence. Landsc Urban Plan 97(3):147–155

Brun M, Di Pietro F, Bonthoux S (2017) Residents' perceptions and valuations of urban wastelands are influenced by vegetation structure. Urban for Urban Green. https://doi.org/10.1016/j.ufug.2017.01.005

Buglife (2009) Planning for Brownfield Biodiversity: A best practice guide. The Invertebrate Conservation Trust, Peterborough/UK

COM (COM-European Commission) (2013) Green Infrastructure (GI)—enhancing Europe's natural capital. European Commission 249 final. Brussels

Costanza R, d'Argee R, de Groot R, Farber S, Grasso M, Hannon B, Limburg K, Naeem S, O'Neill RV, Paruelo J, Raskin RG, Sutton P, van den Belt M (1997) The value of the world's ecosystem services and natural capital. Nature 387:253–260

Cortinovis C, Geneletti D (2018) Mapping and assessing ecosystem services to support urban planning: a case study on brownfield regeneration in Trento, Italy. One Ecosyst 3:e25477. https://doi.org/10.3897/oneeco.3.e25477

De Valck J, Beames A, Liekens I, Bettens M, Seuntjens P, Broekx S (2019) Valuing urban ecosystem services in sustainable brownfield redevelopment. Ecosyst Serv 35:139–149

Dettmar J (1995) Industriebedingte Lebensräume in Europa [Biotops caused by industries in Europe]. Schriftenreihe Für Vegetationskunde 27:111–118

Dettmar J (2005) Forests for shrinking cities? The project "Industrial Forest of the Ruhr". In: Kowarik I, Körner S (eds) Wild urban woodlands. New perspectives for urban forestry. Springer, Heidelberg, New York, Berlin, pp 263–276

DG Environment (Directorate-General Environment of the European Commission) (2012) The multifunctionality of green infrastructure. In-depth report. Science for Environment Policy. Bristol, Science Communication Unit, the University of the West of England (UWE)

Doick KJ (2010) Learning lessons in monitoring brownfield land regeneration to greenspace through logic modelling. In: Fox HR, Moore HM (eds) Restoration and recovery: regenerating land and communities, pp 16–25

Doick KJ, Hutchings T (2007) Greenspace establishment on brownfield land: the site selection and investigation process. Of Forest Research, UK Forestry Commission

Elmqvist T, Fragkias M, Goodness J, Güneralp B, Marcotullio PJ, McDonald R, Parnell S, Schewenius M, Sendstad M, Seto KC, Wilkinson C (eds) (2013) Urbanization, biodiversity and ecosystem services: challenges and opportunities. A global assessment. A part of the cities and biodiversity outlook project. Springer, Dordrecht, Heidelberg, New York, London

Gill SE, Handley JF, Ennos AR, Pauleit S (2007) Adapting cities for climate change: the role of the green infrastructure. Built Environ 33(1):115–133

Gómez F, Jabaloyes J, Montero L, De Vicente V, Valcuende M (2011) Green areas, the most significant indicator of the sustainability of cities: research on their utility for urban planning. J Urban Plan Dev 137(3):311–328

Gómez-Baggethun E, Gren Å, Barton DN, Langemeyer J, McPhearson T, O'Farrell P, Andersson E, Hamstead Z, Kremer P (2013) Urban ecosystem services. In: Elmqvist T, Fragkias M, Goodness J, Güneralp B, Marcotullio PJ, McDonald R, Parnell S, Schewenius M, Sendstad M, Seto KC, Wilkinson C (eds) Urbanization, biodiversity and ecosystem services: challenges and opportunities. A global assessment. a part of the cities and biodiversity outlook project. Springer, Dordrecht, Heidelberg, New York, London, pp 175–251

Hannig M (2006) Wie viel „Wildnis" ist erwünscht? Zur Akzeptanz von Sukzession auf städtischen und stadtnahen Flächen [How much "wilderness" is welcome? On acceptance of succession on urban and city near areas]. Stadt + Grün 1/2006:36–42

Hansen R, Heidebach M, Kuchler F, Pauleit S (2012) Brachflächen im Spannungsfeld zwischen Naturschutz und baulicher Wiedernutzung [Wastelands in the area of conflict between nature conservation and reuse for building]. BfN-Skripten 324, Bonn, Bundesamt für Naturschutz (BfN)

Herbst H, Herbst V (2006) The development of an evaluation method using a geographic information system to determine the importance of wasteland sites as urban wildlife areas. Landsc Urban Plan 77(1–2):178–195

Hobbs RJ, Arico S, Aronson J, Baron JS, Bridgewater P, Cramer VA, Epstein PR, Ewel JJ, Klink CA, Lugo AE et al (2006) Novel ecosystems: theoretical and management aspects of the new ecological world order. Glob Ecol Biogeogr 15:1–7

Hofmann M, Westermann JR, Kowarik I, van der Meer E (2012) Perception of parks and urban derelict land by landscape planners and residents. Urban for Urban Green 11(3):303–312

Hohn U, Jürgens C, Otto KH, Prey G, Piniek S, Schmitt T (2007) Industriewälder als Baustein innovativer Flächenentwicklung in postindustriellen Stadtlandschaften. Ansätze zu einer integrative wissenschaftlichen Betrachtung am Beispiel des Ruhrgebiets. [Industrial forests as a component of innovative land development in post-industrial urban landscapes. Approaches to an integrative scientific consideration using the example of the Ruhr area]. CONTUREC 2:53–67

Kattwinkel M, Biedermann R, Kleyer M (2011) Temporary conservation for urban biodiversity. Biol Cons 144:2335–2343

Keil A (2005) Use and perception of post-industrial urban landscapes in the Ruhr. In: Kowarik I, Körner S (eds) Wild urban woodlands. New perspectives for urban forestry. Springer, Berlin, Heidelberg, New York, pp 117–130

Koch F, Schlink U, Bilke L, Helbig C (2019) How to reconcile the ecosystem service of regulating the microclimate with urban planning projects on brownfields? The case study Bayerischer Bahnhof in Leipzig, Germany. In: Schröter M et al (eds) Atlas of ecosystem services. Springer International Publishing AG, part of Springer Nature, pp 147–156

Konopka T, Wüstendorfer W (1995) Zur Wertschätzung städtischer Brachen durch die Stadtbevölkerung. Eine empirische Untersuchung von Anwohnern und Anwohnerinnen in Nürnberg. [On the valuation of urban wastelands by city dwellers. An empirical study of residents in Nuremberg]. Stadt + Grün 11/95:763–771

Kowarik I (2011) Novel urban ecosystems, biodiversity, and conservation. Environ Pollut 159(8–9):1974–1983

Kowarik I (2013) Cities and wilderness. A new perspective. Int J Wilderness 19(3):32–36

Kremer P, Hamstead ZA, McPherson (2013) A social-ecological assessment of vacant lots in New York City. Landscape Urban Plann.120:218–233

Lafortezza R, Corry RC, Sanesi G, Brown RD (2008) Visual preference and ecological assessments for designed alternative brownfield rehabilitations. J Environ Manag 89(3):257–269

Lehmann I, Mathey J, Rößler S, Bräuer A, Goldberg V (2014) Urban vegetation structure types as a methodological approach for identifying ecosystem services—application to the analysis of micro-climatic effects. Ecol Ind 42:58–72

LH DD (Landeshauptstadt Dresden) (2008) Internal information on urban wastelands in Dresden for urban development planning on AGS. Brachflächen und Baulücken in der Dresdner Stadtentwicklungsplanung [Wastelands and gaps between buildings in Dresden's urban development planning]. Internal Report for Planning Department of LH DD

LH DD (Landeshauptstadt Dresden), Umweltamt (2018) Landschaftsplan der Landeshauptstadt Dresden, Mai 2018. Beschluss des Stadtrates vom 17. Mai 2018. [Landscape plan of the City of Dresden, May 2018. Resolution of the city council of 17 May 2018]. https://www.dresden.de/media/pdf/umwelt/LP_Erlaeuterungstext_LP2018n.pdf. Accessed 16 Ap 2020

LH DD (Landeshauptstadt Dresden) (2019) 1.2.3 Bevölkerung laut Melderegister 1989 bis 2018. [population according to register of residents]. https://www.dresden.de/media/pdf/statistik/Statistik_1201_10_E2013-1981_Melderegister.pdf. Accessed 16 Ap 2020

Mathey J, Rink D (2010) Urban wastelands—a chance for biodiversity in cities? Ecological aspects, social perceptions and acceptance of wilderness by residents. In: Müller N, Werner P, Kerlcey J (eds) Urban biodiversity and design. Conservation science and practice series. Wiley-Blackwell, Oxford, pp 406–424

Mathey J, Rink D (2020) Greening brownfields in urban redevelopment. In: Meyers R (ed) Encyclopedia of sustainability science and technology. Springer, New York. https://doi.org/10.1007/978-1-4939-2493-6_211-5

Mathey J, Kochan B, Stutzriemer S (2003) Biodiversität auf städtischen Brachflächen? Planerische Aspekte naturverträglicher Folgenutzungen [Biodiversity on urban wastelands? Planning aspects of nature-compatible after-use]. Bayerisches Landesamt für Umweltschutz, Evangelische Akademie Tutzing (eds) StadtNatur – Bedeutung der Stadt für die Natur und der Natur für die Stadt, Augsburg, pp 47–57

Mathey J, Rößler S, Banse J, Lehmann I (2014) How can urban brownfields contribute to climate adaptation and human well-being in cities? In: Korn H, Stadler J, Bonn A, Bockmühl K, Macgregor N (eds) Proceedings of the European conference "Climate change and nature conservation in Europe—an ecological, policy and economic perspective", Bonn, pp 114–115

Mathey J, Rößler S, Banse J, Lehmann I, Bräuer A (2015) Brownfields as an element of green infrastructure for implementing ecosystem services into urban areas. J Urban Plan Dev 141(3):A4015001-1-13

Mathey J, Artmann M, Bastian O, Rößler S (2017) Dresden: The Leitbild of a compact city in an ecological network. In: Grunewald K, Li J, Xie G, Kümper-Schlake L (eds) Towards green cities—urban biodiversity and ecosystem services in China and Germany. Cities and Nature, Springer

Mathey J, Arndt T, Banse J, Rink D (2018) Public perception of spontaneous vegetation on brownfields in urban areas. Results from surveys in Dresden and Leipzig (Germany). Urban for Urban Green 29:384–392

Muratet A, Machon N, Jiguet F, Porcher E (2007) The role of urban structures in the distribution of wastland flora in the greater Paris area, France. Ecosystems 10(4):661–671

Naumann S, McKenna D, Kaphengst T, Mav P, Rayment M (2011) Design, implementation and cost elements of Green Infrastructure projects. Final report to the European Commission. DG Environment, Berlin, Brussels

Németh J, Langhorst J (2014) Rethinking urban transformation: temporary uses for vacant land. Cities 40(b):143–150

Niemelä J, Saarela SR, Södermann T, Kopperoinen L, Yli-Pelkonen V, Väre S, Kotze DJ (2010) Using the ecosystem services approach for better planning and conservation of urban green spaces: a Finland case study. Biodivers Conserv 19(11):3225–3243

Pediaditi K, Doick KJ, Moffat AJ (2010) Monitoring and evaluation practice for brownfield, regeneration to greenspace initiatives. A meta-evaluation of assessment and monitoring tools. Landsc Urban Plan 97(1):22–36

Plappert M-L (2013) Ökosystemdienstleistungen urbaner Brachflächen und potentieller Freiraumnutzungen – Bewertung aus Sicht einer nachhaltigen Stadtentwicklung [Ecosystem services of urban brownfields and potential open space uses—evaluation from the perspective of sustainable urban development]. Masterarbeit an der Fakultät Umweltwissenschaften. Masterstudiengang „Raumentwicklung und Naturressourcenmanagement an der Technischen Universität Dresden

Rall EL, Haase D (2011) Creative intervention in a dynamic city: a sustainability assessment of an interim use strategy for brownfields in Leipzig, Germany. Landsc Urban Plan 100:189–201

Rebele F (2013) Differential succession towards woodland along a nutrient gradient. Appl Veg Sci 16:365–378

Rebele F, Dettmar J (1996) Industriebrachen. Ökologie und Management [Industrial brownfields. Ecology and management]. Eugen Ulmer, Stuttgart

Riley CB, Perry KI, Ard K, Gardiner MM (2018) Asset or liability? Ecological and sociological tradeoffs of urban spontaneous vegetation on vacant land in shrinking cities. Sustainability 10:2139

Rink D, Arndt T (2016) Investigating perception of green structure configuration for afforestation in urban brownfield development by visual methods—a case study in Leipzig, Germany. Urban for Urban Green 15:65–74

Rößler S, Mathey J (2014) Potenziale urbaner Brachflächen für den Umgang mit dem Klimawandel in der Stadt [Potentials of urban wastelands for dealing with climate change in the city]. In: Wende W, Rößler S, Krüger T (eds) Grundlagen für eine klimawandelangepasste Stadt- und Freiraumplanung [Basics for urban and open space planning adapted to climate change]. REGKLAM-Publikationsreihe, 6, Berlin, Rhombos, pp 57–75

Rößler S, Mathey J (2018) Brachfläche, Konversionsfläche [Definition of the terms wasteland and conversion area]. In: ARL – Akademie für Raumforschung und Landesplanung (ed) Handwörterbuch der Stadt- und Raumentwicklung [Hand dictionary of urban and spatial development]. Verlag: Akademie für Raumforschung und Landesplanung, Hannover, pp 293–307

Rößler S, Böhme E, Arndt T (2019a) Biologische Vielfalt und Naturschutz im Förderprogramm Stadtumbau [Biological diversity and nature conservation in the urban redevelopment funding program]. Natur und Landschaft 94 2019/12:517–523

Rößler S, Böhme E, Mathey J, Knippschild R (2019b) Biologische Vielfalt und Stadterneuerung [Biodiversity and urban redevelopment]. In: Altrock U, Kurth D, Kunze R, Schmidt H, Schmitt G (eds) Programmatik der Stadterneuerung. Jahrbuch Stadterneuerung; 2019 [Program of urban renewal. Yearbook Urban Renewal; 2019]. Springer VS, Wiesbaden, pp 171–190

Rupprecht CDD, Byrne JA, Ueda H, Lo AY (2015) "It's real, not fake like a park": Residents' perception and use of informal urban green-space in Brisbane, Australia and Sapporo, Japan. Landsc Urban Plan 143:205–218

Schadek U, Strauss B, Biedermann R, Kleyer M (2009) Plant species richness, vegetation structure and soil resources of urban brownfield sites linked to successional age. Urban Ecosyst 12:115–126

Schemel HJ, Reidl K, Blinkert B (2005) Naturerfahrungsräume im besiedelten Bereich. Ergebnisse eines interdisziplinären Forschungsprojektes [Nature experience places in populated areas. Results of an interdisciplinary research project]. Naturschutz und Landschaftsplanung 37(1):5–14

Sukopp H (2005) Welche Biodiversität soll in Siedlungen erhalten werden? [Which biodiversity should be conserved in urban areas?]. CONTUREC 1:15–18

Tian Y, Jim C, Tao Y (2012) Challenges and strategies for greening the compact city of Hong Kong. J Urban Plan Dev 138(2):101–109

Tzoulas K, Korpela K, Venn S, Yli-Pelkonen V, Kazmierczak A, Niemelä J, James P (2007) Promoting ecosystem and human health in urban areas using Green Infrastructure: a literature review. Landsc Urban Plan 81(3):167–178

Unt A-L, Bell S (2014) The impact of small-scale design interventions on the behavior patterns of the users of an urban wasteland. Urban for Urban Green 13(1):121–135

UrbanNBS-Team (eds) (2020) Mehr biologische Vielfalt in Städten und Gemeinden – Eine Arbeitshilfe zur Erstellung kommunaler Biodiversitäts-strategien [More biological diversity in cities and communities—a work tool for the development of municipal biodiversity strategies]. DUH, Radolfzell

Völksen G (2005) Vom Brachgelände zum „Ruderalpark". Der Leinepark in Göttingen: eine Alternative zu herkömmlichen Parkanlagen. [From wasteland site to a "ruderal park". The Leinepark in Göttingen, Germany: an alternative to traditional parks]. Stadt und Grün 1/2005:30–33

Washbourne CL, Goddard MA, Le Provost G, Manning DAC, Manning P (2020) Trade-offs and synergies in the ecosystem service demand of urban brownfield stakeholders. Ecosyst Serv 42:101074. https://doi.org/10.1016/j.ecoser.2020.101074

Wende W, Wojtkiewicz W, Marschall I, Heiland S, Lipp T, Reinke M, Schaal P, Schmidt C (2012) Putting the plan into practice—implementation of proposals for measures of local landscape plans. Landsc Res 37(4):483–500

Werner P (2019) Wissensstand urbane Biodiversität [State of knowledge on urban biodiversity]. Projektbericht urbane Biodiversität im Verbundprojekt „Städtische Grünstrukturen für biologische Vielfalt – Integrierte Strategien und Maßnahmen zum Schutz und zur Förderung der Biodiversität in Städten"

Wirth P, Chang J, Syrbe R-U, Wende W, Hu T (2018) Green infrastructure: a planning concept for the urban transformation of former coalmining cities. Int J Coal Sci Technol 5(1):78–91

Wittig R (2010) Biodiversity of urban-industrial areas and its evaluation—a critical review. In: Müller N, Werner P, Kerlcey J (eds) Urban biodiversity and design. Wiley-Blackwell, Oxford, pp 37–55

Wittig R, Zucchi H (eds) (1993) Städtische Brachflächen und ihre Bedeutung aus der Sicht von Ökologie, Umwelterziehung und Planung [Urban wastelands and their importance from the perspective of ecology, environmental education and planning]. Geobotanische Kolloquien 9, Frankfurt

Zhong Q, Zhang L, Zhu Y, Konijnendijk van den Bosch C, Han J, Zhang G, Li Y (2020) A conceptual framework for ex-ante valuation of ecosystem services of brownfield greening from a systematic perspective. Ecosyst Health Sustain. https://doi.org/10.1080/20964129.2020.1743206

Dr. rer. nat. Juliane Mathey studied biology and since 1992 has been working at the Leibniz Institute of Ecological Urban and Regional Development (IOER), Dresden (Germany). There she is head and a co-worker of various research projects dealing with urban wastelands, urban green spaces, and adaptation to climate change in a national and European context. Her research interests lie in the field of urban regeneration and development of urban wastelands with focus on urban biodiversity, ecosystem services, concepts for green infrastructure as well as on human perception and acceptance of wastelands.

Dr.-Ing. Stefanie Rößler graduated in landscape architecture at Technische Universität Dresden (Germany). She has been working since 2003 at the Leibniz Institute of Ecological Urban and Regional Development (IOER). Her research interest is in urban green space development regarding challenges of demographic change, climate change, and biodiversity development. As a senior researcher she is focusing on formal and informal planning instruments and processes in urban development.

From Isolated Wastelands to Informal Open Spaces Connected to a Metropolitan Park System

Analyzing the Fluctuation of the Urban Pressure on Semi-natural Sites Depending on Their Location Within Five Green Network Projects for Brussels (Belgium)

Séréna Vanbutsele

Abstract In a period of growing urbanization and growing concern for environmental resilience, any design framework for sustainable cities includes a strategy on open spaces. One of the most common strategies is the metropolitan park system. It interconnects open spaces to form a green infrastructure on a metropolitan scale. This strategy embodies the landscape inversion approach where ecological and landscape processes, through open spaces, shape urban development. Meanwhile, urban wastelands are gaining increasing attention. When transformed into informal open spaces, they are increasingly recognized for their ecological and social values. However, wastelands keep being highly coveted for urbanization. In addition, little has been said about the important role of wastelands in the formation of green continuities on a metropolitan scale. This chapter foregrounds the question of wastelands by focusing on the intensity of the urban pressure on three semi-natural spaces in Brussels compared with their location within or without five green network projects for Brussels. The case of semi-natural spaces in Brussels is analyzed, as these spaces were previous urban wastelands that have been progressively transformed into informal open spaces. The analysis shows that the more a site is integrated into a park system, the lower is the expectation to build on it, and vice versa. It is important to specify that the integration of a site into a park system does not involve the loss of its informal characteristics. Additionally, case studies highlight the key role of wastelands in developing cross-boundary strategies and linking the core city with its hinterland.

Keywords Landscape inversion · Green infrastructure · Open space · Ecology · Planning

S. Vanbutsele (✉)
Haute école d'ingénierie et d'architecture, Fribourg, Switzerland
e-mail: Serena.Vanbutsele@hefr.ch

F. Di Pietro and A. Robert (eds.), *Urban Wastelands*, Cities and Nature,
https://doi.org/10.1007/978-3-030-74882-1_14

1 Introduction and Organization of the Chapter

In this contribution, the question of wastelands is related to the search for green continuities. What could be the role of wastelands in relation to a metropolitan park system? Could the location of a wasteland—included, close or disconnected—to a park system influence the real estate pressure on this space? And, could the presence of wastelands reinforce or weaken a park system project?

In order to do so, the chapter is organized into five parts.

The first part defines the concept of open space and exposes different theoretical approaches. Each of these theoretical trends considers open spaces from a particular point of view and with different functions that range from mineral public space designing the urban fabric to the support of green infrastructures, including more prestigious spaces representative of the bourgeoisie. All these theoretical approaches are gathered into the concept of *landscape inversion* as they all share a common approach to open spaces as a crucial element for planning.

The second part exposes the question of wastelands as a particular type of open space and the question of green continuities such as a park system. In order to clarify the concept of the park system, the example of the Boston metropolitan park system is described as it is renowned as an inspirational example of such green infrastructures.

The third part deepens the case study of semi-natural spaces in Brussels. Firstly, it shows how semi-natural spaces in Brussels have become informal open spaces with their own specificities. Secondly, it shows the pressure the real estate market exerts on semi-natural sites leading to different scenarios—from the extreme urbanization of a site to its preservation. Thirdly, it exposes five open space strategies applied in Brussels from the end of the nineteenth century to the present. Each of these strategies embodies the concept of a park system. The semi-natural spaces are foregrounded in each of these strategies.

The fourth part exposes the correlation between the level of urban pressure on semi-natural sites and their inclusion or disconnection with park system projects.

Finally, the fifth part discusses three major issues in planning: considering the urbanization or preservation of a wasteland (1); acknowledging the informal specificity of a wasteland (2); and planning beyond the administrative boundaries of a city. These three issues could be considerably influenced by integrating wastelands into a park system.

2 Planning Open Spaces Through the *Landscape Inversion* Approach

2.1 *From Open Space…*

In urban planning, the term *open space* refers to a large diversity of unbuilt spaces. It is a type of land use dominated by natural land use (Maruani and Amit-Cohen

2007; Banzo 2015). Open spaces are characterized by a close relation with natural soil (Legenne et al. 2010). They are usually defined in complementarity to the urban fabric that represents a higher level of intervention in the ecosystem, altering the landscape and interfering with natural processes (Maruani and Amit-Cohen 2007). There is a large variety of open spaces: green spaces, wooded areas, agricultural or natural spaces, parks and gardens, sport fields… They are public or private. They result either from production (forest, agriculture), from planning (parks), from protection (protected natural areas) or from abandonment (Banzo 2015). Wasteland and brownfield sites are part of this last category. A brownfield site is specifically related to an abandoned former industrial site, raising questions of soil decontamination, for example, while wasteland, at the core of this book, refers more generally to any abandoned vacant site where vegetation develops spontaneously (Bonthoux et al. 2014). In a period of growing urbanization and growing concern for sustainability and environmental resilience, open spaces are key resources in land use planning (van der Valk and van Dijk 2009). As open spaces support natural systems, they provide numerous ecosystem services. They provide water supply, one-day recreation, flood control, farmland, wetland benefits, soil erosion/sedimentation protection, biodiversity, waste absorption/breakdown, aesthetics or inspiration (Forman 2010, p. 317), soft mobility support, etc.

Maruani and Amit-Cohen classified open spaces into two major categories that may interfere with each other: the provision of recreation and other services to society and the conservation of natural values (Maruani and Amit-Cohen 2007). If social and cultural values of open spaces have always included attitudes toward nature and the desire for contact with it, contemporary understanding of ecology offers new insights into ways to serve both human needs and the broader ecological framework of urban open space structures (Ward Thompson 2002). Nowadays, there is a growing concern directed at informal urban greenspaces (Rupprecht and Byrne 2014). The informal uses and uncertainty of those spaces make them an experimental field for group practices and innovative approaches (Denef et al. 2015). They are open from a spatial point of view as well as a social point of view. Therefore, open spaces in general and even more so wastelands and informal open spaces appear to be places of opportunity.

2.2 *… To Landscape Inversion*

Different planning approaches highlight the crucial importance of open spaces. Among these approaches, the *culturalist model* (Choay 1965) and *new urbanism* (Dupuis 2009) focus on urban form and urban morphology. Unlike the previous definition of open space, for them the concept of open space is not strongly related to the notion of a landscape supporting natural dynamics and the omnipresence of vegetation but, rather, open spaces are considered for the way their spatial openness contrasts with their built environment. In these approaches, the concept of open space usually refers to public spaces as composed essentially of minerals such as a square, a plaza, and a street. The authors argue that the urban fabric results from the combination of three entities: street networks (1), demarcated plots of land (2) and the location of buildings (3). The organization of these three entities generates a specific alternation of open spaces and built spaces characterizing the urban fabric. Since open spaces contribute to a global scenography of the urban fabric, the way buildings are located and related to each other appears to be more crucial than the design of the buildings themselves (Sitte 1889; Rowe and Koetter 1978; Panerai 2002; Salat 2011). For Camillo Sitte, a building has to contribute to a coherent combination rather than stand alone for itself (1889).

During the nineteenth century, following urban expansion and population growth, a tradition of vegetal urbanism arose (Stefulesco 1993). Numerous occidental cities became aware of the importance of their image where open spaces played a significant role. The period witnessed the emergence of the hygienist movement, which also emphasized the crucial role of open spaces and greenery. Parks, squares, green boulevards and promenades, shape the upcoming urbanization of new districts. Open spaces, and especially public parks, are prestigious spaces depicting the urban fabric. Landscape architects and horticulturalists played a key role in organizing urban spaces (Marot 1995). In many cases, public parks are interrelated with other open spaces, such as streets, avenues, riverbanks, natural reserves…, contributing to upscale planning of open spaces. Major open space models are developed: park systems (such in the Emerald Necklace Park System in Boston at the end of the nineteenth century), green belts (mostly applied in England as a response to the uncontrolled growth of cities at the end of the nineteenth and the beginning of the twentieth centuries) and green fingers (such as in Copenhagen's five-finger plan initiated in 1947). These models define the shape of open spaces, which in turn impacts the spatial arrangement of the adjacent built-up zones (Maruani and Amit-Cohen 2007).

Since the 2000s, the practices and know-how of landscape architects, actively used in the nineteenth century but partly forgotten during the twentieth century, have been re-introduced in planning and urban practices and extended with expertise on ecological planning (McHarg 1969) and landscape ecology [Forman and Godron 1986]. Multiple designations embody this trend: *landscape alternative* (*alternative du paysage)* (Marot 1995), *sub-urbanism* (Marot 2003), *landscape urbanism* (Waldheim 2005) and *ecological urbanism* (Mostafavi et al. 2010). We gather those trends into the concept of *landscape inversion* describing an approach where open spaces are structural elements; backbones for the development of the territory (Vanbutsele 2017) (see Box 1). Nowadays, the principles of *landscape inversion* are commonly shared and debated among academics and practitioners—architects, planners and landscape architects. However, in practice, the conventional approach to urban development, driven by economic sense, population projections, built infrastructures and architectural objects, still defines the prevailing trends and takes over the *landscape inversion* approach.

Box 1: Landscape inversion approach

We coined the term "inversion paysagère" translated into English as "landscape inversion" referring to Elena Cogato Lanza who describes the inversion of the urban landscape (l'inversion du paysage urbain) as an attempt theoretically to reverse the relations between built areas and green places, center and outskirts, city and countryside (Cogato Lanza 2005).

The landscape inversion reverses the functionalist paradigm that conceives urban planning as a perpetual adaptation of the territory to the needs of the society and the economy. On the contrary, the landscape inversion approach considers landscape and natural resources as a matrix from which emerges the project of territory. In that perspective, open spaces are structural elements for the development of the territory.

The landscape inversion approach refers to multiple scales, from pocket parks included in a dense urban fabric to regional strategies such as the metropolitan park system.

According to Kongjian, there is a positive and negative approach—and landscape inversion belongs to the latter:

> The conventional ('positive') approach to urban development in which urban growth is defined by built, gray infrastructure comprised of roads and pipes that provide services for the urban development. The negative approach considers the green and unbuilt ecological infrastructure that provides ecosystem services and acts as a framework to define urban growth and urban forms across all scales (Kongjian 2010, p. 59).

We detect four characteristics of the landscape inversion approach (Vanbutsele 2017, p. 127-133): (i) the conditions for a project should emerge from the site and not the opposite; (ii) as territorial development is dynamic and not static, the process is more relevant for planning than a static master plan. "This emphasis on urban processes is not meant to exclude spatial form but rather seeks to construct a dialectical understanding of how it relates to the processes that flow through, manifest, and sustain it" (Corner 2006, p. 28); (iii) the need for an interdisciplinary approach that would also include citizen participation, bringing collective intelligence; (iv) a higher focus on relationships between objects, leading to special attention to transition, interfaces, edges and in-between spaces.

The landscape inversion approach is closely related to landscape urbanism and ecological urbanism. Those concepts describe "the aspirations of an urban practice informed by environmental issues and imbued with the sensibilities associated with landscape" (Sordi 2014, p. 6). Nevertheless, those concepts emerge in the context of North American urbanism with a strong link to North American schools, especially the University of Pennsylvania and the Harvard Graduate School of Design. The concept of landscape inversion intends to include contributions from landscape and ecological urbanism and extends it to European approaches such as the landscape alternative (alternative du paysage) and sub-urbanism developed by Sebastien Marot (1995 and 2003). Those concepts place the site at the core of the architectural, urban and landscape practices, reversing the traditional hierarchy between site and program.

3 Two Major Challenges: Planning Green Continuities and Considering Wastelands

3.1 Planning Continuity: The Park System Model

According to the *landscape inversion* approach, natural ground, landscape and ecological processes are the lens through which to develop the territory. In this perspective, connecting open spaces appears as a priority in order to form a natural network. This green infrastructure is thus the backbone for urban development. The concept is defined through different terminologies: park system, green and blue networks, a green way, green fingers, green infrastructures…These models share the idea that open spaces are interrelated in a given geographical area according to fundamental ecological features, especially topography (ridges and valleys) and hydrography (rivers and streams). The open space model defines the shape and spatial arrangement of the adjacent built-up zone.

A park system approach had already emerged at the end of the nineteenth century when the concept was used to solve urban problems such as congestion and sanitation. Boston's Emerald Necklace (1880–1890) and the Metropolitan Park System are commonly recognized as pioneer examples of the park system. They comprise landscape architecture, engineering and urban planning, prefiguring the current search for green infrastructures in planning. The Emerald Necklace and the Metropolitan Park System of Boston embody essential potentialities of the *landscape inversion* approach and landscape urbanism: "*the ability to shift scales, to locate urban fabrics in their regional and biotic contexts and to design relationships between dynamic environmental processes and urban form*" (Corner 2006, p. 24).

The Emerald Necklace park system, designed by Frederick Law Olmsted and Charles Eliot, structures Boston (USA) in connecting different types of green spaces (parks, urban forest and arboretum). This continuity of open spaces stretches over more than 10 km and covers 445 ha. The goal of this project is manifold:

- the project framed the upcoming densification of the city,
- the green network follows the natural typologies of the territory and river banks. Its location stimulates better water management and flood resilience, and
- the system provides a recreational place with esthetic value.

Designed more than a century ago, this infrastructure has evolved though it still structures the city today. It is a destination for more than one million visitors each year (Emerald Necklace conservancy 2020).

The Emerald Necklace park system is a major element of another project at a larger scale: the Metropolitan Park System for Boston designed by Charles Eliot (Fig. 1). This project covers an entire metropolitan area: 16 km around the core city.

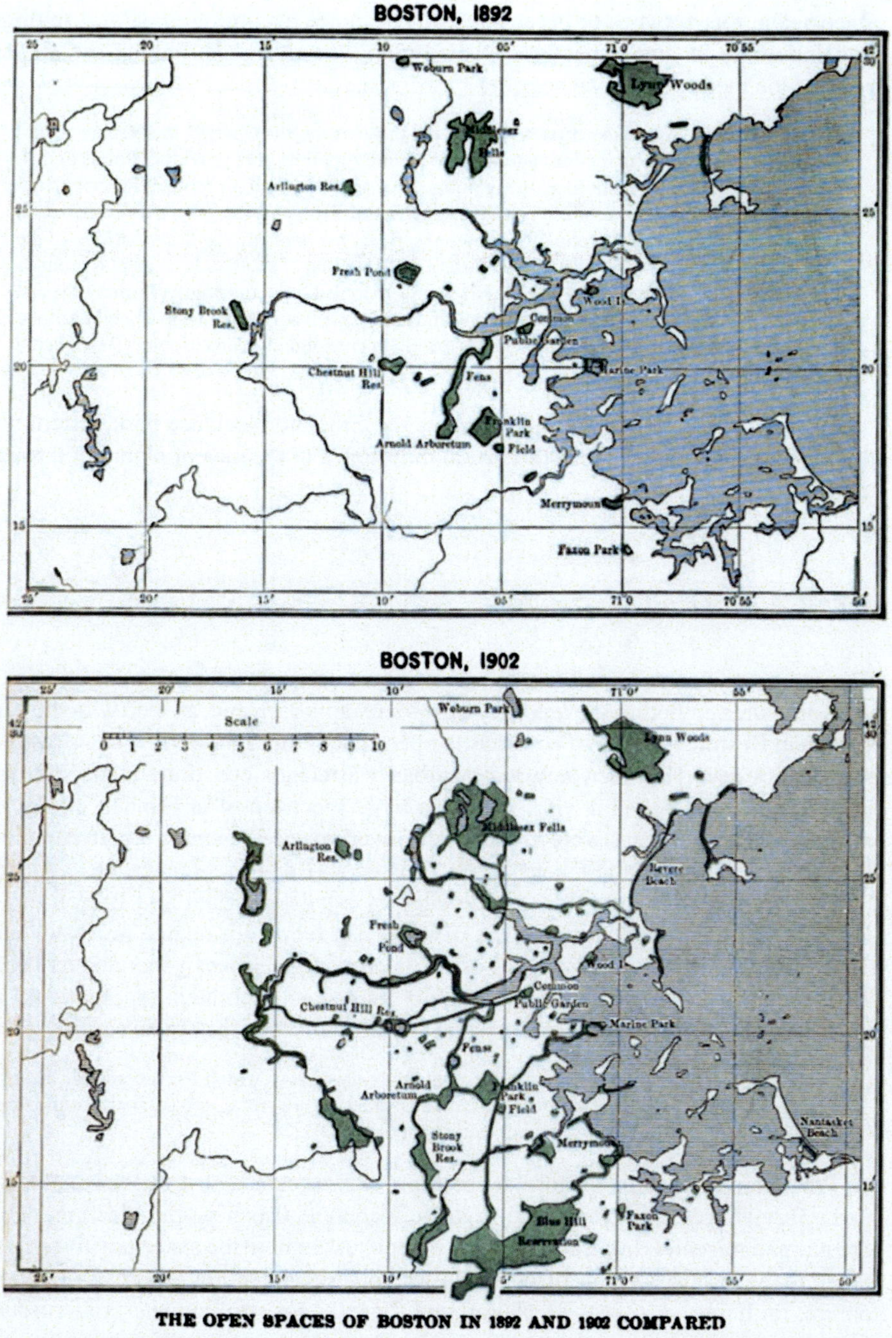

Fig. 1 Comparison of the public open spaces of Boston in 1892 and 1901. This comparison highlights how much of the Metropolitan Park System of Greater Boston had been implemented in 9 years. In Eliot 1903, p. 738

Once again, this network of green spaces is designed according to natural settings. The continuity of green spaces is located along the riverbeds, linking the highest point of the metropole to the seaside.

> Eliot proposes the following: first, to protect the three rivers [the Charles, the Neponset, and the Mystic] that are the main drainage ways of the metropolitan area providing flood control; second, to protect the seashore assuring public accessibility; third, to protect the tops of the highest hills for public accessibility and view sheds; fourth, to provide for an even distribution of large reserves ten miles around the center to serve as anchors guiding the expansion of the city; fifth, and contrasting in scale, to include a constellation of small municipal playgrounds and parks. Later, parkways and boulevards would be added for connectivity, forming the last layer of a complex system in which geology and landscape from a regional to a neighborhood scale served as the structuring element for the city's consolidation and growth (Berrizbeitia 2014, p. 43).

The sum of those characteristics makes the Emerald Necklace park system and the Metropolitan Park System of Boston milestones in the idea of planning through open spaces.

3.2 *Considering Wastelands*

The attempt to connect open spaces and structure urban development through green infrastructures embodies the *landscape inversion* approach and, as seen with the park system in Boston, the idea of connecting open spaces through such green infrastructures is not new. However, nowadays, urban wastelands constitute a specific type of open spaces to contend with. Indeed, a large spectrum of unplanned, neglected and derelict spaces that were traditionally overlooked is gaining great attention. Loose space (Franck and Stevens 2006), drosscape (Berger 2006), terrain vague (de Solà-Morales 1995), interstices at the edge of a pale (Mariani and Barron 2014) and other kinds of plots at the fringe of the urban fabric are related to wastelands. This terminology includes all types of waste and empty spaces generated by urban and economic development. These *voids* produced an alternative or generic urban landscape that is increasingly seen as the key element for urban redevelopment.

> Voids have become a lens to look at the contemporary city, to which we can add all these spaces that in fact aren't void, but that have lost their architectural, urban, social, cultural function, and therefore their meaning (Sordi 2016, p. 91).

When transformed into informal open spaces, wastelands host various unplanned and informal uses. Some could be negatively connoted such as drug dealing, delinquency and mischief, littering …, while other uses connote the space positively. The latter is the case when the place is known for experimenting and learning about nature, for being an adventure playground for kids, for strolling and crisscrossing, for the development of biodiversity, as a shortcut for inhabitants from one district to another… The specificities of wastelands and informal open spaces are potentially valuable from social, ecological and spatial points of view.

The social potential of a wasteland is related to the experience and perception of its livability. Already throughout the nineteenth and twentieth centuries, wastelands and derelict lands within and at the fringe of the urban fabric were used as informal green spaces. They were crucial in shaping the citizens' perception of the urban landscape and the quality of life in the city (Tritsmans 2014).

From a planning perspective, all kinds of spontaneous and alternative (temporary) activities that take place in urban locations awaiting official redevelopment are able to contribute significantly to the future redevelopment of these places. They generate initiatives, creativity and innovation. They are characterized by an open, nurturing and fertile nature (Ward Thompson 2002; De Smet 2013; Oswalt et al. 2013).

From an ecological point of view, unplanned landscapes contribute to biodiversity conservation (Bonthoux et al. 2014). Clément points out that if some spaces are seen as *void* from an urban perspective, they are actually *full* from a biological perspective (Clément 1991). As the urban habitat is heavily planned by conventional urban policies and the rural habitat by intensive farming and rural development, unplanned landscapes offer a refuge for wildlife that is threatened anywhere else (Shoard 2002).

From an ecological perspective, Brun and co-authors (2015) demonstrate the importance of derelict spaces as components of regional conservation networks such as the green and blue corridors in France. Although it is critical to safeguard large protected open and natural spaces from development, these areas alone are not sufficient for biodiversity conservation. Smaller green spaces such as derelict spaces spread over the urban fabric considerably contribute in forming ecological continuities (Brun et al. 2015).

However, from a planning perspective, little has been said about the contribution of wastelands at a metropolitan scale. Traditional planning approaches keep considering wastelands as a local matter, either for built redevelopment or for conventional green spaces. On the one hand, if they are considered as urban plots suitable for redevelopment, their urbanization will favor the densification of the city and prevent urban sprawl by developing the city inside itself. On the other hand, if they are considered as conventional green spaces like public parks or even natural reserves, they play an indispensable role in the livability and biodiversity of a dense city. Their messy and unregulated appearance generally favors changes as they appear as places *where something better has to be done*. Nevertheless, in both scenarios—consisting either in building them or in keeping them open—the unregulated character of the wasteland is lost. Moreover, when their transformation into a new district or a tidy green space serves as a strategy of city marketing and the improvement of the image of the city, this phenomenon may lead to gentrification and the eviction of a part of the citizens. Even if the ambiguity, informality and malleability of wastelands seem valuable, planners are poorly equipped to deal with those specificities, and wastelands are usually left vulnerable to urban pressure.

4 The Role of Wastelands in a Metropolitan Park System: The Case of Semi-natural Spaces and Green Network Projects in Brussels

In this contribution, the question of wastelands is related to the search for green continuities. What could be the role of wastelands in relation to a metropolitan park system? Could the location of a wasteland—included within, close to or disconnected from—to a park system influence the real estate pressure on this space? And, could the presence of wastelands reinforce or weaken a park system project? To do so, semi-natural spaces in Brussels have been analyzed as a case study. They offer an example of former wastelands that have partly evolved into informal open spaces and that have been partly urbanized. The real-estate pressure occurring on three semi-natural sites has been analyzed and the results have been compared with several proposals of green networks: the plan Besme (1863), the green–blue network (1995), the structuring green spaces from the Sustainable Regional Development Plan (2014) and its pre-version (2012) and the proposal for a metropolitan park system (2017).

4.1 Semi-natural Spaces in Brussels: Former Wastelands Turned into Informal Open Spaces

During the second half of the twentieth century, in Brussels-Capital Region, some spaces spontaneously evolved into semi-natural spaces because of the withdrawal of human activities, such as extraction or farming, and the delay of redevelopment projects. Therefore, semi-natural spaces embody the major characteristics discussed in this article. As nature is taking over human activities, semi-natural spaces are a form of nature in the city, while they are also considered as wasteland highly coveted for urban development. This blurry situation allows an amalgam of informal uses and appropriations by citizens and nature. They share common characteristics with informal open spaces "*such as uncertainty with regard to land tenure, conservation status, maintenance regimes, use, regulation and legitimacy*" (Rupprecht and Byrne 2014, p. 598).

We identified 16 semi-natural spaces spread over 336 ha, all located at the fringe of Brussels-Capital Region (Vanbutsele 2017, p. 84). Until 2001, semi-natural spaces were assigned as urban land stock for development. They appeared as white spaces on official plans. The color white represented their availability for major urban projects such as highways, railways, public equipment or housing. Nevertheless, urban projects were delayed for bureaucratic and financial reasons as well as due to public opposition. The history of semi-natural spaces in Brussels is highly related to the implication of inhabitants and non-governmental associations advocating in favor of the preservation of natural resources and green spaces as a major factor in the quality of life in the city. Some semi-natural spaces have witnessed a history of urban struggles. Citizens were fighting to preserve an open space where landowners and municipalities were projecting building blocks, large infrastructures like prisons, railroads or highways (Vanbutsele 2018). Nature and local communities took over those spaces, giving rise to informal uses like crisscrossing, playing, motocross, gardening, harvesting, drug dealing, dwelling, sightseeing, botanical surveying… Since 2001, semi-natural spaces have been officially assigned as "*green zones*" in the official land use plan. In theory, this recognition secures semi-natural spaces as open spaces. In practice, these spaces are not maintained by regional authorities, they have no clear boundaries and there remain ambiguities concerning ownership. Several landowners may share a small piece of land, thereby freezing any project. This is, for example, the case at the Scheutbos where more than 200 owners share a piece of land of less than eight hectares (Fig. 2).

Those characteristics—land dedicated for urbanization for a period but nowadays officially recognized as a green zone, located at the fringe areas of the city, no official–regional management and confusing ownership—complicate the classification of semi-natural spaces. They are commonly seen as derelict spaces, spaces waiting for urbanization, wastelands, … For example, until 2010, several semi-natural spaces were commonly called the *far west* by local residents alluding to the lawless character of the place (Ost et al. 1993).

In Brussels, those spaces are in high demand for urbanization. Since the 2000s, after 30 years of decline, Brussels' population has been growing. Between 2007 and 2012, the population growth was so great that it was qualified as a demographic boom. Current predictions keep forecasting an increase in inhabitants for the coming years (Institut Bruxellois de statistique et d'analyse 2019). The public authorities seek places to accommodate this population growth. Projects for housing, equipment, new infrastructures and new public spaces tend to bloom wherever land is available. This context of growing demand is correlated with a strict limitation of the territory of the Brussels-Capital Region. Due to institutional reasons, Brussels' administrative boundaries are locked in between the Wallonia and Flemish regions. The planning of Brussels-Capital Region is thus strictly limited to 162 km^2 even if essential concerns of the city go beyond administrative boundaries (workers living outside the city commute every day to Brussels, most of the food is produced outside the Region, as well as drinking water, energy…). The growing demand for land correlated with territorial insulation leads to increased pressure from real estate developers. In Brussels, any square meter is in high demand. In this context, open spaces that are not strongly embedded, such as wastelands or informal open spaces, are weakened. They are coveted as land stock for future urban development, recreational parks or areas of natural conservation. In any case, the unregulated character of the space is lost. Authors have, however, pointed out the quality of the unplanned character of semi-natural spaces (Ost et al. 1993; Ibge-Bim and Cooparch-R.U. 2004; Vanbutsele 2017). For Ost and co-authors (1993), the hybrid and messy character of semi-natural spaces is positive because it allows particular uses that cannot take place anywhere else. Semi-natural spaces have an open and unfinished character. The challenge of planning is to respect these specificities and not to transform semi-natural spaces into built areas or into well-maintained, traditional green spaces (Ost et al. 1993, p. 12 and 24). To this end, Ost and co-authors suggest fostering interactions between

Fig. 2 Example of the high fragmentation of ownership—semi-natural site of the Scheutbos. The south part of the site is divided into multiple cadastral parcels, whereas this fragmentation is not visible in the field as shown in the picture (see Fig. 3). Author: Vanbutsele

multiple stakeholders to preserve the diversity of uses and functions of semi-natural spaces (Ost et al. 1993). In order to preserve the fertile fuzziness of semi-natural spaces, Vanbutsele (2017) suggests improving the morphological anchoring of the sites. She highlights that plots located at the edges of semi-natural spaces are on the first line of urbanization. Edges have a high potential for improving the interaction between nature and the city (Vanbutsele 2017).

Fig. 3 Case 1: Semi-natural site of the Scheutbos—September 2013 © Vanbutsele

4.2 Urban Pressure on Three Semi-natural Sites

As the institutional and urban context of Brussels leads to an increased demand for the transformation of spaces such as semi-natural sites, the urban pressure occurring on three semi-natural sites has been analyzed (Boxes 1, 2 and 3). As illustrated in Figs. 3, 4 and 5, the three sites represent the diversity of situations of semi-natural spaces in terms of habitats, sizes, historical evolution and urban pressure. Their sizes vary from 10 to 70 ha. Some are wooded and others are covered with grassland or occupied by urban gardens. Some sites are former quarries, farmlands or dumps.

Box 2: Case 1—Scheutbos—low urban pressure

The Scheutbos is located on the western border of the Brussels-Capital Region. The site covers 57 ha. It still contains agricultural activities, and traces of rural and traditional landscapes such as hedges, aligned willows and orchards, like the landscape of the adjacent Pajottenland. The site is covered with wooden patches, marshland, grassland, urban allotments and an urban park of six hectares. Projects for its urbanization have existed since 1947, but the site, originally of 60 ha, remains open, with the exception of 2 ha constructed during the 1980s. In the 1990s, while the site was recognized as a land stock for future urbanization, civic opposition obtained the recognition of part of the site as a natural reserve. The history of the site, with multiple episodes about its official recognition, has led to a certain confusion. Some landowners who are also developers are expecting a change in the law to allow them to increase the value of their land by building upon it. Other landowners are not even aware of what is happening on their land, while some non-governmental organizations barely know the ownership of the land they maintain. Today, the site is entirely recognized by the public authorities as a green space, although there are informal discussions for building on some plots at the edge of the site (2.4 ha out of the 57 ha). Non-governmental organizations are particularly alert about these proposals and react quickly to prevent the urbanization of the site (Vanbutsele 2019).

Box 3: Case 2—Val du Bois des Béguines—mitigated urban pressure

The Val du Bois des Béguines is located on the northern border of the Brussels-Capital Region. There is no official delimitation of the site. The core of the site spreads over 6.5 ha. It is surrounded by numerous open spaces covering more than 70 ha. The majority of the site is secured as a green zone (59 ha), while some plots are assigned as buildable land (11 ha). The site has a multifunctional character marked by an intermingled pattern of uses. The site was intended to be urbanized since the beginning of the twentieth century, while it was used for agriculture. The municipality acquired most of the land of the site in that perspective. In 2014, an urban forest was planted in the core of the site. This project intends to reinforce the ecological value of the place while maintaining public accessibility. Some sports fields are located close to a swamp zone. The ecological value of this swamp strongly depends on the upper part of the site with the urban forest. Until 2013, a private company set up an adventure trail in trees for commercial purposes with leisure activities. Some plots are still used for food production by a farmer as well as citizens and the surroundings of the site are being developed for residential construction (Vanbutsele et al. 2018).

Box 4: Case 3—Val d'Or—strong urban pressure

The Val d'Or is located on the eastern border of the Brussels-Capital Region (Belgium). The site was formerly a quarry, which evolved into a landfill. In the 1970s, 40 ha of the place were assigned as land stock. From the 1960s till today, urbanization has been intense in the neighborhood of the site and today only 8 hectares out of 40 remain open. The site is partly wooded and partly covered with low-lying vegetation. There are numerous official and unofficial paths that crisscross the site. In the 1990s, non-governmental organizations fought to conserve the open space, but in the 2000s the land use plan assigned the majority of the zone as buildable land including the most interesting part from a biological point of view (Vanbutsele 2017, pp. 227–234). In the 2010s, negotiations took place between the regional authorities and owner-developers to exchange some plots inside the area.

The urban pressure occurring on each site is inferred by a threefold analysis: the stakeholder's analysis is compared to a morphological survey and a study of historical and administrative maps.

Fig. 4 Case 2: Semi-natural site of the Val du Bois des Béguines—July 2013 © Vanbutsele

Box 5: Methodology in the analysis of the case studies

Three types of stakeholders (landowners (1), municipal and regional public authorities (2), non-governmental organizations and users of the space (3)) have been approached through 14 interviews. The analysis traces the story of encounters in favor of the protection of the semi-natural site as well as practices and uses taking place on the site. The analysis points out how stakeholders have witnessed the transformation of the site and how they predict or would like to transform the future of the site.

The results of the stakeholder analysis were compared to the findings of a morphological survey of physical dispositions of the edges of semi-natural sites. Most of the edges appear hermetic, such as hedge, wall, mesh, backyards, garages behind houses. The anchoring of these semi-natural sites within the urban fabric is poor in terms of public spaces and does not favor interactions.

The results of the stakeholder analysis were compared to the findings of a morphological survey of physical dispositions of the edges of semi-natural sites. Most of the edges appear hermetic, such as hedge, wall, mesh, backyards, garages behind houses. The anchoring of these semi-natural sites within the urban fabric is poor in terms of public spaces and does not favor interactions.

The stakeholder analysis and the morphological survey were also compared with the historical evolution of semi-natural sites on maps. Official land use plans and cadastral maps were analyzed to trace the administrative evolution of semi-natural sites. This analysis identifies the land still available for urban redevelopment and plots that are under more pressure for urbanization projects than others (Vanbutsele 2019).

Fig. 5 Case 3: Semi-natural site of the Val d'Or—July 2013 © Vanbutsele

This threefold analysis highlights three ongoing scenarios on semi-natural sites: a laissez-faire scenario, an alternative trend scenario and a complete urbanization scenario.

1. **Laissez-faire scenario**. The most common trend partly urbanizes semi-natural sites. Urbanization continues to occur, especially on the edges of the semi-natural site. This trend reduces the size of the site while keeping a small urban green space in the center.
2. **Alternative scenario**. The second trend maintains the site as an open space in its entirety. The site persists as an entire open space. It remains within its current boundaries. Urbanization inside the site is blocked.
3. **Extreme urbanization scenario**. This trend urbanizes the semi-natural site so that it would completely disappear over time. This trend has not been encountered for current semi-natural spaces.

These three scenarios are supported by different stakeholders. They pressure the authorities to apply their favorite scenario. Thus, for example, non-governmental organizations for the protection of the environment lobby to keep the Scheutbos site in its entirety, while landowners—who are also developers—apply pressure to build on it, at least in part.

More interestingly, there is a discrepancy between an official scenario and a *most likely to happen* scenario. The urban pressure is inferred by this discrepancy. The official scenario is the future officially assigned to the site by public authorities through land use planning. The *most likely to happen* scenario is the trend detected through the threefold analysis (interviews with stakeholders, urban morphology and structure of plots).

In the case of the Scheutbos, the official result matches the *alternative scenario.* It is officially recognized as a green space in its entirety. But some plots located on its edges are informally coveted to be built upon. This scenario is discussed in the backstage of planning inducing a form of pressure in favor of constructing on the site (Fig. 6). On the other hand, the Val du Bois des Béguines is officially recognized as a smaller space than what it currently is, but informal discussions have generated pressure to recognize the site as a bigger green space. Finally, at the Val d'Or, the official scenario partly urbanizes the site, but the urbanization of the site has been so intense and rapid that one wonders if the *extreme urbanization scenario* is not occurring (Fig. 6). In other words, the Scheutbos experiences the lowest urban pressure, the Val d'Or experiences the strongest urban pressure and the Val du Bois des Béguines experiences a mitigated urban pressure.

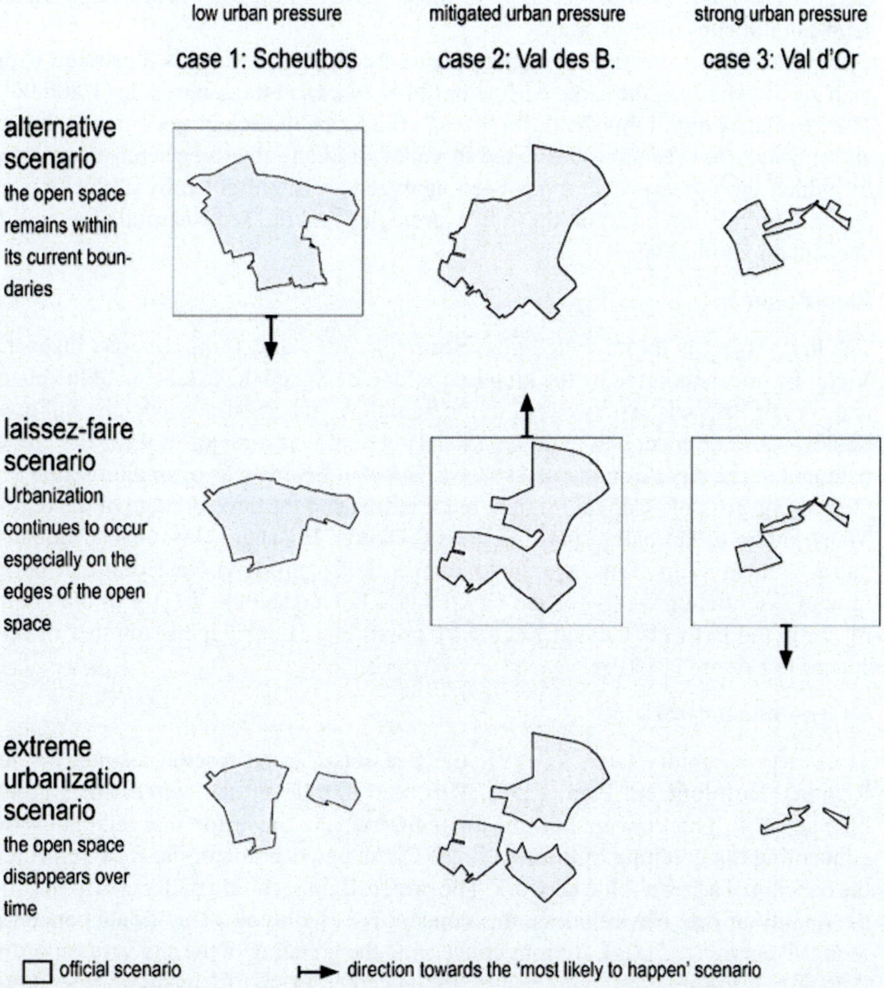

Fig. 6 Three scenarios showing three levels of urban pressure occurring on the semi-natural sites of the Scheutbos, the Val des Béguines and the Val d'Or, Author: Vanbutsele

Having determined the intensity of the urban pressure on the three semi-natural sites, we will now determine whether these three sites contribute or not to a metropolitan park system.

4.3 *Five Proposals for a Park System in Brussels*

Brussels does not have an official project for a metropolitan park system. However, different strategies of open spaces have been elaborated through time. Five of them have been considered in this research: the plan Besme (1863), the green–blue network (1995), the structuring green space from the Sustainable Regional Development Plan (2014) and its pre-version (2012) and the proposal for a metropolitan park system (2017). The analysis looks at the five maps representing each project and locates semi-natural sites on each map.

In what follows, the five strategies are presented. Each of them is illustrated with a map always showing the same perimeter (24 X 24 km) at the same scale (1/200,000). This regularity highlights the differences in the scope in each project. For each map, all semi-natural sites are represented in yellow–red and the three semi-natural sites for which the urban pressure has been analyzed are highlighted by a black square frame. The relation between these five strategies and the semi-natural sites will be deepened in point five.

Plan Besme

The first project is the so-called plan Besme. In 1863 and 1866, the road inspector, Victor Besme, supported by the king Leopold II, designed the extension of the historical core of Brussels: a network of monumental streets, tree-lined avenues and green boulevards to connect new public parks. This plan was a major tool for the embellishment of the city through open spaces. The plan Besme was of smaller scope than the present Brussels-Capital Region, but it influenced the development of the region. Many public urban parks (park de Forest, Duden, Elisabeth, Josaphat, Cinquantenaire, Woluwe, Tervuren, Jagersveld, étangs de Boitsfort…) created at that period remain essential elements for the city today. They contribute greatly to the quality of life in the Brussels-Capital Region by providing a considerable number of open spaces in a dense city (Fig. 7).

Green–blue network

More than a century later, in 1995, the Brussels-Capital Region adapted its first Regional Development Plan (RDP). This plan introduced the concept of a green–blue network. This concept became the main frame of any effort and regional action concerning the greening of the city. In 2002, the new version of the RDP reinforced the concept of a green–blue network. The project thus became a major tool to improve the quality of life. Nevertheless, this concept is based upon a traditional concentric model (Vanempten 2009). Actions concerning the greening of the city vary according to their place in relation to the center. In the center, efforts focus on re-greening; in

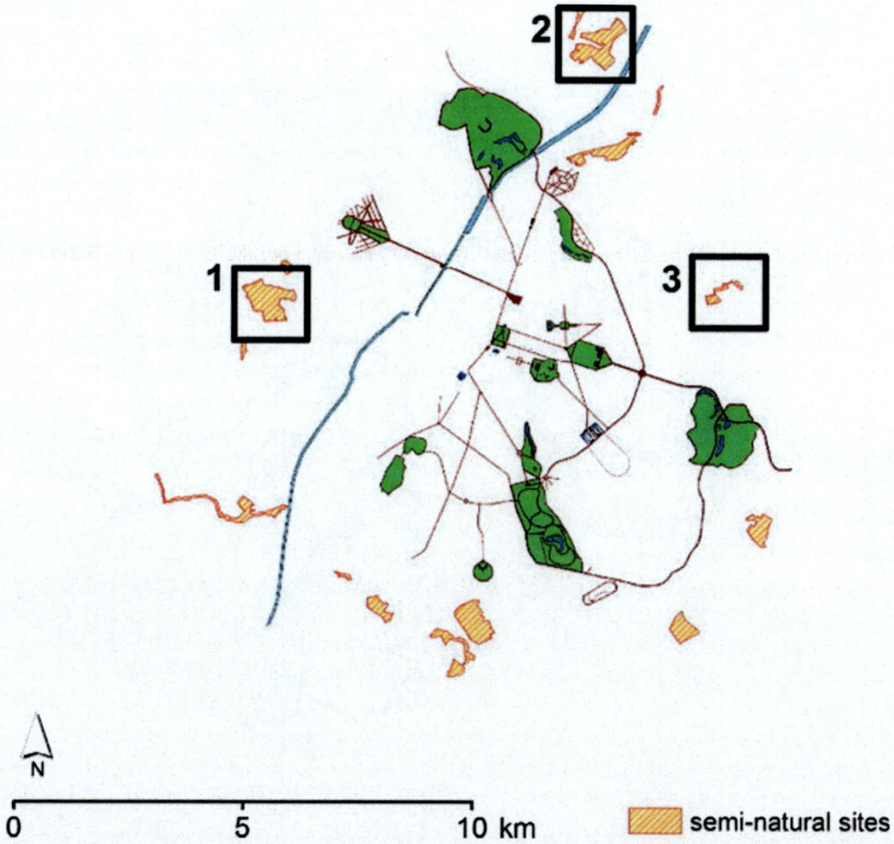

Fig. 7 Plan Besme and the semi-natural sites. Author: Vanbutsele from Billen and Duvosquel 2000

the direct periphery, the aim is to preserve residential and scenic characteristics. The first achievement of this strategy is a ring of 60 km called *green promenade* for pedestrians and bikers. This circuit connects major green spaces to the periphery (Fig. 8).

Structuring green spaces from the pre-version of the Sustainable Regional Development Plan (2012)

For the past 10 years, the need to elaborate a coherent and larger vision of Brussels and its hinterlands has become more urgent. Open spaces and landscapes appear to be a potential tool to start the elaboration of a coherent vision on both sides of the administrative boundaries. Cross-border strategies of open spaces are claimed to be an essential part of any design framework for a sustainable city. The radio-concentric vision of a green and blue network evolves into a cross-border green model more integrated into the territorial structure. Models like the green fingers or the park system are increasingly debated (Maruani and Amit-Cohen 2007). Authors

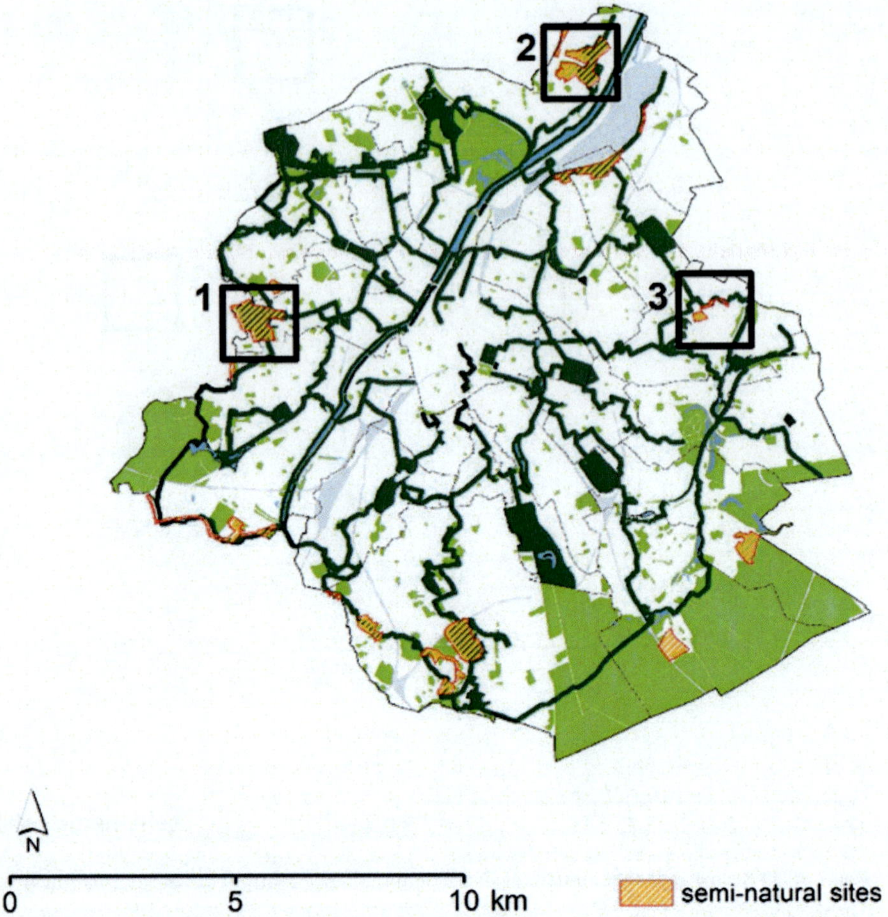

Fig. 8 Green–blue network (1995 and 2002) and the semi-natural sites. Author: Vanbutsele from Région de Bruxelles-Capitale 2002

highlight the interest of a green network that would cross boundaries and be aligned with the topographical structure of the territory (Vanderstraeten et al. 2009; Aerts 2012; Roland 2012). Brussels is structured by many watersheds. This particularity offers a high potential considering water management issues, biodiversity, climate regulation, biomass, soft mobility, recreation and landscape identity (Vanderstraeten et al. 2009, p. 1–2, 51N4E et al. 2011). A network of central and peripheral open spaces is also an opportunity to link dense urban fabrics and their inner urban parks with regional landscapes as the Brabant plateau and the Dijle valley (Vanempten 2009). From 2012, when the Brussels-Capital Region prepared a new Sustainable Regional Development Plan, strategies for a network of open spaces became part of any debate about Brussels. Three teams of experts, architects, landscape architects and planners, were in charge of designing three visions for the development of the

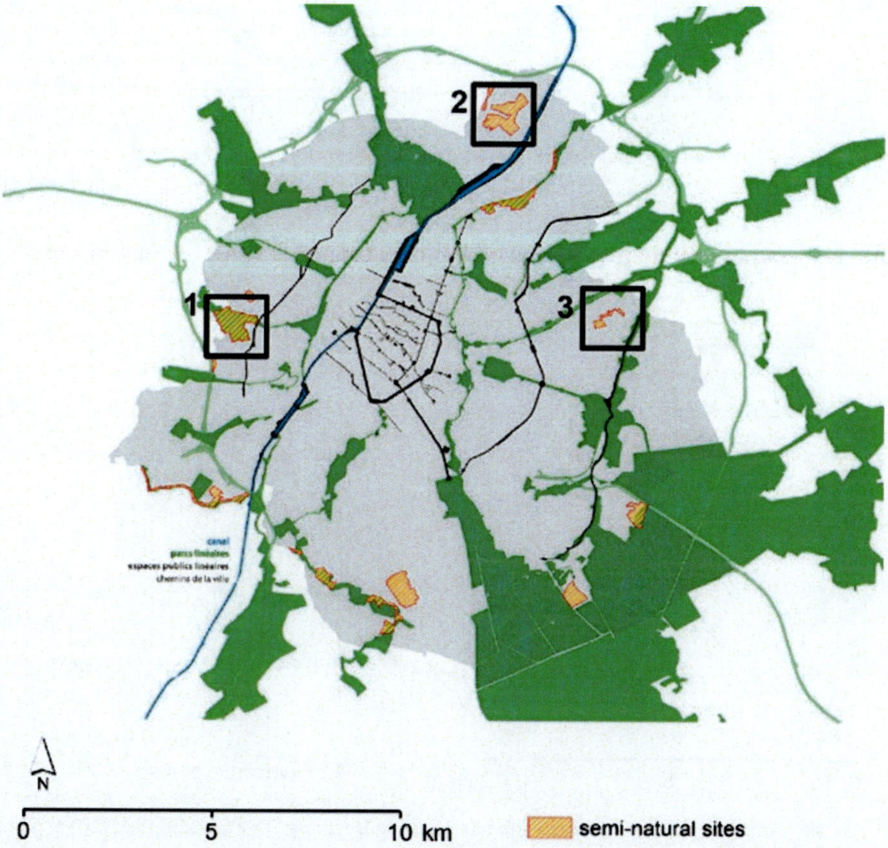

Fig. 9 Structuring green spaces from the pre-version of the Sustainable Regional Development Plan (2012) and the semi-natural sites. Author: Vanbutsele from Aerts 2012

Brussels metropolitan area. Some visions addressed strategies of open spaces that form structural landscapes across the regional boundaries (51N4E et al. 2011). Those proposals have been included in the summary map below showing the continuities of structural green spaces (Fig. 9).

Structuring green spaces from the Sustainable Regional Development Plan (2014)

The final document of the SRDP reaffirms the priority for the green and blue network in connection with the hinterland of Brussels.

> Beyond the administrative boundaries of the Region, the green network should be linked with existing open spaces outside the Brussels Region in order to insure hydrographic continuities, ecological network, soft mobility network and landscape coherence (Région de Bruxelles-Capitale 2014, p. 84) (Fig. 10).

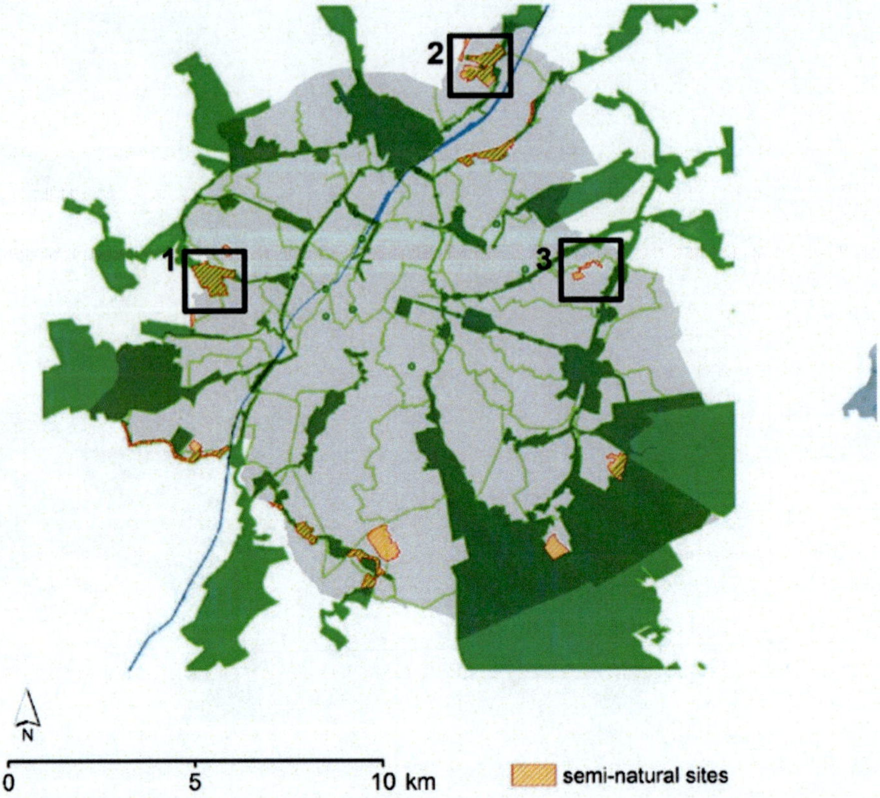

Fig. 10 Structuring green space from the Sustainable Regional Development Plan (2014) and the semi-natural sites. Author: Vanbutsele from Région de Bruxelles-Capitale 2014

Proposal for a metropolitan park system

In 2017, Vanbutsele brought together the previous proposals and extended them into a map suggesting the future park system for the Brussels metropolitan area.[1] The map shows an extension of the plan Besme connected to regional landscape continuities. Green continuities mainly follow hydrography and the valley bottom (Vanbutsele 2017, pp. 187–198) (Fig. 11).

[1] This proposal for a park system of the Brussels metropolitan area has been named in French *"charpente paysagère"* [Vanbutsele 2017].

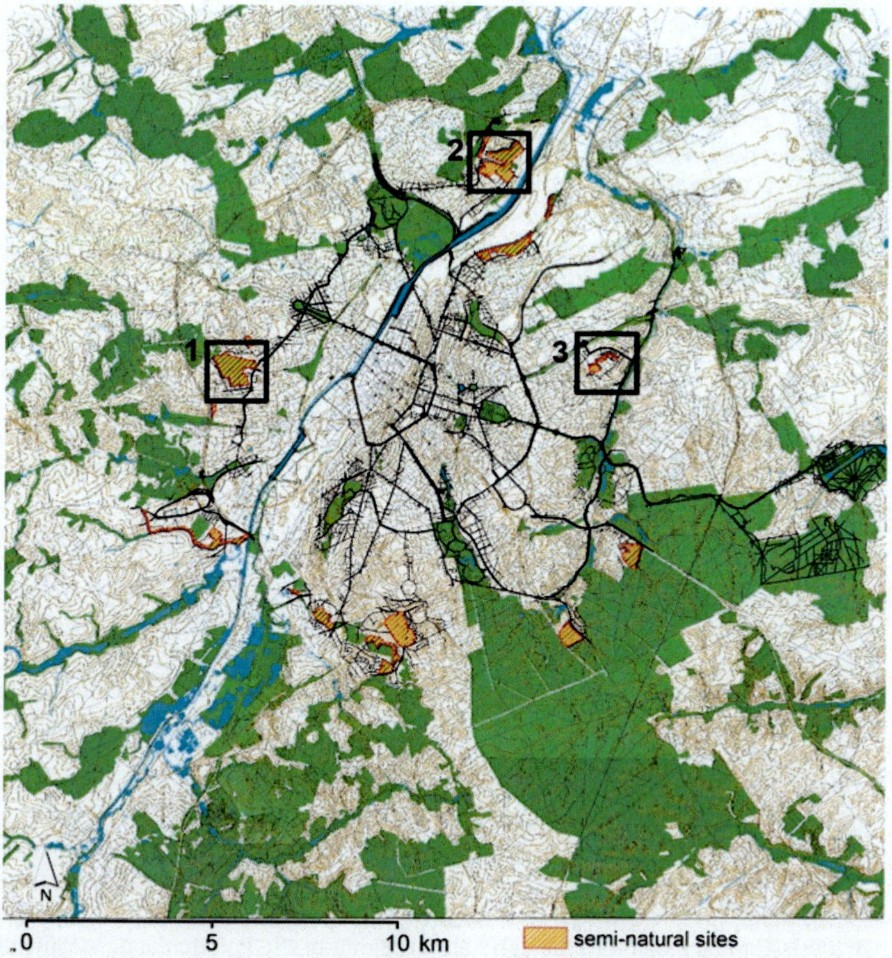

Fig. 11 Proposal for a metropolitan park system in Brussels (2017) and the semi-natural sites. From Vanbutsele 2017, pp. 196–197

5 Correlation Between Urban Pressure and the Inclusion of Semi-natural Sites in the Park System Projects

Now that we have described the five open space strategies and mapped them along with the semi-natural sites, we now discuss the different roles that semi-natural spaces play within the five analyzed projects.

In the nineteenth century, semi-natural spaces were completely disconnected from the network of open spaces designed by Besme. Although this plan is already an extension of the city, semi-natural spaces are too peripheral and are not included in the scope of the plan.

At the end of the twentieth century, with the Regional Development Plan and the green–blue network, the geographical scope of Brussels planning included the semi-natural spaces. They are still peripheral elements, at the edge of the Brussels-Capital Region. Connections between semi-natural spaces and the historical core are weak, but semi-natural spaces are relatively well connected to other green spaces on the outskirts. Among the semi-natural spaces, some clearly appear to be essential links of the green–blue network, while others are absent.

From 2012, some semi-natural spaces became parts of important green continuities allowing one to cross the administrative boundaries, while others still remain totally absent from the green network.

In the proposal for a metropolitan park system, semi-natural spaces are not peripheral spaces anymore but rather spaces at the heart of the metropolis. In that proposal, semi-natural spaces are critical links joining the central open space network of Besme to major landscape continuities on a metropolitan scale. The proposal for a metropolitan park system switches the scale of the reflection as well as the role of semi-natural spaces from the periphery to the center.

Depending on the strategy for open spaces and their representation, the contribution of semi-natural spaces varies. Some semi-natural sites constantly appear as essential elements of the green network, while others remain absent from any representation. Those spaces are smaller than the former, usually smaller than 10 ha. Some authors suggest that those smaller landlocked spaces should be complementary to the park system. They offer a singular scenery. Wooded areas or agricultural plots are closely surrounded by buildings. Those smaller spaces are also key resources for greening urban fabrics (Desvigne 2009, p. 23).

Moreover, the analysis suggests that the contribution of a semi-natural site seems to be correlated with the urban pressure occurring on the site. Among the three cases, case 1, the Scheutbos, is clearly affirmed as an essential link for every green network project. Since 1995, the Scheutbos clearly appears in each representation of open space strategies. The site with its rural and traditional landscape literally fastens the Brussels-Capital Region to the agricultural region of the Pajottenland. Meanwhile, the Scheutbos has the clearest status and experiences the least urban pressure. The entire site is officially assigned as a green zone. There is some pressure to urbanize some plots located on the edges of the site. This pressure is mainly due to landowners who are also developers. However, that pressure occurs on a minor area in terms of surfaces (2.4 ha out of the 57 ha). In addition, non-governmental organizations and administrations are lobbying strongly to prevent urbanization of the edges of the Scheutbos and to keep it within its current boundaries.

On the other hand, case 3, the Val d'Or, has been strongly urbanized during the last decades. Previous mobilizations from citizens and non-governmental organizations have left room for negotiations between landowners who are mainly public authorities and developers. The remaining 8 ha of open spaces out of the original 40 ha are a leftover of the urbanization. While the Val d'Or suffers from the highest urban pressure, it also provides the weakest contribution to all proposals of green networks. It never appears in any strategy of open spaces.

Finally, case 2, Val du Bois des Béguines, is in an intermediate situation. It is a fragmented site consisting of a sum of different open spaces mainly—but not exclusively—belonging to the municipality. Over time, the official designation of the site has evolved from buildable space, to land reserve, to green space. The site currently includes more than 70 ha of open spaces, though that includes 11 ha that are officially buildable. Their urbanization would lead to a drastic reduction and fragmentation of the open space. However, non-governmental organizations are working to maintain the site in its entirety. Moreover, the municipality is developing an urban forest project reinforcing the green status of the place. As the status of the site is unclear with a very fluctuating recognition, the contribution of this space to a park system project is also unclear. Depending on the project, the Val du Bois des Béguines is sometimes absent from and sometimes part of the open spaces strategy.

6 Discussion and Conclusion: From Periphery to Center

Semi-natural spaces in Brussels have evolved from former wastelands to informal open spaces. Confronted with growing urbanization, they will still evolve in the coming years. Planners, inhabitants, developers and citizens will be confronted with the perpetual choice of either building on them or keeping them open. Faced with this ineluctable choice, planners may use the relations these open spaces (potentially) have with a metropolitan park system to come to a decision.

Indeed, the case studies suggest that there is a correlation between the urban pressure on a semi-natural site and its contribution to a metropolitan park system. The more a site takes part in a continuity of green spaces, the higher its chances are of being preserved as open space. In other words, the more a site belongs to the park system, the lower the expectations to build on it. In contrast, sites that are not clearly identified as essential links for the park system tend to face higher urban pressure. In order to preserve an informal open space, connecting it with metropolitan green continuities could thus be essential.

Moreover, considering the belonging of wastelands to a metropolitan park system could also help to preserve its informal character. When traditional planning approaches tend to consider wastelands either for building redevelopment or for conventional open public space, the unregulated character of the space is lost, although these unconventional characteristics are increasingly recognized and valued. Considering wastelands as parts of a chain of green spaces, their loose character could be complementary to other open spaces with a higher level of intervention (e.g. recreational green spaces) or with a more restrictive public access in order to enhance biodiversity.

From a strategic point of view, the case of semi-natural spaces in Brussels highlights the importance of the location of wastelands. Precisely because semi-natural spaces were wastelands located at the edges of the dense city, they underwent less urban pressure than other wastelands located in the city center. This situation has favored the evolution of semi-natural spaces into informal open spaces rather than being urbanized. And precisely because they are located on the fringe of the city, they are also at the core of the metropolitan area, and they can become essential links between the city and its outskirts. In Brussels, semi-natural spaces may contribute in an essential way to a coherent vision of Brussels and its hinterland. More generally speaking, wastelands turned into informal open spaces could become a strong link between dense urban fabric and peripheral fabric. Informal open spaces as part of a metropolitan park system could become a powerful tool to plan a city beyond its administrative boundaries and articulate multiple scales.

Acknowledgements This research has been partly founded by the UCLouvain (BE) and University of Geneva (CH).

References

51N4E, l'AUC, Bureau Bas Smets, Mint, Chôros, Urban Solutions (2011) Double Brussels small metropolis global metropolis. Elaboration d'une vision territoriale métropolitaine à l'horizon 2040 pour Bruxelles dans le cadre du Plan Régional de Développement Durable [Elaboration of a territorial vision at the metropolitan scale for 2040 for Brussels in the context of the sustainable regional development plan], Bruxelles

Aerts J (2012) Bruxelles, vers une métropole polycentrique—vers des territoires structurants [Brussels, towards a polycentric metropolis—towards structural territories]. In: Atelier paysage 2012: Natures et Paysages de la métropole, UcLouvain, LOCI, Louvain-La-Neuve

Banzo M (2015) L'espace ouvert pour recomposer avec la matérialité de l'espace urbain [The open space to reconsider urban space materiality]. Articulo—J Urban Res 6 [Special issue 6 la fabrique des espaces ouverts]. https://doi.org/10.4000/articulo.2708

Berger A (2006) Drosscape: wasting land in urban America. Princeton Architectural Press, New York

Berrizbeitia A (2014) Between deep and ephemeral time: representations of geology and temporality in Charles Eliot's Metropolitan Park System, Boston (1892–1893). Studies in the history of gardens et designed landscapes, 38–51

Billen C, Duvosquel J-M (2000) Bruxelles [Brussels]. collection L'esprit des villes d'Europe. Fonds Mercator, Anvers

Bonthoux S, Brun M, Di Pietro F, Greulich S, Bouche-Pillon S (2014) How can wastelands promote biodiversity in cities? a review. Landsc Urban Plan 132:79–88

Brun M (2015) Biodiversité végétale et délaissés dans l'aménagement urbain. Contribution potentielle des délaissés urbains aux continuités écologiques [Plant biodiversity and derelict spaces in urban planning. Potential contribution of urban derelict spaces to ecological continuities]. PhD dissertation, Université François-Rabelais de Tours

Choay F (1965) L'urbanisme, utopies et réalités—Une anthologie [Urbanism, utopias and realities—an anthology]. Collection Points -Essais. Editions du Seuil, Paris

Clément G (1991) Le jardin en mouvement [The garden in movement]. Pandora, Paris

Cogato Lanza E (2005) Le territoire inversé [The inversion of the territory]. In: Versteegh P (ed) Méandres. Presses Polytechniques et Universitaires Romandes, Lausanne, Penser le paysage urbain, pp 117–141

Corner J (2006) Terra Fluxus. In: Waldheim C (ed) The landscape urbanism reader. Princeton Architectural Press, New York, pp 21–33

De Smet A (2013) The role of temporary use in urban (re)development: examples from Brussels. Brussels Studies 72. https://www.brusselsstudies.be

de Solà-Morales I (1995) Terrain Vague. In: Davidson C (ed) Anyplace. MIT Press, Cambridge
Denef J, de Lestrange R, Vanbutsele S (2015) Les espaces ouverts, lieux communs de la recherche et de l'action territoriale [Open spaces, common places of research and territorial action]. In: Hanin Y (ed) Cinquante ans d'action territoriale. Presses universitaires de Louvain, Louvain-la-neuve
Desvigne M (2009) Epaissir les lisières [Thicken ecotones]. In Livret de chantier 2: Naissances et renaissance de mille et un bonheurs parisiens. Consultation internationale de recherche et de développement sur le grand pari de l'agglomération parisienne Paris. https://ciudadyterritorio.files.wordpress.com/2017/04/nouvel-arep-mcd_livret_chantier_2.pdf
Dupuis B (2009) Le mouvement du New Urbanism et le paysage urbain. La circulation d'une doctrine urbanistique [The *New Urbanism* movement and the *urban* landscape: the *circulation* of a town-planning *doctrine*]. Articulo—J Urban Res, Special issue 2. https://doi.org/10.4000/articulo.1133
Eliot C (1903) Charles eliot landscape architect a lover of nature and of his kind who trained himself for a new profession practised it happily and through it wrought much good. The Riverside Press, Cambridge
Emerald Necklace conservancy (2020) Welcome to the Emerald Necklace. https://www.emeraldnecklace.org/park-overview/
Forman RTT (2010) Urban ecology and the arrangement of nature in urban regions. In: Mostafavi M, Doherty G (eds) Ecological urbanism. Harvard University Graduate School of Design, Lars Müller, pp 312–323
Forman RTT, Godron M (1986) Landscape ecology. Wiley, New York
Franck KA, Stevens Q (2006) Loose space: possibility and diversity in urban life. Routledge, London, New York
Ibge-Bim, Cooparch-RU (2004) Plan de gestion des sites semi-naturels [Management plan for semi-natural spaces]. Bruxelles
Institut Bruxellois de statistique et d'analyse (2019) D'ici 2025, la population bruxelloise augmentera, mais moins que prévu [By 2025, the population of Brussels will increase, but less than expected]. https://ibsa.brussels/publications/titres/a-la-une/avril-2019-d-ici-2025-la-population-bruxelloise-augmentera-mais-moins-que-prevu#.XpCE9XLgqUk
Kongjian Y (2010) Five traditions for landscape urbanism thinking. Topos 71
Legenne C, Arbelbide LL, Thibault C, Tricaud P-M, Vivien J-F (2010) Comment traiter les fronts urbains? Les carnets pratiques [How to deal with urban fronts? practical note book]. Institut d'Aménagement et d'Urbanisme Ile-de-France, Paris
Mariani M, Barron P (2014) Terrain vague: interstices at the edge of the pale. Routledge, London, New York
Marot S (1995) L'alternative du paysage [Landscape alternative]. Le Visiteur: Ville, Territoire, Paysage, Architecture : Revue Trimestrielle De Critique Des Situations Construites 1:54–81
Marot S (2003) Sub-urbanism and the art of memory. Architectural Association, London
Maruani T, Amit-Cohen I (2007) Open space planning models: a review of approaches and methods. Landsc Urban Plan 81(1–2):1–13. https://doi.org/10.1016/j.landurbplan.2007.01.003
McHarg I (1969) Design with nature. Published for the American Museum of Natural History by the Natural History Press, Garden City, New York
Metropolitan Park Commission (1893) Report of the Board of Metropolitan Park Commissioners. Boston
Mostafavi M, Doherty G (2010) Ecological urbanism. Harvard University Graduate School of Design, Boston, Lars Müller
Ost F, Remy J, Campenhoudt L (1993) Entre ville et nature, les sites semi-naturels: approches sociologique et juridique des sites bruxellois [Between city and nature, semi-natural sites: sociological and legal approaches of Brussels sites]. Facultés universitaires Saint-Louis, Bruxelles
Oswalt P, Overmeyer K, Misselwitz P (2013) Urban catalyst: the power of temporary use. DOM Publishers, Berlin
Panerai P (2002) Analyse urbaine [urban analysis]. Editions Parenthèses, Marseille

Région de Bruxelles-Capitale (2002) Plan régional de développement [PRD] [Regional Development Plan]. Gouvernement de la Région de Bruxelles-Capitale, Bruxelles

Région de Bruxelles-Capitale (2014) Projet de plan régional de développement durable [PRDD] [Sustainable Regional Development Plan]. Bruxelles

Roland LC (2012) When you can't see the city for the trees. A joint analysis of the Sonian Forest and urban reality. Brussels Studies 60. https://www.brusselsstudies.be

Rowe C, Koetter F (1978) Collage city. the MIT Press, Boston

Rupprecht CDD, Byrne JA (2014) Informal urban greenspace: a typology and trilingual systematic review of its role for urban residents and trends in the literature. Urban for Urban Green 13(4):597–611

Salat S (2011) Les villes et les formes: Sur l'urbanisme durable [Cities and forms: about sustainable urbanism]. Editions Hermann, Paris

Shoard M (2002) Edgelands. In: Jenkins J (ed) Remaking the landscape: the changing face of Britain. London, Profile Books, pp 117–146

Sitte C (1889) Städtebau nach seinen künstlerischen Grundsätzen. English edition: city planning according to artistic principles (often translated as The Art of Building Cities). Éditions du Seuil (1996), Paris

Sordi J (2014) Beyond urbanism. Babel, Trento

Sordi J (2016) Ordinary voids, special openings. In: Mosé R, Schröder J (eds) Towards a pro-active manifesto. Arcane Rome, pp 91–97

Stefulesco C (1993) L'urbanisme végétal [Vegetal urbanism]. Institut pour le développement forestier, Paris

Tritsmans B (2014) Versatile green: an alternative perspective on urban green space in late nineteenth-century Antwerp. Urban History. https://doi.org/10.1017/S0963926814000509

van der Valk A, van Dijk T (2009) Regional planning for open space. Routledge, London

Vanbutsele S (2017) Du ménagement de l'espace ouvert au dé-ménagement de la ville: exploration des sites semi-naturels pour une densification qualitative de Bruxelles [From open space preservation to city remodeling: exploration of semi-natural sites in favor of a qualitative densification of Brussels]. PhD dissertation. Presses universitaires de Louvain, Louvain-la-Neuve

Vanbutsele S (2018) Des réserves foncières aux zones vertes : projets et contre-projets d'urbanisation contribuant à la formation des sites semi-naturels bruxellois [From land reserve to green area: projects and counter-projects contributing to the formation of semi-natural sites in Brussels]. In: Mantziaras P, Viganò P (eds) Projeter des horizons d'attente : urbanisme de l'espoir, Lausanne, actes de la 3e Journée d'étude Bernardo Secchi, École polytechnique fédérale, 20 septembre 2017. Metispresses, Genève

Vanbutsele S (2019) Les bords des sites semi-naturels bruxellois : analyse et pistes d'actions [The edges of semi-natural sites in Brussels: analysis and proposal for actions]. In: Zastavni D, Leduc D (eds) Recherche et enseignement en architecture. Influences et complémentarités [Research and teaching in architecture. Influences and complementarities]. Actes de la journée de la recherche et des doctorants 2015 (LOCI, UCLouvain). Presses universitaires de Louvain, Louvain-la-Neuve

Vanbutsele S, De Smet A, Dewaelheyns V, Tristmans B (2018) Exploring semi-natural space in the Val des Béguines. In: Valérie D, Hans L, Gulinck H (eds) Challenging the boxes, interfaces in landscape and land use. Gompel&Svacina, Oud-Turnhout

Vanderstraeten P, De Corte S, Deprez B, Thielemans B (2009) États généraux de Bruxelles [General assemblies of Brussels]. Note de synthèse n°4, Brussels Studies, Bruxelles

Vanempten E (2009) Challenging urbanism: the rurban reality of the Brussels metropolitan area. In: De Meulder B, Ryckewaert M, Shannon K (eds) Proceedings of international urbanism & urbanisation PhD seminar: transcending the discipline. Urbanism & urbanization as receptors of multiple practices, discourses and realities, Leuven, 2009/10/01–2009/10/03

Waldheim C (2005) The landscape urbanism reader. Princeton Architectural Press, New York

Ward Thompson C (2002) Urban open space in the 21st century. Landsc Urban Plan 60:59–72

Séréna Vanbutsele is Teacher and Research Assistant at the University of Geneva (Switzerland), School of Social Sciences/Environmental Governance and Territorial Development Institute. Trained as an architect and planner, she has a PhD in Urban Planning from UCLouvain, Belgium. Her expertise lies at the interface of urban and landscape planning. She questions the paradigm of the compact city and the process of densification through the transformation of unplanned open spaces such as wastelands, informal green spaces, semi-natural sites, and urban fringes.

Becoming Urban Wastelands

The Evolutions of the Catchment Areas of the 'Traditional' Water Management System of Tamil Nadu (India)

Laura Verdelli, Geeva Chandana Balasubramanian, and Rukkumany R. Harishankar

Abstract The city of Chennai still showed, on a map of 1908, large open water reservoirs and a multitude of waterbodies. They were the last remnants of the 'traditional' water management system that existed in the villages of Tamil Nadu. These reservoirs, as water storage structures, are the visible elements but only a part of a much larger system. The efficiency of this system is based on a range of practices related to land use planning. The lands surrounding the reservoirs belonged to the village community and participated fully in the territorial system. The system was based on an organization around the tanks made of community-owned lands and managed by the village authorities. However, these common lands are perceived (since the colonial period) only as uncultivated lands and are indicated on cartography as 'wastelands.' This, coupled with societal and institutional changes, the evolution of agricultural methods, the gradual transition to individual pumps, and the increase in land pressure, leads to their gradual functional abandonment. These lands are often ignored by public policies and territorial planning, and are frequently (more or less illegally) occupied and built on. We will focus on the process that has driven this change in perspective, which transforms lands that were once useful to the community into 'wastelands' to wildly urbanize.

Keywords Water storage · Monsoon · Common lands · Reservoir · Water management system

L. Verdelli (✉)
UMR CITERES-CNRS, Université de Tours, Tours, France
e-mail: laura.verdelli@univ-tours.fr

G. C. Balasubramanian
Department of Architecture and Interior Design, SRM University, Chennai, India
e-mail: ar.geevachandana@gmail.com

R. R. Harishankar
Department of Architecture, SAP School of Architecture and Planning, Anna University, Chennai, India
e-mail: rhrukkumany@annauniv.edu

F. Di Pietro and A. Robert (eds.), *Urban Wastelands*, Cities and Nature,
https://doi.org/10.1007/978-3-030-74882-1_15

1 Introduction

Water resource management is essential for human survival, especially in geographic areas where water is not consistently and abundantly available, and where societies are directly dependent on their ability to conserve, store, and distribute water and to make a rational and frugal use of it. This is particularly true in regions where the rain falls very intensely but for short periods (Agarwal and Narain 1997), as in the state of Tamil Nadu—India (Fig. 1). Seen from the sky just after the monsoon, the state of Tamil Nadu appears to be a landscape filled with waterbodies (Weitz 2005). This corresponds to the presence of reservoirs (commonly called ‘water tanks’ or simply ‘tanks’), used by man to retain water.

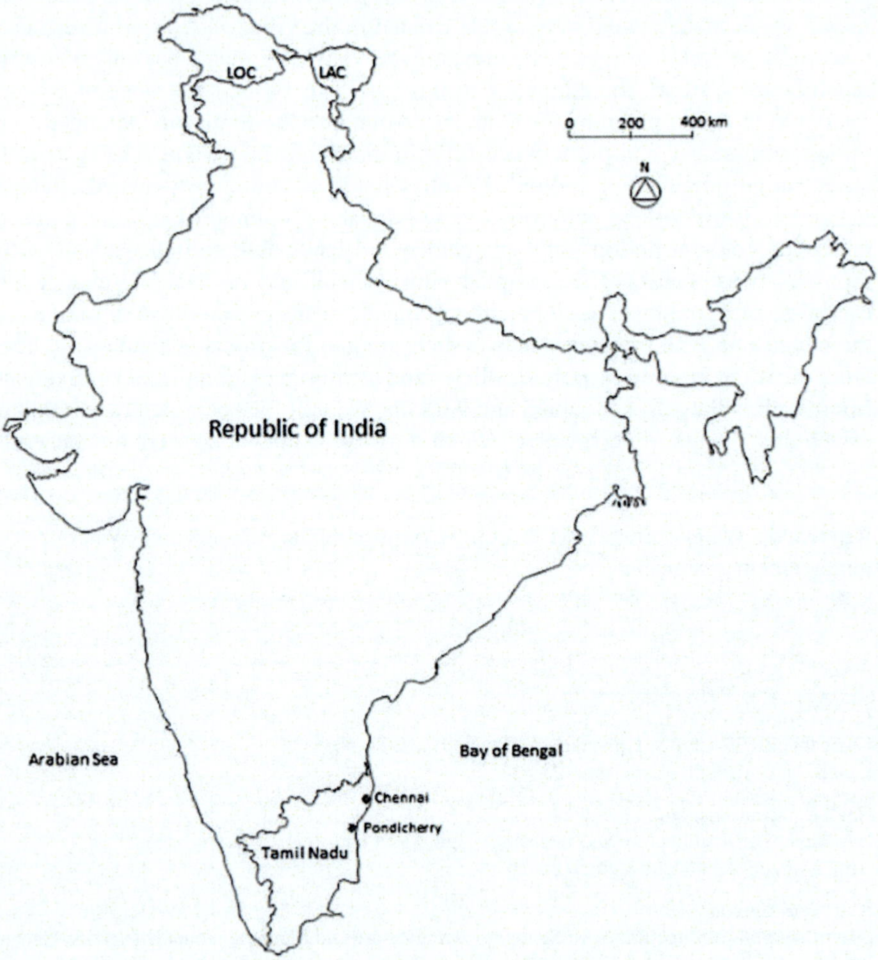

Fig. 1 Map of the Republic of India showing the location of Tamil Nadu, Chennai and Pondicherry

These reservoirs, as water storage structures, are the visible elements but only a part of a much larger system made of the presence of lands used as catchment area. The status of these lands, which up to British rule belonged to the village community and participated fully in the territorial system, is what sparked our interest. What this article tries to achieve, putting together all the different parts through the existing knowledge and completing the scheme via on-site observations and discussions with various stakeholders, is to highlight the role played by the lands owned by the community. These lands passed from uncultivated useful parts of the water management system to 'wastelands' and, during the last three decades, to urbanized lands, thus proposing an unusual example of the trajectory of wastelands. If most urban wastelands in Europe come from the abandonment of industrial buildings, in this part of India the legal definition of wastelands was applied to lands that looked abandoned (or had never been exploited) and, once officially considered as unused, became 'urbanization friendly'.

The authors wish to report on this evolution in the area between Chennai (state capital of Tamil Nadu) and Pondicherry (Union Territory of Puducherry) which comprises two very dense urban areas and a gradient from urban to rural in between the two (allowing us to observe the complete cycle of transformation: commonly used, unused, encroached, and built-up). Our goal is to question, at least partially (in time and space), the reasons that led these lands to become wastelands and that encouraged their disappearance as 'free' lands. Actually, their transformations are less physical or ecological than they are conceptual, dependent entirely on assumptions embedded in our ideas about different kinds of landscapes and the kinds of associations that are evoked when we encounter and use terms like 'wilderness' and 'wasteland' (Di Palma 2014). We try to analyze the dynamics behind the almost radical disappearance of the common territorial system that was organized around the water management system, assuming that change in the nomenclature that called unproductive lands 'wastelands' is one of the factors that led to their evolution.

Box 1: Method

The main assumption of the present article comes from the further interpretation of the fieldwork of a research project that focused specifically on the place of water. The project fieldwork, conducted between 2013 and 2019, was spread over several years and was conducted both by researchers and Master/Ph.D. students (University of Tours, France), focusing on analysis of the water and land management system (in particular, via the Ph.D., in progress, by Rukkumany R. Harishankar); analysis of the representations of urban water (in particular, via the Ph.D. by Karine Hochart); available statistical data (on the evolution of land use and urbanization pressure); and in situ observations and interviews with local stakeholders (including inhabitants and farmers). Rukkumany R. Harishankar's Ph.D., in particular, analyzes the land allocation system in four villages in Tamil Nadu at the junction between agricultural areas and the urban areas of Chennai: Mannur, Mevalarkuppam, Valarpuram, and Thandalam (district of Kancheepuram). From the statements of village receipts excerpted from the registers of the offices of Village Administrative Agents established on a parcel basis over four decades (1986–2016), the thesis is able to show the changes in land use.

Apart from the two Ph.D.s, the field data were collected by Master students from the Department of Spatial Planning and Environment—Polytechnic Engineering College (EPU-DAE), during field research work and internships on water-related topics. Seven Master theses and nine internships put together sets of interviews and questionnaires with inhabitants living on the ancient *puramboke* lands (common lands), with public and private water suppliers as well as cartographic diachronic analysis of land use (on GIS basis, using existing maps, satellite land cover, and Google Maps/Google Earth images), which we were able to consult so as to follow the intuition that underpins the present article.

In urban areas, the landscape has so significantly evolved that it has become extremely challenging to recognize whether common lands, often including waterbodies, even existed. To understand what the system was made of, its interests, and its functioning, it was necessary to slowly recreate a figure of it, assembling literature references, remnants in rural areas, and some urban vestiges so as to have an understanding of it, as reconstructed in what follows.

2 Contemporary Progressive Urban Pressure on Unbuilt Lands

The dissolution of the community organizations that administered the water management systems after Indian independence contributed to the decline of a part of territorial organization. Formerly common lands have become either private or state-owned. Perceived as uncultivated lands, what remains of them is today ignored by public policies and territorial planning, and frequently more or less illegally occupied and built on. Their classification as 'wastelands' has allowed real estate developers to monopolize them, and to install urban projects whose size varies from small residential units to large public interest equipment. Official interventions were directly framed by the government authorities, as landowners. In the meantime, Indian society has also undergone a significant evolution of its structure, which has tended to move away from collective functioning toward more individual ways of functioning.

In addition, public policies and contemporary strategic territorial planning do not define specific action plans for these lands, which, in the absence of protection, are occupied, to the point sometimes of eliminating the surfaces of the tanks themselves, and are built on, either via the transfer by the State for public interest, or via a more or less informal privatization (officially these lands cannot be sold, but some kind of land property titles accompany the transaction for the ownership of a house—during the interviews the authors were unable to verify the legal status of these documents).

The surviving tanks in the city suffer from eutrophication, encroachments, narrowing of the connection channels, garbage and sewage disposal, siltation, saline water intrusion, and other issues. They have turned into urban wastelands that are losing their identity and value. Chennai has a thriving population with fast-growing needs. This accounts for tremendous pressure on waterbodies for urban infrastructure and real estate projects that reduce the rainwater carriage capacity of the few existing waterways. Due to the shortage of available land, the space occupied by water has

Fig. 2 The Aya Aiya Kulam, in the surroundings of Pondicherry, in November 2018 at the maximum level of water just after the northeast monsoon—Picture courtesy of Frédéric Landy

even been perceived as the only space to locate some transportation facilities like the Chennai's Mass Rapid Transit System (MRTS), which runs over the Buckingham Canal.[1] The ownership of waterbodies is scattered among various government departments, thus complicating the situation even more. To counter this issue, the Protection of Tanks and Eviction of Encroachment Act came into effect from October 2007. Unfortunately, this law has suffered from a lack of implementation since.

The importance of urbanization erased almost all the traces of these systems and only the knowledge of their previous existence allows us to sketch out their supposed extent. The only areas where it is possible to prove the disappearance using what little historical mapping is available and some remnants in place names (i.e., Tank Road) are those covered by water. This phenomenon is even more important since seasonal variations can hide the existence of a water reservoir. Being located often in a small land depression without rigid boundaries, as shown in Figs. 2 and 3, a water tank can be totally invisible during the dry season and only appear as such once filled with water.

Depending on their location, the attested flood risk and the time lapse when they were urbanized (the phenomenon having started in the 1960s) we can find: informal slums, ordinary lower middle-class districts (unplanned as well as planned), and/or wealthy residential buildings (Verdelli 2019), or government infrastructures. The

[1] Parallel to the Bay of Bengal, in a north-south direction, over a length of about 420 km, the Buckingham Canal, built during the British period for transport purposes (navigation ended in 1975), connects the State of Andhra Pradesh to the State of Tamil Nadu.

Fig. 3 The same Aya Aiya Kulam, in the surroundings of Pondicherry, in September 2018 at the end of the dry season—Picture courtesy of Frédéric Landy

latter started to become fancy when the urban space available for construction in dense urban areas shrunk to almost zero (the rates of urban growth and rural–urban migration in India being high, rapid, and difficult to control) and in the most recent decade when the presence of water started being valued and appreciated by the upper-class population (Hochart 2019).

Today, different districts proudly house education institutions, office, and industrial complexes on reclaimed lakes, marshlands, and creeks (Fig. 4). Waterbodies have turned into garbage dumps, catchment areas have been filled up to accommodate urbanization, and even the memory of the system functioning has been lost.

Fig. 4 A Google Maps' capture showing the Indira Gandhi Medical College and the Kathirkamam Hospital built on the main water area of the so-called Kanagan Lake in Pondicherry, dividing the former water space (inside the blue line) into a smaller lake (inside the white line—in the east), a built-up area (in the middle), and a 'wasteland' ready to be further urbanized (in the west). It is also possible to see the area already partly urbanized around the blue perimeter, where there survive unbuilt parts of the former catchment area. September 2019

3 Origins of the Mutation of the *Puramboke* Lands into Wastelands

3.1 Common Land for Common Uses

The land on which the extensive system of protection around the reservoirs was located and the reservoirs themselves were part of common lands, belonging to the villagers. Called *puramboke* in Tamil Nadu, these common lands were important for the functioning of the network that supplied the reservoirs. Since ancient times special purposes were assigned to these lands, recognizing their contribution to the preservation of the region's ecological balance. This included rivers, river banks, irrigation tanks, grazing and pasture lands, marshlands, backwaters, and salt pans. Since there was no single ownership, crops were not grown in these common lands. The ruling dynasties offered these lands for public use as the region suffered alternatively from seasonal droughts and heavy rains. These lands were widely used for water conservation purposes, and some were exploited for seasonal cropping to generate revenue for public use and also used for the grazing of domestic animals. Still the distribution of rights and of access in local irrigation systems was not necessarily equitable (Mollinga 2010). But the *puramboke* lands were transformed over the years.

3.2 Identification of Puramboke Lands as Wastelands Under British Rule

The concept of wastelands in India appeared during the British period and included all lands that were not under cultivation through the process of revenue 'settlement,' a term applied to the method of assessing the land revenue demand, and linked to property regimes. Under British rule, state authorization of private property rights in land prevailed, resulting in separation of public and private lands. With the system of land classification post-1858, when the administration of Indian territories passed on from the English East India Company to the British Crown, the colonial administration distinguished between cultivated and uncultivated lands. *Puramboke* lands not being under cultivation were termed 'wastelands' as they did not provide revenue to the state. Starting from this moment, *puramboke* lands were officially recorded and mapped as wastelands, were declared to belong to the state, and were taken over by the revenue department. This classification, primarily done to reorganize land classification and structure revenue administration, totally failed to identify and acknowledge the 'traditional'[2] practices associated with the *puramboke* lands, which were included under the revenue settlement. An order of the Court of Directors of the

[2] The authors chose to use the word 'traditional', even if not exactly adapted, since it is still the least inadequate in describing these systems. By this term, the authors mean systems stemming from tradition, not static but evolving over a long period of time, reflecting the sociocultural, economic,

East India Company in 1856 offered title to such lands to local landlords and to the general public, and rules for selling 'wastelands' were published in 1864 (Kasturi 2008). The precincts policy was extended to India, and large portions of common lands became the property of the British Crown (Ghotge 2011).

Before British rule, common land resources were under the control of villages. A large part of the country's natural resources was available to the rural population. Either relying on the remnants of the feudal systems (like the *zamindari* or the *inamdari*[3] ones), or directly on cultivators for revenue collection, the British system extended state control over these resources and provoked a substantial decline of uncultivated lands available to villagers (Ghotge 2011). The actual identification of wastelands and forestlands took place through settlement. Extensive land survey and settlement operations were carried out throughout the country to streamline the land revenue collection system. These operations resulted in creation of detailed village records and often demarcation of cultivated lands and wastelands/forestlands on maps (Saigal 2011).

3.3 *Post-independence Government Policies: Evolution of Urban Wastelands*

Since the British land classification system, uncultivated lands have been considered to be government property. In most Indian states, all the lands not under cultivation were acquired by the state as a result of intermediary abolition (Indian Planning Commission 1966). After independence, the 1927 Forest Act continued with few changes. Common lands were taken under government control and the rights provided to the local communities were slowly withdrawn, thus alienating the people further. The Land Acquisition Act of 1894 gave the authorities power to take the common lands. The complete exclusion of the public from common lands quickly and irreversibly deteriorated the traditions and purposes associated with them.

Even though governmental policies made a large contribution to this major depletion, lack of interest and of a sense of collective belonging also drove common lands out of daily use. On the other hand, and at the same time, anyone became entitled to have access to privately owned land, even if this democratization did not really question the social inequalities of Indian society. In this situation, the government authorities, which influenced planning bodies, acquired via self-attribution these common lands for public purposes (often meaning 'constructions').

and political structure of society and influenced by many factors like the environmental, societal, and technical/technological conditions.

[3] The *Zamindar* and the *Inamdar* were landowners. They ruled on both cultivated and uncultivated lands and reserved the right to collect taxes from peasants on behalf of the powers that be. They would allocate to tenants the cultivation of the land, and would authorize tenants to use uncultivated land, graze their cattle and have access to water resources. The *zamindari* system, which ended in India in 1951, had never been massively present in the South of the subcontinent, while the *inamdari* was primarily present in the states of Maharashtra, Karnataka, and Gujarat.

4 Gradual Abandonment of the Water System that Lost its Identity and Values

Irrigation systems through reservoirs bring together various components that connect the reservoirs themselves to users through a complex dynamic (Weitz 2005; Mukundan 2005). A number of studies on different aspects of this system have focused on: tanks, their construction, their structure (Ratnavel and Gomathinayagam 2002); the social institutions that maintain the system (Aubriot and Prabhakar 2011; Mukundan 2005; Sreenivasan 2002; Mosse 1997); associated rituals (Weitz 2005); the socioeconomic benefits derived from the system (Ariza et al. 2011; Pandey 2000); and the presence and benefits of the vegetation associated with the reservoirs (Pandey 2000). Likewise, studies have been devoted to the creation of land and administrative systems created by the British and the changes brought about by them, notably in Tamil Nadu (Yanagisawa 2011; Mizushima 1996). However, very few studies focus on the wide identification and reconstruction of the entire system of water management that set together, in terms of space, catchment areas (based on *puramboke* lands) and waterbodies. When the system worked in the past, reservoirs fulfilled several roles and served as a bridge between anthropic and natural functioning. Today, a large part of these systems is perceived to be in serious decline, even if some advantages and links survive. If we think of the presence of tanks as closely linked to that of other territorial components, we realize that the latter, although less visually identifiable, were just as necessary for overall functioning.

Actually, a plurality of factors led to the present situation where what once were common lands were first abandoned and then evolved either into agricultural areas or into urbanized areas.

4.1 A First Step: The Shrinking of 'Unproductive Lands' Starting from the Nineteenth Century

Significant changes in the administrative system, first brought about by the British, then by the Indian government (Aubriot and Prabhakar 2011), have had a definite impact on the perception, classification, and functioning of *puramboke* lands.

The allocation of common land followed a customary pattern (Singh 2013), which was apparently discouraged by the British administration (from 1757–1858 to 1947) when it took over the administration of the villages. In addition, the transition, unintentionally or intentionally, led to a certain homogenization of the nuances described above in the classification of *puramboke* lands, which became the basis of the British land registers. This incomprehension of the functioning of the global system, together with societal evolution (which led to the decline of the communal society and the transfer of ownership of the common land resources from the *panchayats* to the state), as well as the change in the 'contract' between the state and its citizens (Mosse 1997), led to a transformation that is definitive. Common lands now belong to a single category: that of not producing income, translated as wastelands (Singh 2013; Marikkani 2012), which put together 'uncultivated land' and 'abandoned land.'

These changes have removed the customary protection of the land, making it vulnerable to encroachment and conversion for other purposes. The classification of these common lands as uncultivated lands also meant that they could be allocated to other needs, as if unproductive lands were inconceivable and ignoring that unproductive does not mean useless. In the nineteenth century, a large-scale conversion of common lands in the villages started, as well as in the twentieth century, where lands belonging to common property gradually decreased in Tamil Nadu (Marikkani 2012). The conversion of common lands to private land for agriculture has significantly changed their functions. The land, both of catchment areas and of tanks themselves, was 1. leveled; 2. waterproofed; 3. covered by surface stagnation of rainwater which no longer has an outlet, during the rainy season.

In the area under urban pressure around Chennai and Pondicherry, the changes in the 'traditional' water management system are largely linked to the evolution of land use itself, since it changed from 'agricultural' to 'industrial' first and then to 'residential.' There is a general increase in land used for non-agricultural purposes. As a result, reservoirs have become superfluous in terms of use as water supplies and in terms of needs perceived by the populations, both for agriculture and for domestic uses.

4.2 The Conversion to Private Agriculture or to Built-Up Areas in the Last Three Decades and the End of Collective Benefits

Encroachment on the tank bed occurs nowadays due to pressure on land, meaning that farmers with fields in the foreshore areas encroach on part of the upper lands of the tank bed for agricultural purposes. This encroachment is allowed and organized in some tanks in the summer season, when the tank is empty, so that this fertile land is exploited by landless people, or by landowner farmers paying a fee to the village. In other tanks, encroachment is committed illegally by landowners with lands adjacent to the tank bed, without permission or the consensus of the rest of the village. Due to the increase of urbanization pressure, the lands formerly participating in the water management system are particularly vulnerable: pollution of water, decreased water storage capacity, risk of breaching of the bund creating dangerous floods, shortage of drinking water, etc.

As quoted by the newspaper The Hindu: The area of the waterbodies in Chennai City and its suburbs shrank from nearly 12.6 km^2 in 1893 to about 3.2 km^2 in 2017, mainly due to urbanization. A study by the Department of Geology, Anna University, based on a city map of 1893, has revealed that there were nearly 60 large waterbodies in the core of what was then Madras. The study traced the shrinking and vanishing waterbodies through a series of city maps. In 2017, the number of waterbodies, both large and small, came down to 28 (Lakshmi 2018). Between the 1950s and the 1970s, most of the waterbodies disappeared progressively, swallowed up by settlements. The encroachments of the waterbodies depleted the volume of water harvested and the loss of groundwater recharge as more waterbodies vanished over the decades. It is estimated that if the average depth of lakes and ponds was 2 m and filled twice a year, the volume of surface water stored dipped from 1.335 million cubic feet (mcft) in 1893 to 339 mcft in 2017 (Lakshmi 2018). The disappearance of waterbodies and open spaces also means a drastic drop in groundwater recharge potential and increases saltwater intrusion in the coastal belt (Central Ground Water Body 2017). Indeed, the surface water tables are heavily exploited for urban domestic uses via mechanical extraction. Their refilling was historically secured by the tanks system, providing consistent volumes of harvested rainwater able to slowly infiltrate the soil. Today, the volumes of underground water extracted are increasing, and less and less of the refilling is done via the tanks system, thus allowing ocean saltwater to infiltrate the water tables instead.

Because the Public Works Department used to maintain tank memoirs with hydraulic details of tanks during the British period, it is possible to trace minor flood events and to link their geographical extension to a vanished waterbody in the neighborhood.

5 Conclusion

It is generally accepted that the old water management systems in Tamil Nadu are today highly deteriorated. The word *puramboke* has come to signify worthlessness to the general public. Being named wastelands by the British rulers turned them into wastelands over the years. Their depletion is mainly due to disregard for the importance of their role, to the evolution of the state-society contract, and to the will to transform all land into revenue-producing land. This means that over the years hospitable *puramboke* lands have been subject to ecological destruction. The technical systems that made them up, centered around the tanks, were the visible and easily identifiable part of a much larger and less tangible system. The fact of identifying the tank with the system led to progressive ignorance of the rest of the system, to disregard for its different components and therefore to its disappearance as a complete and complex sociotechnical system, which was strongly linked to social and community considerations.

What appears as somehow original in the trajectory of these urban wastelands is the fact that they have never been built on before, and that they did not originate in brownfields of deindustrialized areas, even in big metropolitan areas. The urban phenomenon in India is so specific and so rapid, compared with Europe, that the 'wastelands' treated in the present article look like what in Europe are wastelands of agricultural origin where the sprawl of small and medium-sized towns progressively swallows agricultural areas as the urban front progresses.

Today, if you ask any architecture or spatial planning student at Anna University (Chennai) what are the *puramboke*, he/she will answer 'wastelands'! And of course, he/she is right, today they have turned into 'wastelands' but it was not always so.

Acknowledgements
We warmly thank:
Frédéric Landy (Director of the IFP French Institute of Pondicherry) for use of the pictures of the Aya Aiya Kulam.
Ayesha Ahmad, Saloni Shrestha, and Julie Lausin for their help with the English proofreading and with the logical framework and coherence of the article.

References

Agarwal A, Narain S (1997) Dying wisdom: rise, fall and potential of India's traditional Water Harvesting Systems, CSE, New Delhi
Ariza P, Galán E, Serrano T, Reyes V (2011) Water tanks as ecosystems. Local ecosystemic perception for integral management of water tanks in Tamil Nadu, South India. In: Aubriot O (ed) Tank and well irrigation crisis: spatial and social challenges of tank and well irrigation in Puducherry and Villupuram Districts (South India). Concept Publishing House, Delhi, pp 303–327
Aubriot O, Prabhakar P (2011) Water institutions and the revival of tanks in South India: what is at stake locally? Water Altern 4(3):325–346
Central Ground Water Body (2017) Report on aquifer mapping and ground water management—Chennai aquifer system, Government of India, Tamil Nadu. Chennai
Di Palma V (2014) Wasteland: a history. Yale University Press, New Haven and London
Ghotge N (2011) How grazing lands became 'waste' lands, Info change Agenda 'enclosure of the commons' 21:16–20
Hochart K (2019) L'Adyar n'est pas un long fleuve tranquille. Politisation de l'habiter au prisme de l'inondation à Chennai, Inde du Sud [The Adyar is not a long quiet river. Politicization of living in it through the prism of flooding in Chennai, South India]. PhD, University of Tours, Tours and Anna University, Chennai
Implementation of Land Reforms—A Review by the Land Reforms Implementation Committee of the National Development Council (1966) Planning Commission, Government of India, New Delhi
Kasturi K (2008) Whose land is 'wasteland'?, Info change Agenda 'Battles over land 11:48–50
Lakshmi K (2018) The vanishing waterbodies of Chennai. The Hindu 01(04):2018
The Land Acquisition Act (1894). Act no. 1 of 1894, Dated 2nd February 1894
Marikkani M (2012) Managing of common property resources by Gramma panchayat—a case study. Indian Streams Res J 2(4):1–4
Ministry of Environment, Forest and Climate Change (1927) The Indian forest act. An act to consolidate the law relating to forests, the transit of forest-produce and the duty leviable on timber and other forest-produce 21(09):1927
Mizushima T (1996) Mirasi system and local society in Pre-colonial South India. In: Robb P, Sugihara K, Yanagisawa H (eds) Local Agarian societies in colonial India–Japanese perspectives. Routledge, Taylor and Francis Group: London, UK, pp. 77–146
Mollinga PP (2010) Boundary concepts for interdisciplinary analysis of irrigation water management in South Asia. In: ZEF working paper series 64, Center for Development Research (ZEF), University of Bonn
Mosse D (1997) The symbolic making of a common property resource: history, ecology and locality in a tank irrigated landscape in South India. Dev Change 28(3):467–504

Mukundan TM (2005) Eri systems of South India—traditional water harvesting. PPST Bull 16:1–37
Pandey DN (2000) Sacred water and sanctified vegetation: tanks and trees in India, constituting the commons: crafting sustainable commons in the New Millennium. In: 8th Biennial conference of the international association for the study of common property. Bloomington
Reyes V, Aubriot O, Ariza-Montobbio P, Galán E, Serrano-Tovar T, Martinez J (2011) Local perception of the multifunctionality of water tanks in two villages of Tamil Nadu. South India, Soc Natl Res Int J 24(5):485–499
Ratnavel SM, Gomathinayagam P (2002) Village tanks of South Asia. Madurai, DHAN foundation
Saigal S (2011) Greening the 'Wastelands': evolving discourse on wastelands and its impact on community rights in India. In: Paper presented at the 13th Biennial conference of the international association for the study on commons
Singh S (2013) Common lands made 'Waste lands'—making of the 'Wastelands' into common lands. In: 14th Global conference of the international association for the study of the commons, June 03–07, Kitafuji, Japan
Sreenivasan R (2002) Water management and water managers in traditional tank systems. Madurai, DHAN foundation
Tamil Nadu Protection of Tanks and Eviction of Encroachment Rules (2007) Published vide notification No. G.O. (Ms.) No. 320, Public Works (W2), dated 28.09.2007
Verdelli L (2019) Integrating flood risk in urban and architectural projects. In: Bremner L (ed) Monsoon assemblages, monsoon waters. London, UK, 2019, pp. 245–257
Weitz B (2005) Water reservoirs in South India—an anthropological approach. PhD, Maxmillans university, München
Yanagisawa H (2011) Village common land. Manure, fodder, and intensive agricultural practices in Tamil Nadu from the mid-Nineteenth century. Rev Agrarian Stud 1(1):23–42

Laura Verdelli is Associate Professor at the University of Tours (France), Planning & Environment Department, UMR CITERES. She has a degree in Architecture (Politecnico di Milano, Italy), in Landscape Architecture (Fondazione Minoprio, Como, Italy), in Spatial Planning (PhD in Urban and Regional Planning in Tours, France), and in Cultural Heritage and Museology (PhD in Coimbra, Portugal). She focuses her research on the recent reappropriation of unbuilt lands along water in urban and peri-urban areas guided by heritagization processes in the Mediterranean and in India.

Geeva Chandana Balasubramanian is Assistant Professor at School of Architecture and Interior Design, SRM Institute of Science and Technology, Chennai (India). She is an architect with a Master's Degree in landscape architecture from the School of Architecture & Planning, Anna University, Chennai. During her Master's Degree she spent a semester on an exchange program in the Spatial Planning and Environment Department of the University of Tours, France. Her research focuses on urban landscape design and spatial planning, evolution of urban wastelands, revitalization of underutilized spaces in urban areas, and the efficacy of green infrastructure treatments in urban areas.

Rukkumany R. Harishankar is an Associate Professor and co-ordinates the Master's Program in Landscape Architecture at the School of Architecture and Planning, Anna University, Chennai (India). She did her architecture degree followed by a Master's Degree in landscape architecture at CEPT, Ahmedabad. She is on the verge of completing her PhD on assessing the impact of urbanization on peri-urban water management systems. Her research specifically explores the social, economic, and ecological role of traditional water management systems and how they are affected by land use changes and specifically by urbanization in the peri-urban areas of Chennai.

Unscripted Spaces. Urban Green Space and *Terrains Vagues* in Historical Perspective. Antwerp (Belgium) c. 1900

Bart Tritsmans

Abstract This article studies urban green spaces and *terrains vagues* in late nineteenth- and early twentieth-century Antwerp. The research compares the perspective of citizens with the official policy on urban public green spaces. The article reveals the importance of these "unscripted spaces" for the urban population as opposed to the negative connotation that is generally assigned to them and in contrast with the official policy about open spaces at the turn of the century.

Keywords Urban green space · Wastelands · Appropriation · Perception · Popular education · Urban history

> These strange places exist outside the city's effective circuits and productive structures. [...] In short, they are foreign to the urban system, mentally exterior in the physical interior of the city, its negative image, as much a critique as a possible alternative (de Solà-Morales 1995, pp. 119–120).

Terrains vagues have long been considered as negligible and even reprehensible elements in the history of cities. This article reevaluates this wrongful status by studying the valuable potential of urban wastelands in changing urban societies by means of the case study of Antwerp around the turn of the twentieth century. The main objective of this article is to unravel the different meanings and perceptions of urban green spaces and terrains vagues. The research reveals the importance of *terrains vagues* in times of urban transformation in the late nineteenth and early twentieth centuries by studying urban spaces from the perspective of citizens and by comparing the uses and perceptions with the official policy on urban green spaces. The term "unscripted spaces" was chosen to reflect the free, uncontrolled, and versatile nature of *terrains vagues* as opposed to the negative connotation that is generally assigned to them and in contrast with the official policy on open spaces at the turn of the century.

To reveal the tensions between the daily uses and perceptions of urban space and the official approximation, this article will look into the urban planning context

B. Tritsmans (✉)
History Department, University of Antwerp, Antwerp, Belgium
e-mail: Bart.Tritsmans@uantwerpen.be

F. Di Pietro and A. Robert (eds.), *Urban Wastelands*, Cities and Nature,
https://doi.org/10.1007/978-3-030-74882-1_16

and the government policies toward urban green space and wastelands and compare them with their everyday uses and users. The article consists of an exploration of the meaning of *terrains vagues* in urban history, a study of the government policy on urban green spaces and the citizens' perspective on urban green spaces and wastelands and the appropriation of these spaces. This article will show the multifaceted character and vital importance of wastelands in research on the history of urban space.

1 *Terrains Vagues* from a Historical Perspective

Unscripted spaces—or loose spaces (Franck and Stevens 2007)—are characterized by a range of potential meanings and functions in the urban environment, and they are constantly in flux. *Terrains vagues* are generally defined as the city's negative (de Solà-Morales 1995). Although the term implies vacancy as well as freedom, and absence as well as possibility, *terrains vagues* have often been allocated an unfavorable image—as abandoned, polluted, useless, and even dangerous sites (Barron and Mariani 2001, 2008; Qviström 2007). Nevertheless, the *terrain vague* has recently been going through a phase of revaluation. In 1985, Rem Koolhaas wrote his visionary essay "*Eloge du terrain vague*," a homage to the added value of *terrains vagues.* A decade later, Ignasi de Solà-Morales wrote his leading article, with which he established the use of the more neutral term *terrain vague* in reference to the element of fascination that characterizes these spaces.

de Solà-Morales (1995) indicated the impossibility of capturing the meaning of *terrain vagues* in a single English word or phrase. The Latin origin of the word *vague, vacuus,* means empty and unoccupied, and free and available at the same time. The second underlying meaning of the word *vague* originates in the Latin *vagus:* imprecise, blurred, uncertain. The term *terrain vague,* therefore, captures the complex meaning of these spaces: empty and unused, but also free, promising, and full of possibilities. Luc Lévesque argued that the paradoxical nature of this forgotten oxymoron determines the appeal of the French expression *terrain vague* (Lésveque 1999).

Terrains vagues—also called urban wastelands (Nabarro and Richards 1980), urban wilds (Lynch 1960), spaces of no more and not yet (Parrinello 2013), etc.—are not only characterized by a wide-ranging and often divergent terminology, but also by a kaleidoscopic array of typologies, users, and uses. Urban *terrains vagues* consist of a large number of varieties including construction sites, industrial wastelands, demilitarized ramparts, building lots, urban dumps, etc. Their accumulation forms a vast area of open space in the historical city. Although recent landscape studies, ecology research, and urban planning initiatives have addressed the value of *terrains vagues* as informal open space (Kivell and Hatfield 1998; Stilgoe 2005), the topic has remained largely unexposed to historical research.

Lésveque (1999, pp. 29) stressed the transient, temporary character of *terrains vagues*: "Interface entre la fin d'un monde et le début d'un autre, le terrain vague n'a décidément aucune 'valeur historique.' (...) C'est un espace fuyant qu'on ne peut évidemment pas tenter de 'conserver' en le fixant." Considering their ephemeral character and changing nature, attempts to categorize or to more clearly conceptualize *terrains vagues* would be a pointless exercise (Barron and Mariani 2001). However, retracing the evolving connotations of the term and analyzing the fluctuating perceptions from the perspective of a range of actors in the urban society can reveal the complex meaning of the *terrain vague* in urban societies at the turn of the twentieth century.

Since the transition to the creation of modern urban environments in the second half of the nineteenth century, the reconfiguration of the urban environment has constantly raised issues on open space. Congestion, housing conditions, green spaces, sanitation, and livability *avant-la-lettre* were profoundly discussed by municipal and national governments, urban planners, and philanthropists (Meller 1995). Despite persistent and increasingly professionalized spatial planning, and a growing urge to dispose of improper elements in the urban environment, cities were dotted with all kinds of—sometimes very large and manifest—*terrains vagues.* To explain the value of *terrains vagues* for citizens, and to reveal their controversial status, it is worth investigating how these spaces forced their way into urban policy and planning and whether *terrains vagues* were in some cases considered as an asset more than a nuisance.

In his book, *The Feel of the City,* Nicolas Kenny argued that turn-of-the-century bourgeois commentators typically portrayed cities in their totality, as entities in which flaws—such as the existence and uncharted use of *terrains vagues*—were overlooked in favor of a coherent whole. Needless to say, these panoramas of forward-moving, modern cities were but a utopian fiction (Kenny 2014). According to Latour and Hermant, the city has to be approached from the ground up, from the patchwork of individual spaces that vivifies it (Latour and Hermant 1998). Swyngedouw and Kaïka refer to Henri Lefebvre's work to remind us of what a city really is: "something akin to a vast and variegated whirlpool replete with all the ambivalence of a space full of opportunities, playfulness, and liberating potential, while being entwined with spaces of oppression, exclusion, and marginalization" (Swyngedouw and Kaïka 2000, pp. 570). In the light of Lefebvre's production of space-triad—the constant interaction of "spatial practice, representation of space and representational space" (Lefebvre 1991)—*terrains vagues* represent subversive forms of those three elements (Barron and Mariani 2001). They could be seen as "le talon d'Achille de la cité productiviste" (Lésveque 1999). Because, while planned open space has always been essential to city planning, diametrically opposed to the orderly voids, *terrains vagues* embody uncompromising discontinuity (Lésveque 1999). A pertinent remark Lévesque made on the role of urban wastelands is whether their interruptive character could be considered beneficial to an environment in which everything is increasingly planned out and pinched into the preconceived image of the modern city (Lésveque 1999). Precisely this tension between planned, controlled official urban spaces and spontaneous, unofficial, and controversial spaces lies at the heart of this article.

The daily uses and perceptions of various groups in the urban society play a pivotal role in the study of *terrains vagues.* De Certeau's influential volume, *l'Invention du*

Quotidien (1980), and Jonathan Raban's *The Soft City* (1976) inspired a new generation of historians to focus on everyday life and perceptions. New approaches, such as the city of the mind (Taverne and Wagenaar 1993; Bertels et al. 2013) and the city of the senses (Corbin 1998; Kenny 2014), were explored. Since *terrains vagues* as such are rarely part of official policies, and are often regarded as residual and ambiguous, they generally reveal themselves as spaces of which a range of people make creative, alternative, and controversial use (Tritsmans et al. 2018; Edensor 1995; Barron and Mariani 2001; Tritsmans 2015). Therefore, from a historical perspective, these unplanned, alternative spaces reverberate the heterotopic qualities, the social complexity, and the spontaneity of the modernizing city like no other (Groth and Corijn 2005; Tritsmans 2015). Although research on unofficial urban spaces is delicate—the daily use of a city is generally considered too trivial to write about, and spontaneous spaces and wastelands do not often appear in official documents besides as potential building land—it is crucial for an adequate and full understanding of urban space. During the nineteenth and twentieth centuries, a multitude of *terrains vagues* featured in growing cities on both sides of the Atlantic, providing citizens with much needed, informal open spaces (Parrinello 2013).

Whereas urban *terrains vagues* recently gained attention in (urban) planning and landscape studies, the theme is decidedly novel in the field of (urban) history. Recently, it was repeatedly suggested that wastelands should be the topic of historical research (Clark and Jauhiainen 2006; Clark et al. 2009; Brantz and Dümpelmann 2011; Tritsmans and Stynen 2015). However, until now, one could at most catch a glimpse of the *terrain vague*: hidden in the general history of a city or as an oblique aspect in historical studies on related topics such as the history of pollution (Massart-Guilbaud 2010), the recent history of environmental awareness in an urban context (Lachmund 2013), or the historical perspective on youth and urban space (Schildt and Siegfried 2005; Laakkonen 2011). Under the impulse of urban green space studies (among other research topics), *terrains vagues*—as unofficial but crucial urban spaces—are now gaining ground within urban history (Talja 2014; Tritsmans 2015). In the following paragraphs, I will trace the history of *terrains vagues* in the city of Antwerp during the late nineteenth and early twentieth centuries by studying how the urban population discovered and appropriated wastelands in and at the edge of the city. To understand the need of the population to explore unscripted spaces, I will compare the official green space policy with the daily uses and needs of the urban population.

The first part of the article is based on a thorough search of the written debates that were held by the city council, published as *Gemeenteblad*. The archives of the head engineer and the archives of the parks and public gardens service have been studied and revealed correspondence with the city government, city dwellers (often as representatives of sports or cultural associations) and the police, newspaper clippings, drawings, and reports. The archival documents shed light on the conflicts and negotiations about the use of public spaces. The perception and the everyday use of urban space, and, most importantly, the use of urban wastelands, were studied through contemporary publications. The perspective of citizens is based on the study of a range of written sources such as petitions, appeals, and letters from citizens to the city authorities, diaries, travelers' and citizens' accounts, local magazines, newspapers, and folkloristic literature. Visual sources such as maps and photographs were consulted to complement the written accounts.

2 Antwerp at the Turn of the Century: An Official View on Urban Green Space

A lot has been written about the impact of Haussmannization on the development of cities in the second half of the nineteenth century. The great attention paid to Parisian urban development in terms of the history of urban planning and architecture is not surprising, since Napoleon III and Georges-Eugène Haussmann carried out groundbreaking urbanist transformations in the mid-nineteenth century. They were also the first to integrate a green space model in their plans for the transformation of the urban fabric (Papayanis 2009). In 1863, César Daly wrote in *Revue de l'Architecture* that the most admirable interventions in Paris during the Second Republic were the new green zones. "Paris, de carrière de pierre, de moellon et de grès qu'elle était, se transforme en bouquet" (Daly 1863, pp. 249). Haussmannization marked the transition to modern urban living environments like no other. Other cities followed the example of the urbanization plans and the transformation of medieval urban tissue into rational urban spaces.

The transformation that Antwerp underwent during the second half of the nineteenth century was not comparable to the drastic interventions Haussmann carried out. Haussmann's plans demolished medieval neighborhoods to construct 137 km of boulevards. In his expansion plan for the city of Antwerp, Theodoor Van Bever retained most of the existing, organically grown neighborhoods and connected them to the newly built neighborhoods by means of new traffic axes (Fig. 1).

However, the major infrastructure works often left a deep impression on the urban population. With the new urban expansion, the city council wanted to tackle the problem of the green space shortage. Antwerp had hardly any green spaces within its walls: the zoo that was opened to the public in 1843, but for which an expensive entrance ticket had to be bought, the Promenade du glacis, a promenade that was built on the city walls from 1845 onward, and the neglected Warande Park, which was located outside the ramparts. In the literature, one can find mention of a study conducted in the early twentieth century which concludes that Antwerp was the European city with the smallest amount of public green spaces in Europe. Although this study has yet to be found, and was most likely made up to prove a point, the lack of public green space is apparent. The demolition of the Spanish ramparts between 1859 and 1881 illustrates the city government's pursuit of a modern and green image, and the efforts to profile itself as an international metropolis by constructing a new prestigious district and urban park according to the latest fashion (Tritsmans 2016). A considerable part of the city council discussions about urban expansion was devoted to the green zones that would be implanted in the new district.

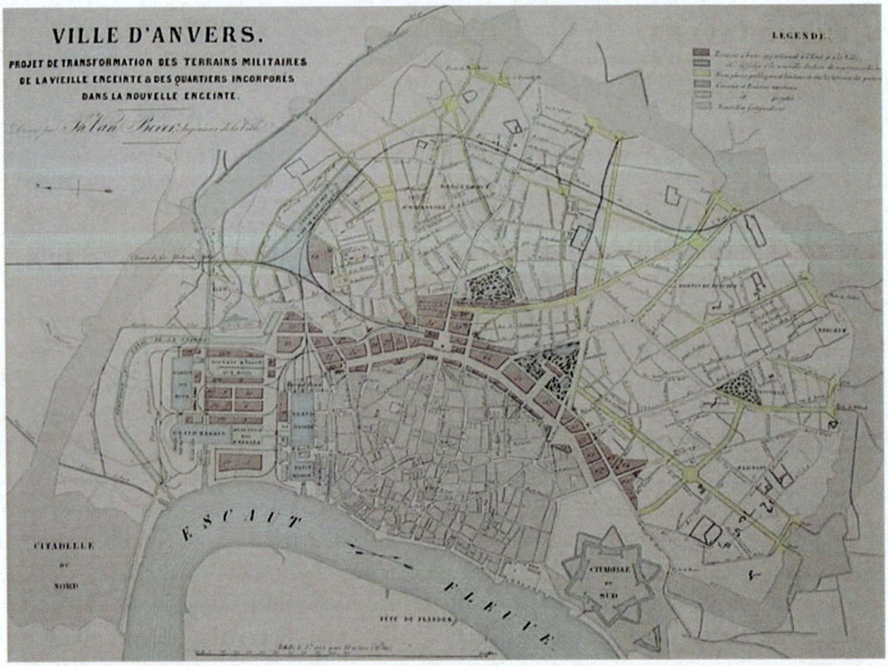

Fig. 1 Urban extension plan for Antwerp by city engineer Theodoor Van Bever, 1863. *Source* City Archive Antwerp (SAA), 12#488

Interurban contacts became more intensive, and the (inter)national urban competition encouraged city governments to adapt the urban space to new trends. This evolution is also noticeable in the specific, relatively new domain of urban green space (Clark and Jauhiainen 2006). Urban competition became an increasingly important element in the discussions. Although international prestige was an important objective, and inspiration was eagerly drawn from leading green projects in other European cities, the specificity of Antwerp remained a powerful factor. Urban green space is a topic that concerned the urban population. The urban extension plans and the transformations in the late nineteenth-century urban fabric often incited protest, but it was specifically the topic of urban green space that caused fierce debates during council meetings and in public opinion.

The disappearance of the *Promenade du glacis*, the popular promenade on the sixteenth-century ramparts, led to protests from the urban population such as several petitions, upheaval during city council meetings, etc. The opinion of the population was taken into account, but the expertise of urban planners and landscape architects prevailed. The changing urban environment and the loss of open space in the city led to debates between the urban government and the population. Besides the fact that the creation of urban green spaces was mostly decided without consulting the urban population, the strict rules and regulations that were issued by the urban government and which further restrained and guided the daily use of urban space by the city dwellers regularly caused protests, upheaval, and debates in council meetings.

3 Managing Public Green Space in the Bourgeois City

The tension over the desired use and users of urban green spaces reflects the aspiration of the city government to create a bourgeois city. The Parc des Buttes-Chaumont in Paris (opened in 1867) showed that the construction of a park in a district that was considered a dangerous, derelict place could be used as an element in the bourgeoisization of the area (Strohmayer 2006). Improving the quality of life in the district, which was located in the periphery of the city and which attracted many newcomers and people from low social classes, would improve the district and educate the residents. Strohmayer states that "the design of the Buttes subtly encourages those stereotypical forms of behavior so beloved by the bourgeoisie" (Strohmayer 2006, p. 576).

In the Antwerp city parks, visitors were expected to behave according to certain standards. The park users were closely monitored by a large number of guards and police officers, and the aversion to popular entertainment became evident in the regulations for public green spaces issued in 1870, which referred to forms of leisure that were preferably kept out of urban public spaces.[1] The city council clearly wanted to build in a quality control on the activities that took place in the public parks. Kiosks for organizing concerts became a fixture in the nineteenth-century parks, but the character of the performances was strictly monitored. Other activities in urban green spaces were also subject to rules. For example, moving with cattle, wheelbarrows, (dog) carriages, and pedal cars was prohibited on the walkways and in the parks. The park rules also prohibited walking on the lawns, damaging plants, trees, fences, statues and benches, violation of modesty, fishing, bathing, and even gathering and playing in the park.[2] Protecting the quiet, bourgeois character of the urban parks and promenades was high on the agenda (Stynen 2010; Tritsmans 2016).

During discussions by the city council, little attention was paid to recreation; a bourgeois vision aimed at international prestige dominated the debates. In other European cities, recreation was increasingly common in late nineteenth-century green spaces. In Stockholm, when the Lindhagen Plan was implemented in 1879–1880, much attention was paid to the construction of new parks that were suitable for all kinds of activities, and not just for strolling and looking at the flowers and other park visitors: "visitors could sit down on the grass wherever they wanted, drink coffee and eat from their picnic hampers, play games and during the winter season go skating and skiing" (Nolin 2006, p. 115). Recreation would also play an increasingly important role in cities such as London, Paris, Berlin, Helsinki, and Saint Petersburg during the second half of the nineteenth century (Clark and Jauhiainen 2006).

[1] Stad Antwerpen, *Gemeenteblad,* 1870:2, 315.

[2] Stad Antwerpen, *Gemeenteblad,* 1870:1, 316–317.

Between 1894 and 1899, James Whitton of Glasgow, in his role as Superintendent of Parks, made several trips through Europe to make a comparative study of the Scottish, British, and continental European parks. He noted that the British were at the forefront of providing sports infrastructure in the parks (Maver 1998). In the Netherlands, Germany, and Belgium, recreational and especially sports infrastructure was almost completely missing. It should come as no surprise that this lack of recreational opportunities regularly led to discussions and conflicts. At the turn of the century, there were conflicts and negotiations between the Antwerp City Government and different groups in the urban population.

Bouncing ball games challenged the use of urban (green) space. Squares, parks, and vacant lots in various parts of the city were provided with courts. They were often improvised and created without the intervention of the city council. The council was reluctant to allow games in public spaces since they were popular and attracted many working-class spectators and posed a potential danger to security in public spaces. Sports clubs repeatedly requested that the city government allow them to create temporary courts on undeveloped building land, and to create official courts in public spaces. But the city government was hesitant, since in their opinion the games posed a threat to passers-by.[3] An argument that also played a part in the discussion was the low-threshold character of the ball game. The popular game was mainly played in the squares in the busy, crowded neighborhoods of the city, making it a potential threat to the bourgeois image of the city. Only after a 30-year struggle by several sports clubs did the city government take action, and official courts were designed and constructed by the city engineer in various places in the city.[4] The idea that recreation deserved a place in the densely populated city center took time to mature.

An innovation that caused a stir in the city council and in public spaces was the invention of the pedal by a French carriage smith in 1864, which initiated the craze of the pedal cars (later called bicycles) in Antwerp from 1869 (Den Hollander 2006). The maladaptive handling of recreation in public spaces by the city officials is reflected in the debates about the pedal car. Although they were a luxury, and only wealthy citizens could afford one, pedal cars had the potential to seriously disrupt the peaceful atmosphere in the public green areas. The pedal car drivers were considered a danger to the walkers on the trails in the park, to the horses in the avenues, and especially to the plantations. During the council discussion about the bicycle regulations in the city park, the majority of the council members were in favor of strict regulations.[5] It was repeatedly indicated that cyclists were a danger to walkers in the park, but that the plantations, in particular, suffered severely.[6]

---

[3] City Archive Antwerp (SAA), 571#32, Archives of the head engineer, 1886.

[4] City Archive Antwerp (SAA), 571#32, Archives of the head engineer, Plans for courts designed by engineer Lemeunier, 1904–1905.

[5] Stad Antwerpen, *Gemeenteblad,* 1869:2, 54–65.

[6] Ibid., 54–65.

4 Popular Education as an Answer to Vandalism

Innocent forms of misconduct, such as dangerous cycling and ball games in public spaces, should not be surprising, given the strict regulations that forbade almost any form of recreation in all public green areas in the city. Margo De Koster and Herbert Reinke showed that imposed codes of conduct were always contested by city dwellers who appropriated urban spaces in a different way. In addition to a sense of freedom and fun, the counter-reaction, according to De Koster and Reinke, could also be seen as an expression of frustration about social exclusion (De Koster and Reinke 2012). One of the manifestations of social discontent and a deviating sense of norms, which was a thorn in the side of the city government, was vandalism.

Shortly after the park regulations were issued by the Antwerp government in 1870, recurring violations and vandalism resulted in the recruitment of a team of guards, who patrolled the park's trails day and night.[7] Vandalism, especially in public green spaces, was regularly discussed by the Antwerp City Council. The urban government was increasingly worried by the rise in the vandalism that new green spaces suffered from. Some council members argued that trees, flowers, fences, and furniture were regularly damaged, especially in working-class neighborhoods. It is remarkable that the needs of the working-class population were absent from the discussions about the creation of urban green spaces in Antwerp. In nineteenth-century England, there was a strong belief that urban greenery besides art could positively influence workers' morals. For example, the 1833 London Select Committee on Public Walks report stated that walking in a regulated public green space would contribute to a more orderly society and social emulation (Hannikainen 2014). In David Reeder's words, the central idea was that "nature, like art, was thought to have a morally beneficial influence as well as recuperative powers" (Reeder 2006, p. 43). The power of greenery as an element in popular education received a lot of attention in the international literature on urban green spaces in the late nineteenth century. These elements were largely disregarded in the discussions that arose in Antwerp.

A specific type of green space that was perceived by the city government as an instrument for popular education was the square. The concept was developed in the UK, where squares originated as planted semi-public spaces in between building blocks from the late seventeenth century on. Henri Lawrence stated that "the squares reflected and responded to changes in ideas about the use and control of land, about nature and its role in urban life, and about the shape and form of the city" (Lawrence 1993, pp. 90–91). In the late nineteenth century, the phenomenon spread across the continent as an ingredient of Haussmannization: Haussmann laid out 20 squares in Paris to inject greenery into densely built-up areas (Lawrence 1993). As a result of the great success of the squares in Paris, "une exposition permanente," and Brussels, "de vraies perles" (Stynen 2010), the Antwerp City Council developed a markedly positive attitude toward the new small status symbols. Thus, the focus on the decorative and prestige-oriented functions of public green spaces was reinforced. In the

[7] SAA, 571#7, Letters between the city architect and the city council about security in the city park, June 1869.

Fig. 2 Gustave Hermans, The square in front of the south station is fenced for the public, Antwerp, ca. 1900. *Source* SAA, 40#964

late nineteenth and early twentieth centuries, numerous squares were created; some contemporary sources speak of a wave of "squaremania."[8] The city government discovered the square as a means to inject into the urban fabric small green spaces, which were inaccessible to the population and were easy to maintain. Many urban planners and architects attributed urban greenery a beneficial function, but nature also stood for the uncivilized, dark, untamed wilderness that needed to be controlled by man (Swyngedouw and Kaïka 2000).

The city government's efforts to civilize urban space were often counterproductive. As Stynen aptly put it in his research, green projects were often built on an overly naive image of society (Stynen 2010). In particular, the idea that the construction of squares in working-class neighborhoods would improve the environment and educate the lower classes proved to be a misjudgment. Although the mayor emphasized that all strata of society suffered from innate destructiveness, it is striking that the squares mentioned during the council discussions due to recurring vandalism were all located in densely populated working-class neighborhoods.[9]

Many of the squares in working-class neighborhoods were in bad shape, although according to an alderman they received the same maintenance as all other squares in the city. The majority of the squares in Antwerp were fenced off (Fig. 2), but the destruction of plants, fences, and furniture was worse in densely populated neighborhoods, it was argued. With this remark, he indirectly alluded to the idea that squares

[8] Stad Antwerpen, *Gemeenteblad*, 1872:2, 350–359.

[9] Stad Antwerpen, *Gemeenteblad*, 1895:1, 186.

were not suitable in densely populated neighborhoods after all. In the vicinity of one of the squares, where various social housing complexes were built in the second half of the nineteenth century, "the nature of the people" regularly caused vandalism.[10] On the St. Jans square, "qui constitue une grande ressource pour la population ouvrière entassée dans les rues populeuses qui avoisinent la place St-Jean,"[11] the trees were said to be in bad condition, because vandals destroyed them.[12] The question was raised whether it would be better to replace the green space with a paved square. "Now the space has been given away to the street children and its destruction hurts the heart."[13] According to the mayor, the argument that there was not enough police surveillance was not valid: "Not a week goes by without official reports being drawn up!"[14] However, the strict action of the police was to no avail, "the destructiveness seems innate to man (…) and when the police have their back turned, the destruction begins."[15]

The Stuivenberg Cemetery, which was converted into a public square in 1887, faced a similar problem.[16] The district was known to be a rough neighborhood (Van Houtven 2010). The city engineer and the police were concerned about the attitude of the residents. Plantations were regularly damaged and trees were even torn from the ground. The city officials decided that the square would remain walled (and thus inaccessible) to protect the newly planted trees.[17] In the summer of 1897, a group of vandals climbed over the wall of Stuivenberg Square, where they extracted two young trees, and again in 1898 eleven young trees were deliberately damaged, confirming the city government in their decision to keep the square closed to the population. Only in 1899 was the wall removed and the square opened to the neighborhood.[18]

As a response to the perceived rise in vandalism and the children's lack of education, the city government came up with an idea to educate the population about caring for nature. Council members and police officers mentioned that the education of "the street youth" was deficient.[19] A tree planting ceremony was organized with a view to involving the working-class population, especially children, in the process of planting Stuivenberg Square. The idea behind this initiative was to make sure people felt connected to the trees and the space they themselves co-created. Inspired by

[10] Ibid.

[11] SAA, 571#32, Archives of the Head Engineer, Letter from Willems to the college of aldermen, 11 January 1895.

[12] Stad Antwerpen, *Gemeenteblad,* 1895:1, 185.

[13] Stad Antwerpen, *Gemeenteblad,* 1894:2, 308–309.

[14] Ibid.

[15] Ibid.

[16] SAA, MA#46092B, Archive of the Parks and Public Garden Service, Letter from the head of plantations to the city engineer, 23 January 1888.

[17] SAA, MA#46092B, Archive of the Parks and Public Garden Service, Letter from the head of plantations to the college of aldermen, 5 April 1893.

[18] SAA, MA#46092B, Archive of the Parks and Public Garden Service, Letters about the wall around the Stuivenbergplein, 1893–1899.

[19] SAA, MA#46092B, Archive of the Parks and Public Garden Service, Letter from the aldermen's college to the police, 5 April 1895.

the internationally spread belief that greenery had pedagogical potential, a large tree festival was organized on Stuivenberg Square on Sunday, 11 March 1906. The city authorities invited music companies to perform, they decorated the square with flags and other decorations and gathered up an army of dignitaries and police officers on horseback to give the tree planting the necessary cachet.[20] Six hundred children in trios received the honor of planting one of the 203 trees in the square. The city council aimed to teach the "children from the working-class neighborhood" to respect nature by actively involving them in the tree planting festival.[21]

As deputy mayor and alderman for education, an important office for the Liberal Party, Victor Desguin (1838–1919) addressed the children directly in his opening speech. Desguin was known as an advocate of improved urban education and a founder of special education and popular libraries in the city. He emphasized that the square would become a children's square, which would be planted with trees by children for their leisure, and that it would become a magnificent park in the future.[22] Desguin played on the emotions of the parents by making a comparison between the growth process of a tree and the raising of a child. The good care of the mothers in the audience would make the children strong, and teach them honor and good morality. The positive influence of the school and the proper education of the parents would ensure that the vandalism that ravaged the urban green spaces would be a thing of the past.[23] In spite of the festive atmosphere that prevailed during the tree festival, the position of the city government remained ambiguous. On the one hand, they were ardent in their advocacy of creating new green zones, but, on the other hand, they were dissatisfied with the disrespectful attitude of the urban population.

A closer look on the photograph from 1906 (Fig. 3) reveals the irony of the event. City officials, formally dressed and equipped with hats and canes, are admiring the newly planted trees on the square. Police officers are guarding the event, and in the background, behind what seem to be crush barriers or a rope are dozens of children with their parents watching the ceremony from a distance. The tree planting ceremony did not quite have the effect that the city council had hoped for. A few months after the tree planting day, the Deputy Police Commissioner of the Stuivenberg District reported that two trees and forty bushes had been damaged.[24] They had received information that "a gang of toughs from 9 to 11 years old from Lange Kool- or Veldstraat" were likely responsible for the vandalism.[25]

The city council was reluctant to take a recreational approach to urban greenery and maintained the nineteenth-century bourgeois reflex to use green as an element in the beautification of the city. Although squares increasingly suffered from vandalism,

[20] "La fête municipale des Arbres à Anvers," *Le Samedi*, 17 maart 1906, 1.

[21] SAA, MA#46092B, Archive of the Parks and Public Garden Service, Report about the tree festival by Henri De Bosschere, head of Parks and Public Garden Service, March 1906.

[22] "Discours de M. V. Desguin, échevin, f.f. de bourgemestre," *Le Samedi,* 17 March 1906, 2.

[23] Ibidem.

[24] SAA, MA#46092B, Archive of the Parks and Public Garden Service, Report about the prosecution of an unknown perpetrator by Augustinus De Smet, 21 July 1906.

[25] Stad Antwerpen, *Gemeenteblad,* 1904:2, 425.

Fig. 3 Tree planting festival, Stuivenberg Square, 1906. *Source* SAA, FOTO-OF#13799

and were considered an outdated intervention internationally, they still proved to be a popular way to bring greenery to the city. Before the First World War, they were the most frequently discussed and most implemented form of urban green in Antwerp. While increasing attention was paid internationally to the recreation of urban populations in both large and smaller cities, the Antwerp City Authorities remained convinced of the need for strict regulations and green zones devoid of recreational infrastructure. The growing problem of vandalism, which could not be solved even with an initiative such as the tree plant ceremony, illustrates the dissatisfaction of the urban population with the green politics in Antwerp. The case study of the tree planting ceremony shows the extent to which urban green spaces were not adapted to the use of the urban population; this is where the value of *terrains vagues* comes into focus.

5 Unscripted Spaces: *Terrains Vagues* as Recreational Spaces

Jorgensen and Tylecote have argued that, in research, wastelands have been considered healthy recreational zones, on the one hand, but that they had dangerous, uncontrolled connotations on the other (Jorgensen and Tylecote 2007). Ward-Thompson wrote that it is the tension between these extremes that makes the spaces so popular, and that wastelands are characterized by a wide variety of meanings for different individuals in the city (Ward-Thompson 2002). Based on examples on the border of the Swedish city of Malmö, Qviström showed that throughout the twentieth century there were regular political debates about the transformation of undeveloped land into a park, industrial area, or landfill. He recognized the function of the hybrid spaces as exceptional biotopes, informal playgrounds, shelters for young people, and decor for all kinds of dark activities (Qviström 2007). Qviström and Saltzman indicated the importance of the daily experience of the unofficial green zones in urban space. They concluded that city dwellers had a connection with the wastelands, and that these spaces reveal their main significance when studying the daily use of the city dwellers (Qviström and Saltzman 2006).

Laakkonen approached the daily experience of children in the city of Helsinki from a historical perspective (Laakkonen 2011). The connection between youth, cities, and public space has recently been the subject of research (Schildt and Siegfried 2005). Most studies, however, focus on city government attitudes toward children since it is challenging to reconstruct children's perceptions (Mayer 2009). Laakkonen studied how children experienced their immediate environment in the past, and how they adapted the environment to their own needs. The conclusions of the study highlighted an aspect that is very important in the perception of urban green spaces: the need of children for safe, loving environments on the one hand, and exciting, frightening spaces on the other. Laakkonen indicated that the sources for research into the perception of children are very scarce. His study is largely based on official sources, but he also set up a small-scale experiment with oral history. Laakkonen shows the pitfalls of the perception research very clearly. He argues that people's memories can be influenced by myths about a certain period, or by the memories of others. The world of childhood may never even have really existed, except in someone's personal memory. Collecting memories and drawing conclusions are like making a puzzle: with each piece the picture becomes more complete, and the more material is available, the more representative the result (Laakkonen 2011).

The late nineteenth century and early twentieth century show Antwerp as a city that, within its walls, in addition to densely built-up neighborhoods characterized by poor hygienic conditions and a lack of clean water and fresh air, still had a lot of wastelands and rural elements. During the discussion about setting up playgrounds in Antwerp in 1904, a council member indicated that parents often forbade their children to play on the street, not because the streets were too busy, but because "their children would come into contact with others, who unfortunately had not received enough education. […] At the Stuivenberg cemetery, I will not let my son play in the street; I don't want to speak ill of my neighbors, but unfortunately I must assert that there are many children in my neighborhood whose upbringing has been unfortunately neglected."[26] The city council rejected several requests to create playgrounds, arguing that there were enough vacant lots in the area where children could play.[27] This is a rare mention of *terrains vagues* in official documents. It shows that wastelands were seen as residual spaces where activities could take place which were banned from official urban green spaces: activities that the city council disapproved of or did not wish to invest in.

Available land in the city was mostly considered too valuable to serve as playgrounds. During a council meeting one of the council members proposed to convert Wip Square, an undeveloped area with a bad reputation on the outskirts of the city which was known as an informal recreation area, into an official playground. The other members of the council were reluctant to sacrifice valuable building land for the construction of playgrounds, but with the argument that children had for 10 years claimed the space in the densely built working-class area as a playground, the council member convinced his colleagues.[28] Wip Square was one of the scarce examples of a vacant lot that was transformed into a playground. This case shows a growing awareness for the users of urban greenery, which overall occupied a marginal position in the discourse about the creation and use of urban green spaces.

The mentions of vacant land both within the city and at its fringes pose the question of the everyday use of these spaces, and their value for the population. Based on various sources, accounts by city dwellers in the early twentieth century reveal how spaces were used and appropriated. I studied how citizens bent the rules in official green spaces, and how they discovered a myriad of unofficial, unscripted spaces and appropriated them in their everyday lives (Tritsmans 2015). Wherever possible, the memories of the city dwellers were compared with photographic material, maps, and official sources in order to paint as complete a picture as possible of the scenes that are usually left out of urban history. Taverne aptly stated: "It is striking how little we actually know about the most ordinary things" (Taverne 1978, p. 11). The sources offer the twenty-first-century reader a unique insight into the daily experience of urban public space, and show an alternative approach to the history of urban green space that is usually strongly determined by the official discourses. *Weerspiegeld Antwerpen (Antwerp reflected),* a publication from 1929, forms a unique source for research into the daily use and appropriation of urban spaces by city dwellers. As if the book was tailor-made for historical research into the perception of the city, editor Louis Franck stated in the introduction to *Weerspiegeld Antwerpen*: "The image of a city lives not only in great things and great deeds, in art, in wealth and in beauty! There is the intimacy of the environment and of everyday life…" Franck suspected that some readers would find the book superficial, that they would think "that's what happened to me, and isn't that interesting after all" (Franck 1929, p. X). However, he

[26] Stad Antwerpen, *Gemeenteblad,* 1904:2, 423–424.

[27] SAA, MA#46092B, Archive of the Parks and Public Garden Service, Letter from the college of aldermen to the Parks and Public Garden Service, 26 November 1901.

[28] Stad Antwerpen, *Gemeenteblad,* 1908:2, 703.

was convinced that the more the city changed under the pressure of time, the more it would become important to record it.

It is striking how much attention was paid to urban green spaces when people wrote about their memories and experiences of the city. Most authors in *Weerspiegeld Antwerpen* describe their childhood in the city around the turn of the twentieth century, and although there was hardly any (positive) attention for children in urban spaces during that period, almost all of them describe the public (green) spaces where they played as a child. The descriptions of the city clearly show how green spaces were not always the environments envisaged by the city council. Although the official green spaces were strictly regulated, there was a margin to circumvent the regulations. In addition, the parks, squares, and walking paths reflected a broader social mix than is often assumed in the literature. Bourgeois patterns of behavior only partially penetrated, and a significant part of the appeal of public space was the presence of popular scenes (Tritsmans 2015, 2016).

The most striking element is the importance of *terrains vagues* in the everyday experience of the city. Undeveloped land and meadows within the city limits were eagerly used by children (Fig. 4). Since there was hardly any official space for recreation in the city, children created their own playgrounds. Broeders (1929, p. 92) described the use of the wasteland around Stuivenberg in the early twentieth century: "Thousands of children were swarming and romping there on public holidays [...], whole battalions of soldiers were drilled under the critical eye of old people, idlers

Fig. 4 Léon Keusters, Children playing on a *terrain vague,* Antwerp, late nineteenth century. *Source* SAA, FOTO-OF#7264

and boys." Despite the police checks that were carried out regularly, the vacant lots were a paradise for the city children. "On the wild blades of grass that were called the pasture, caring house mothers bleached their linen. In the middle, where the pasture was bumpy and bold like a dune pan, we, the bad boys, held sway. We dug pits there, built ourselves bulwarks in the ground! - organized shows."

The vacant lots not only served as play areas, but also as places for the gathering of "miserable subjects and tramps and receptacles for all kinds of bad waste. Especially at nightfall the surroundings became a bit 'unheimlich' [creepy], an impression, for young and old, reinforced by the memory of the many assaults, of which the hollow roads and wasteland around the old cemetery had been the scene a few years earlier." Van Hoof (1929, p 82) also wrote about the fallow lands that were used by all kinds of folk figures: "donkeys, mules and a few skinny horses ran around on the wasteland. I believe that all the scissor sharpeners, hawkers and tinkers from all over our country came to buy their four-legged friends at the Dam."

Wastelands were not only to be found in the less well-to-do, peripherally situated neighborhoods. While the *terrains vagues* were mainly associated with playing, mischief, bleaching laundry and grazing donkeys and horses by the inhabitants of the working-class neighborhoods, they were mainly associated with all sorts of shady affairs in bourgeois areas. Qviström used the weed metaphor for this phenomenon: he wrote that activities, plants, buildings, and even people who found themselves in the shadow areas of the city were considered weeds, especially if the fallow areas were near prestigious neighborhoods (Qviström 2007). An example is the bourgeois Zurenborg District where the presence of the *terrains vagues* was considered a blot on the distinguished character of the neighborhood (Schmook 1929, p. 180). Despite the negative connotation of the vacant spaces around Zurenborg, late nineteenth-century photographs of the newly created environment around the Cogels-Osylei show children playing on the wasteland between the bourgeois palaces. In the South, a neighborhood characterized by its star-shaped street pattern and stately buildings, a lot of land remained undeveloped around the new fine arts museum. Horemans (1929, p. 315), who grew up in the neighborhood, remarked that parts of the South "behind its outwardly rigid and neat bourgeois character, hid in their innermost essence a genuine popular image [...]. The vacant area offered young people the opportunity to play and relax; the vacant lots soon became playing meadows, on which the turbulent children of the surrounding area romped and played."

The vacant spaces around the city walls were also a dream playground for the city children (Fig. 5). The authors in *Weerspiegeld Antwerpen* regularly mentioned the city walls as a playground, and their descriptions once again demonstrate the multilayered, accessible, and almost uncontrolled character of the space. Broeders (1929, p. 159) lived in the northern part of the city and described the rampart as a place that lent itself to "swimming, squirting, roaming fields, meadows and canals and doing all kinds of mischief." Broeders' peer Eekeleers (1929, p. 194), who described the ramparts in the south of the city, painted a similar picture of the use of the ramparts.

Although Antwerp was a city with a very limited surface area of public green spaces, an image that remains intact after a study of the official green spaces, the

Fig. 5 Léon Keusters, Children playing on the ramparts around the city, Antwerp, c. 1900. *Source* SAA, FOTO-OF#17

perception of the residents sheds a new light on Antwerp's status as a green-poor city. More than the city council wanted to believe, urban greenery presented itself as a multilayered space (Tritsmans 2015). Much more than was indicated in previous research, the use of public space by city dwellers is of great importance for further historical research into urban space. After all, the perception of the city dwellers strongly contributes to the nuance of the official discourse as it has been the central theme until now. Unofficial green spaces such as *terrains vagues* and (vacant spaces around) city walls play a major role in the perception of urban space, which shows the particular importance that almost every author attributed to urban green space. But it also illustrates how unofficial urban green spaces compensated for the lack of recreational opportunities in the official city parks and squares.

6 Conclusions

During the second half of the nineteenth century, radical urban renewal projects were undertaken, transforming medieval city cores into modern, sophisticated urban centers characterized by urbanistic features like strolling boulevards, city parks, and thoroughfares. These transformations will not only have led to the occurrence of new *terrains vagues*, but the existing wastelands will also have functioned both as alternative, unregulated open spaces in an exceedingly regulated urban environment, as a nostalgic hideaway from modernity, and as free zones for activities that were banned from the bourgeoisie-oriented city centers. *Terrains vagues* reveal their most important meanings when studied in terms of how city dwellers used and perceived them. *Terrains vagues* were a dream playground for children, but they also served as a sports field, strolling boulevard, fish pond, and secret café, as well as a refuge for sexual offenses, questionable business transactions, a construction site for slum dwellings, etc. Earlier research has pointed out that, in times of urban crises or landslide decisions on the transformation of the urban fabric, (temporary) *terrains vagues* constitute both safe havens and free zones for the urban population. The perception of urban space not only shows the particular importance that city dwellers attributed to open spaces, but also the extent to which unofficial urban areas could offset the lack—or the restricted character—of official open space (Tritsmans 2015). These unscripted spaces had a considerable impact on the quality of life and the perception of a city, on the sustainability and the resilience of the urban society, and on the various forms of social interaction, collectivity, and solidarity (Edensor 1995; Barron and Mariani 2001).

Terrains vagues are characterized by a palimpsestic nature. While from an official perspective, these spaces are generally perceived as a free zone for undesirable elements in the urban environment (which on occasion, they were), they also functioned as playgrounds, bleaching greens, sports fields, and vegetable gardens. The ambiguous character of the spaces and the social tensions they raise are reflected in the accounts and experiences written by contemporaries. A recurring dichotomy is the tension between the bourgeois image of the city and the daily use of urban space by the diverse city population. After all, the presence of *terrains vagues* was not restricted to the less well-off, peripherally situated neighborhoods. But while the inhabitants of working-class neighborhoods associated *terrains vagues* primarily with juvenile mischief, bleaching laundry, and grazing donkeys and horses, in bourgeois circles they were associated with all kinds of shady business. This article exposes the versatile uses and perceptions, and the changing value of *terrains vagues* from the perspective of different actors in urban society. *Terrains vagues* assemble an amalgam of meanings, experiences, customs, and images. Moreover, much more than previous studies indicated, the use and perception of unofficial space by the citizens is crucial in the future development of research on public spaces within the field of urban history.

Terrains vagues have had a considerable impact on the perceptions of the urban landscape and the quality of life in the city. Moreover, the value of these previously uncharted *terrains vagues* also reveals the biased nature of urban studies in which they are not taken into account. The urgency of historical research on *terrains vagues* is demonstrated by the case study of the history of urban green spaces in Antwerp around the turn of the twentieth century. The lack of official green spaces and the official vision of the use of urban space can be called into question when contrasted with the perception of contemporary citizens. The city government's attempts to educate and control the urban population in urban green spaces were regularly contested. Unofficial green areas such as fallow (building) land and ramparts proved to be of great significance for the townspeople of fin-de-siècle Antwerp. The perception of urban space shows the extent to which alternative, unofficial urban spaces could offset the lack of official urban green space. In short, research points out that we cannot fully understand cities without taking into account the important role of *terrains vagues*. Further historical research into urban wastelands can contribute to the existing research by acquiring more in-depth knowledge about the significance of unscripted urban spaces through a study of the uses and perceptions of various actors in changing urban societies, by broadening the scope of the spatial dimension in architectural and urban history with an analysis of urban *terrains vagues*, and by expanding and nuancing the existing concepts and knowledge about *terrains vagues* by means of a much needed historical dimension.

References

Barron P, Mariani M (eds) (2001) Terrain vague: interstices at the edge of the pale. Routledge, London

Bertels I, Furnée JH, Sintobin T, Vandevoorde H, van de Schoor R (eds) (2013) Tussen beleving en verbeelding. De stad in de negentiende-eeuwse literatuur. Universitaire Pers, Leuven

Brantz D, Dümpelmann S (eds) (2011) Greening the city. Urban landscapes in the twentieth century. University of Virginia Press, Charlottesville

Broeders J (1929) Het faboert. In: Franck L (ed) Weerspiegeld Antwerpen. Hoe 't vroeger was: onze schrijvers over hun stad. De Sikkel, Antwerpen, pp 153–161

Clark P, Niemi M, Niemelä J (eds) (2009) Sport, recreation and green space in the European City. Finnish Literature Society, Helsinki

Clark P, Jauhiainen J (eds) (2006) The European City and green space. London, Stockholm, Helsinki and St Petersburg, 1850-2000. Ashgate, Aldershot

Corbin A (1998) Village bells: sound and meaning in the nineteenth-century French countryside. Columbia University Press, New York

Daly C (1863) Promenades et plantations. Parcs. Jardins publiques. Squares et boulevards de Paris. Revue De L'architecture 21:245–249

De Solà-Morales I (1995) Terrain vague. In: Davidson C (ed) Anyplace. The MIT Press, Cambridge, pp 118–123

De Koster M, Reinke H (2012) Geheimen van jongeren. De Antwerpse jeugd en haar nachtleven in de vroege twintigste eeuw. Tijdschrift over Cultuur Criminaliteit 2(2):32–46

Den Hollander M (2006) Sport in 't stad: Antwerpen 1830–1914. Universitaire Pers Leuven, Leuven

Edensor T (1995) Industrial ruins: space, aesthetics and materiality. Bloomsbury, Oxford

Eekeleers W (1929) Het Half-Maantje. In: Franck L (ed) Weerspiegeld Antwerpen. Hoe 't vroeger was: onze schrijvers over hun stad. De Sikkel, Antwerpen, pp 194–198

Franck L (1929) Tot inleiding. In: Franck L (ed) Weerspiegeld Antwerpen. Hoe 't vroeger was: onze schrijvers over hun stad. De Sikkel, Antwerpen, pp V–XI

Franck K, Stevens Q (eds) (2007) Loose space: possibility and diversity in public life. Routledge, London

Groth J, Corijn E (2005) Reclaiming urbanity: indeterminate spaces, informal actors and urban agenda setting. Urban Stud 42(3):503–526

Hannikainen M (2014) The greening of London—The development of public green spaces in Camden and Southwark from the mid-1920s to the late 1990s, Unpublished dissertation University of Helsinki, Helsinki

Horemans J (1929) Het Zuidkwartier. In: Weerspiegeld Antwerpen. Hoe 't vroeger was: onze schrijvers over hun stad. De Sikkel, Antwerpen, pp 315–317
Jorgensen A, Tylecote M (2007) Ambivalent landscapes—wilderness in the urban interstices. Landscape Res 32(4):443–462
Kenny N (2014) The feel of the city. Experiences of urban transformation. University of Toronto Press, Toronto
Kivell P, Hatfield S (1998) Derelict land: some positive perspectives. In: Kivell P, Roberts P, Walker GP (eds) Environment, planning and land use. Routledge, London, pp 118–130
Laakkonen S (2011) Asphalt kids and the matrix city: reminiscences of children's urban environmental history. Urban History 38(2):301–323
Lachmund J (2013) Greening Berlin: the co-production of science, politics, and urban nature. The MIT Press, London
Latour B, Hermant E (1998) Paris ville invisible. Institut Synthélabo pour le progrès de la connaissance, Paris
Lawrence HW (1993) The greening of the squares of London: transformation of urban landscapes and ideals. Ann Assoc Am Geogr 83(1):90–118
Lefebvre H (1991) The production of space. Blackwell Publishing, Oxford
Lésveque L (1999) Montréal, l'informe urbanite des terrains vagues: pour un egestion créatrice du mobilier urban. Les annals de la recherché urbaine 85:47–57
Lynch K (1960) The image of the city. The MIT Press, Cambridge
Massart-Guilbaud G (2010) Histoire de la pollution industrielle. France, 1789–1914. Ehess, Paris
Maver I (1998) Glasgow's public parks and the community, 1850–1914: a case study in Scottish civic interventionism. Urban History 25(3):323–347
Mayer I (2009) Children and the quest for purity in the nineteenth century Scottish city. Paedagogica Historica 33(3):801–824
Meller H (1995) Urban renewal and citizenship: the quality of life in British cities, 1890–1990. Urban History 22(1):63–83
Nabarro R, Richards D (1980) Wasteland: a Thames television report. Associated Book Publishers, London
Nolin C (2006) Stockholm's urban parks: meeting places and social contexts from 1860–1930. In: Clark P (ed) The European city and green space. London, Stockholm, Helsinki and St Petersburg, 1850-2000. Ashgate, Aldershot, pp 111–126
Papayanis N (2009) Planning Paris before Haussmann. Johns Hopkins University Press, Baltimore
Parrinello G (2013) The city-territory: large-scale planning and development policies in the aftermath of the 1968 Belice earthquake (Sicily). Plan Perspect Int J History Plan Environ 28(4):571–593
Qviström M (2007) Landscapes out of order: studying the inner urban fringe beyond the rural—urban divide. Human Geography 89(3):269–282
Qviström M (2008) A waste of time? On spatial planning and 'wastelands' at the city edge of Malmö (Sweden). Urban For Urban Green 7:157–169
Qviström M, Saltzman K (2006) Exploring landscape dynamics at the edge of the city: spatial plans and everyday places at the inner urban Fringe of Malmö Sweden. Landscape Res 31(1):21–41
Reeder D (2006) The social construction of green space in London prior to the Second World War. In: Clark P (ed) The European city and green space. London, Stockholm, Helsinki and St Petersburg, 1850-2000. Ashgate, Aldershot, pp 41–67
Schildt A, Siegfried D (eds) (2005) European cities, youth and the public sphere in the twentieth century. Ashgate, Aldershot
Schmook G (1929) Zurenborg. In: Franck L (ed) Weerspiegeld Antwerpen. Hoe 't vroeger was: onze schrijvers over hun stad. De Sikkel, Antwerpen, pp 178–193
Stilgoe J (2005) Landscape and images. University of Virginia Press, Charlottesville
Strohmayer U (2006) Urban design and civic spaces: nature at the Parc des Buttes-Chaumont in Paris. Cultural Geograph 13:557–576

Stynen A (2010) Proeftuinen van burgerlijkheid. Stadsnatuur in negentiende-eeuws België. Unpublished dissertation, University of Leuven, Leuven

Swyngedouw E, Kaïka M (2000) The environment of the city… or the urbanization of nature. In: Bridge G, Watson S (eds) A companion to the city. Blackwell Publishers, London, pp 567–580

Talja S (2014) Sport, recreation and space in urban policy: Helsinki and Dublin from the 1940s to the 1970s. Unpublished Dissertation. University of Helsinki, Helsinki

Taverne E (1978) In 't land van belofte: in de nieue stad. Ideaal en werkelijkheid van de stadsuitleg in de Republiek 1580–1680. Schwartz, Maarssen

Taverne E, Wagenaar C (1993) De 'city of the mind'. In: Taverne E, Visser I (ed) Stedebouw, de geschiedenis van de stad in de Nederlanden van 1500 tot heden. SUN, Nijmegen, pp 330–357

Tritsmans B (2015) Versatile green. An alternative perspective on green space, late nineteenth-century Antwerp. Urban History 42(1):89–112

Tritsmans B (2016) Bomen zijn waardevolle bijkomstigheden. Stedelijk groen in Antwerpen, 1859–1973. Universitaire Pers Leuven, Leuven

Tritsmans B, Stynen A (2015) Stadsgroen onder spanning: debatten over ruimtegebruik negentiende-eeuws Antwerpen. Tijdschrift Voor Geschiedenis 128(1):45–68

Tritsmans B, Dewaelheyns V, De Smet A, Vanbutsele S (2018) Exploring Semi-natural spaces in the val de béguines. In: Dewaelheyns V, Leinfelder H, Gullinck H (eds) Challenging the boxes. Interfaces in landscape and land use. Gompel & Svacina, Oud-Turnhout, pp 132–145

Van Hoof F (1929) De Dam of de parochie van St-Job. In: Franck L (ed) Weerspiegeld Antwerpen. Hoe 't vroeger was: onze schrijvers over hun stad. De Sikkel, Antwerpen, pp 78–86

Van Houtven S (2010) De negentiende-eeuwse stedelijke politiepraktijk door de bril van de buurt de Antwerpse Seefhoek, 1879. In: De Koster M, De Munck B, Greefs H, Willems B, Winter A (eds) Werken aan de stad. Stedelijke actoren en structuren in de Zuidelijke Nederlanden 1500–1900. Vrije Universiteit Brussel, Brussel, pp 269–278

Ward-Thompson C (2002) Urban open space in the 21st century. Landscape Urban Plan 60:59–72

Dr. Bart Tritsmans currently teaches at the History Department of the University of Antwerp (Belgium). He obtained a PhD in history (University of Antwerp) and architectural engineering (Vrije Universiteit Brussel) in 2014 with a study about the changing policies, debates, and uses concerning urban green space. His research focuses on the historical evolution of urban space and wastelands in the nineteenth and twentieth centuries. As head of exhibitions at the Flanders Architecture Institute, he has curated several exhibitions on contemporary and historical topics in architecture and urban space.

Conclusion. Anthropocene Wastelands: From the Margins to the Center

Rémi Beau

Abstract For a long time, wastelands have been geographically and intellectually kept on the margins of society. The dualistic thinking of the relationship between humans and nature permeated modern representations of places where the uncertain or disused sites were held in contempt. However, the deconstruction of the nature–culture dualism opened up a new theoretical landscape that paved the way for the revaluation of wastelands. Combined with the development of urban ecology in a context characterized by the scale of human presence on earth, postmodern ecological thinking tends to place wastelands at the center of attention as places where new ways of living in an unstable world are being developed. From the margins to the center, wastelands challenge our representations of places and forms of life in the Anthropocene.

Keywords Modernity • Postmodernity • Dualism • Nature–culture • Wildness • Multispecies living places • Anthropocene

1 Introduction

Representations of urban wastelands have changed considerably since the last decades of the twentieth century. How have these places, long regarded negatively as uncertain badlands, gradually attracted the attention of ecologists and social scientists? Can brownfields be described as a new form of urban nature, and even more so as desirable nature? Such a reassessment would indicate a cultural transformation in the relationship to nature in the Western world since scientific and philosophical modernity. As French anthropologist Phillipe Descola has established, Western ontology is based on the binary opposition between Nature and Culture and this dualism shapes the social representations of places. In this respect, the two main spatial categories

R. Beau (✉)
Institut d'Ecologie et des Sciences de l'environnement de Paris (IEES Paris), Sorbonne Université, CNRS, UMR 7618 Paris, France

F. Di Pietro and A. Robert (eds.), *Urban Wastelands*, Cities and Nature,
https://doi.org/10.1007/978-3-030-74882-1

of modernity are "the City" and "Nature." They represent the opposite sides of the nature/culture dualism. Each of them is also a center of gravity for modern thinking. On the one hand, European cities were seen as the most favorable environment for the enlightenment of individuals. On the other hand, Nature, defined as the non-human part of the world, has been invested with spiritual and esthetic values since the eighteenth century. So, wild nature and cities are both antithetical and complementary in modern thinking. While the modernity of a city can be measured by its degree of artificiality, the value of nature depends on the absence of human modifications. This cleavage is a cultural source of the contempt expressed for the uncertain places of cities where wild nature occasionally takes over human artifacts. There was no place for wastelands in the binary spatial distribution of modernity.

The hypothesis I will defend here is that the role and value of wastelands as wild and disused places in modern cities could not be recognized until the nature–culture dualism was challenged. Its deconstruction began precisely with the renewal of ecological thinking from the 1970s onwards. At that time, some thinkers, researchers, and activists initiated a postmodern shift in ecological thinking in the sense that they wanted to think about the environmental issues beyond the nature–culture dualism.

Starting from the analysis of negative representations of wastelands in modernity, I will highlight here how theoretical changes brought about by this combination of postmodern thinking and ecological research have contributed to the revaluation of urban wastelands. Following this path, I will analyze how postmodern environmentalism even tends to put urban wastelands at the center of attention. This new centrality will ultimately be challenged in the light of certain criticisms that highlight the social and ecological consequences that can result from such a theoretical and practical move.

2 The Marginalization of Wastelands in Modernity

In one of his most famous books (Latour 2004), French sociologist Bruno Latour offers a provocative interpretation of Modernity as a new worldview, new epistemology, and new political constitution. His main thesis, which he calls "the modern paradox," is that this constitution is both self-destructive and terribly effective, and in a way defeated by its effectiveness. At the heart of this paradox is the dualism that establishes the great divide between nature and society and polarizes the two main sets of practices of modernity: purification and mediation. While the former seeks to separate the things that belong to the order of nature from those that belong to society, the latter precipitates the production of hybrids of nature and culture. From this perspective, the "Moderns" wanted to establish theoretical boundaries between nature and culture while producing objects that constantly cross the boundaries between the natural and the artificial. In concrete terms, the modern constitution has contributed to making the Industrial Revolution possible and at the same time has made it unthinkable.

Examining the history of Western land use planning since the end of the eighteenth century through the perspective of this dualistic theoretical framework gives particular meaning to the relationship to urban wastelands in modern societies. Many ambitious projects for the transformation of European cities were developed in the eighteenth century, but the process of urban modernization really took off in the nineteenth century. Modern town planning emerged as a specific field where the city was largely understood in a functionalist and rationalist way. In France, for example, the transformation of Paris was notably led by Prefect Haussmann (1809–1891) with the objectives of providing citizens with light, water, air, and space. This urban renewal was also motivated by political reasons. In a nutshell, the "embellishment" of Paris was a tool for establishing and maintaining social order against the seditious mood of Parisians (Lefebvre 1968).

Nature was not totally absent from this urban project. Echoing the search for the restoration of social order, the town planners worked to introduce a tamed and orderly nature into cities. However, while trees were planted in the avenues, gardens, and parks created in large cities according to certain hygienic guidelines, there was no room for spontaneous plants or wild animals and no room for unruly nature in functional and modern cities (Mathis and Pépy 2017).

Thus, the rationalization of spatial planning initiated at the time by Western Europe paved the way for very negative representations of empty spaces, underused areas, or abandoned places where spontaneous nature continued to flourish. The modernization of European cities has in fact led to the social construction of these places, which were no longer considered to be anything more than waste, residual elements of modernity. Since then, urban wastelands have mainly been regarded as worthless spaces waiting to be rehabilitated. While these negative judgments were aimed at places, they also included the human and non-human inhabitants of the wastelands. Thus, it can be said that the devaluation of wastelands has since then more broadly qualified those human and non-human beings, who have not been able to find their place in the new spatial order of Western societies.

A Latourian analysis would say that what makes these places so repugnant to modern dualistic thinking is that they remain undefined, no longer belonging to the order of nature and resistant to the order of society (Lizet 1989). On the one hand, wastelands appear to the functionalist view as an underused space since they are not dedicated to a specific function. On the other hand, they bear too many traces of human presence to be recognized as true nature. As the nature writer and professor in literature, John Tallmadge says about the undisciplined urban nature:

> The fact is that urban landscapes are just too mixed up, chaotic, and confused to fit our established notions of beauty and value in nature. Maybe it's not really nature at all, not a real ecosystem, just a bunch of weeds and exotics mixed up with human junk (Tallmadge 2004, p. 43).

Thus, the dualistic framework renders places as wastelands unthinkable for modern people in the sense that they do not correspond to any modern spatial category.

The paradoxical dimension of Modernity, underlined by Latour, really appears when one considers the fact that the production of wastelands is largely a consequence of modern urbanism. Urban wastelands are indeed pure products of Modernity. First of all, uncertain places were not labeled as waste until we considered land mainly from the perspective of productivism and rationalism. Secondly, the spatial transformation of modern cities and the construction of infrastructures such as railway stations, factories, and roads have concretely produced many intermediate or peripheral places that have fallen into the category of urban wastelands. Returning to the transformation of Paris, the Haussmann renovation of the old center had a considerable impact on the periphery of the city, reshaping the boundaries of Paris known as the "zone" (Cannon 2017). The "zone" of Paris, which draws a wider space on the outskirts of the city between the former Wall of the Farmers General and the Thiers Wall, is a typical example of the production of margins of urban life (Merriman 1991). By-products of modernization, these places became objects of contempt and fear for the urban bourgeoisie, who saw in them the association of mixed landscapes with dangerous and seditious people (Lefebvre 1968; Merriman 1991). In the spirit of the Moderns, urban wastelands appear to be reminiscent of the old order. They are hybrid places, made of nature and culture, waiting for the process of purification. In short, urban wastelands materialize the great paradox of Modernity, which is to accelerate the production of places unthinkable for the Moderns.

Criticized by urban planners, the resistance of wasteland to rationalism and functionalism has nevertheless found better allies within another current of thought: Romanticism. As influential as it was, rationalism did not define the only way for Westerners to deal with the world. It is well known that, while the rationalist idea of progress was taking off in Europe, guiding political and industrial revolutions in different ways, a strong feeling for nature emerged at the same time and was expressed in the arts of the eighteenth century in particular. This concern for nature blossomed fully in the nineteenth century in Romantic thought and art, which took a stand against rationalism. Faced with what they saw as the cold modernization of European cities, some thinkers and writers developed a taste for urban wastelands. Among them, Victor Hugo described the Parisian suburbs in "Les Misérables" as fuzzy, but also eminently fascinating places. The French Romantic writer forcefully characterized the hybrid character of the borders of Paris:

> Wandering around musing, in other words dawdling, is a good way to spend time for a philosopher; particularly in that funny, rather ugly semi-rural landscape, with its odd, dual nature, that surrounds certain big cities, notably Paris. To observe the urban outskirts is to observe the amphibian. End of trees, beginning of roofs, end of grass, beginning of pavement, end of furrows, beginning of shops, end of ruts, beginning of passions, end of divine murmuring, beginning of human racket; whence the extraordinary interest (Hugo 2010, p. 963).

However, while the Romantics have made a positive contribution to the requalification of urban wastelands, these discourses have not really changed their status as vestiges of modernization, trapped between nature and society. Indeed, if Romanticism wanted to break with the dualism of rationalist thinking and work toward a monistic reconciliation between the spiritual and the material, human and nature, the natural sites chosen by Romantic thinkers and artists as the best places for an individual to merge with nature were mainly the most remote sites, isolated from society. Virgin forests, deserts, and mountains became the romantic's favorite "spots" (Wordsworth 1995), the places where nature in its pure beauty was most capable of triggering the feeling of the sublime. In such a setting, the urban wildness seemed pale in comparison.

The nature conservation movement, which originated in the second half of the nineteenth century, largely inherited this romantic view of nature (Nash 1967). Proponents of nature conservation have learned from it that nature is at its best when it is removed from all forms of human activity. In this respect, the movement did not participate in the deconstruction of the opposition between humans and nature. Rather it claimed the necessity to preserve the counterpart of human places at a time when industrialization was consuming natural places at a rapid pace. The creation of national parks, which was the most tangible result of the movement, was compatible with the modern spatial distribution of humans and nature. In that sense, nature conservation remains modern until the last quarter of the twentieth century. This modernism explains why environmentalists for a long time paid so little attention to urban nature and even less to urban wastelands.

3 Urban Wastelands in the Anthropocene: The "New Wild" and Multispecies Living Places

The dualistic framework of modernity that dominated spatial and environmental thinking during the nineteenth and twentieth centuries was seriously challenged at the end of the twentieth century, both through a postmodern current of thought's encounter with ecology and the changing global environmental context.

Far from being a unified current, postmodernism could at least be defined by a common intention to overcome modernity. Popularized in reference to developments in architecture opposed to the objectivism and functionalism of the modern movement, the term has spread to many areas of art and thought and generally qualifies a kind of skepticism and irony in the face of the great narrative of Western Modernity (Lyotard 1984). Thus, postmodernism could be minimally described as a critique of the progressive vision of the historical development of societies.

Given this stance, it is easy to conceive that postmodernism has established a strong link with the environmental movement that was reborn in the 1960s. Indeed, the environmental philosophy that emerged in the last quarter of the twentieth century focused on the ecological critique of Modernity and its consequences. In this sense, the current attempted to identify the main theoretical roots of the environmental crisis, which seemed to be nothing more than dualism, rationalism, and faith in progress and technology. Thus, like postmodernism, environmentalism worked to deconstruct the theoretical framework that gave rise to the idea that man's "raison d'être" was the domination of nature.

The deconstruction of modern categories of thought has nevertheless raised an important question for environmentalist thinkers about the idea of nature. Indeed, while the social sciences were ready to give up the idea of nature entirely (Haber 2006), such constructivism seemed to contradict some of the goals of the environmental movement that were still aligned with the idea of protecting nature. This concern gave rise to one of the most important debates within environmental thinking. Known as the "wilderness debate" in the United States (Callicott and Nelson 2008, 1998), this debate challenged the predominance of the conception of wilderness as a space free of human intervention in environmental thought and policy. Contrary to this view, opponents of the classical idea of wilderness have argued that many other forms of nature were valuable, including in places modified by humans. This reassessment of natural places free of the nature–culture dualism could lead to substantial changes in the way we view the places we inhabit. As the eco-feminist philosopher Val Plumwood argued in the wilderness debate:

> Defining our wilderness experience as a quest for the presence of the wild nature, not the absence of humans, creates conceptual space for the interwoven continuum of nature and culture, and for that recognition of the presence of the wild and of the labor of nature we need to make in all our life contexts, both in wilderness and in places closer to home (Plumwood 1998, p. 684).

Described as "postmodern deconstructionist" researchers by their critics (Foreman 2008), authors such as the philosopher Baird Callicott and the historian William Cronon have followed the same path in trying to renew the conception of wilderness in order to think not only of distant spaces but also of nearby places where spontaneous forms of nature can be found. In a widely discussed article, Cronon defined his approach as follows:

> But if we acknowledge the autonomy and otherness of the things and creatures around us—an autonomy our culture has taught us to label with the word 'wild'—then we will at least think carefully about the uses to which we put them, and even ask if we should use them at all (Cronon 1998, p. 495).

This conception of the "wild" has opened up new ground for the search for better ways to interact with nature. While the "wilderness framework" was only oriented toward the great outdoors, the call to take care of the nearby wildness advocated the complementary investigation of inhabited or urban places in search of autonomous natural processes. This research leads precisely toward wastelands.

As a matter of fact, since the 1970s, a few scientists and thinkers in different parts of the Western world have been working to change the way we used to look at urban wastelands. In the scientific field, some botanists have started to study the specific flora that blooms in urban or peri-urban places (Lizet et al. 1999; Sukopp and Hejný 1990). Initiating the development of urban ecology, these studies have contributed to the recognition of the ecological role of ruderal species found in these neglected and intermediate places. The idea that something might have ecological value in this "unofficial countryside" (Mabey 2010) still had a long way to go, but it gradually helped to soften the negative representations of wastelands.

Meanwhile, on the literary level, the reappraisal of wastelands benefits from the reinvention of the figure of the "flâneur" in postmodern literature. Classically framed by Walter Benjamin (Benjamin 1992), in reference to Baudelaire's writings, the flâneur appears as a distinctly modern figure since he was introduced to think about the urban condition in modern cities. For Benjamin, wandering in Paris was a means of escaping the social controls of modern urbanism and a way of rediscovering a kind of experience of the city without intermediaries. On this point, the novelty of the postmodern wanderer is that he aims less at finding a way of thinking about the human condition in a modern city than at exploring the wilderness where he can observe the tangled human and non-human worlds. This wilderness wanderer has become a central figure in the emerging currents of "more-than-human" literature or the "new nature writing" (Smith 2017). Thus, during the last quarter of the twentieth century, in the words of British geographer Matthew Gandy:

> the marginal spaces of Berlin, London, Montreal, and other cities were becoming a significant focus for cultural and scientific attention that reflected a series of developments such as the emergence of new art practices, increasing levels of ecological awareness, and the changing characteristics of cities themselves (Gandy 2013, p. 1301).

The beginning of the twenty-first century has seen this new interest in wastelands not only confirmed but also increased. Research in urban ecology has intensified considerably, giving a significant place to the study of biodiversity on brownfields (Bonthoux et al. 2014; Muratet 2017). As for the social sciences, the development of the "environmental humanities" (van Dooren et al. 2016), including fields such as multispecies ethnography, more-than-human geography, and anthropology beyond humanity, has triggered the publication of many original works concerning wastelands approached as hybrid places where human and non-human beings make up new forms of life (Gandy 2013; Haraway 2016; Lorimer 2015; Tsing 2017a). The real novelty that emerges from these studies is that wastelands are now considered as spaces in their own right and no longer simply defined by what is on either side. Some voices have even been raised in favor of protecting urban wastelands.

Indeed, in the new world of the Anthropocene, characterized by what ecologist Bill McKibben called in the 1980s the "end of nature" (McKibben 2006), humans are everywhere, even in the most remote regions exposed to climate change. Thus, for some environmentalists, the Anthropocene has changed the status of protected areas. They no longer embody the non-human part of the world; they are wild places in a human world. This is the condition of the "new wild" in the Anthropocene. In this framework, the theoretical boundary between wilderness areas and spontaneous urban nature has disappeared. Moreover, if we fail to protect the former, the latter could be the new hope for the future of biodiversity. In his rather provocative book, *The New Wild: why invasive species will be Nature's salvation*, journalist Fred Pearce does not hesitate to call certain urban wastelands new biodiversity hotspots. He writes: "Feral urban Britain turns out to be a wildlife paradise" (Pearce 2016, p. 167) Herein lies the real novelty of the recent reappraisal of wastelands on ecological grounds: the idea that these "badlands" could now embody one of the key issues in nature conservation policies.

This recognition of the "intrinsic" value of wastelands is reinforced by a growing number of works in the social sciences that shed light on the different forms of social life or "assemblages" that have developed in these places. Indeed, a common feature of these studies is their tendency to describe the types of social organization constructed in wastelands no longer as marginal ways of living, but as the most appropriate ways of living in a hybrid and degraded world. In other words, as the entire planet enters a regime of environmental instability, the best way to learn to live in the Anthropocene would be to look at social activities developed in uncertain "historical" places such as wastelands. As anthropologists Anna Tsing and Nils Bubandt put it in the introduction to a volume devoted to what they call "feral dynamics:"

> If we, as a species, want to survive the industrial infrastructures we moderns have made, we as researchers need to understand how more-than-human remaking of engineered landscapes occur (Bubandt and Tsing 2018, p. 3).

Moreover, the thesis has been eloquently defended in Anna Tsing's bestseller *The Mushroom at the End of the World*, where the anthropologist describes the collaborative survival of multispecies collectives within damaged landscapes, which she calls "third nature." This is what Tsing claims:

> Precarity once seemed the fate of the less fortunate. Now it seems that all our lives are precarious—even when, for the moment, our pockets are lined. In contrast to the mid-twentieth century, when poets and philosophers of the global north felt caged by too much stability, now many of us, north and south, confront the condition of trouble without end (Tsing 2017a, p. 21).

Thus, to cope with global precariousness on Earth, we may have to adopt new social values, such as transience, impermanence, mobility, and malleability. Like the matsutake hunters who managed to find new resources in the damaged forests, we would be condemned to adapt to life in the "ruins of capitalism" and to compose resilient lifestyles under these conditions. As the title of a collective book edited by Anna Tsing suggests, we must learn the "arts of living on a damaged planet" (Tsing 2017b). In order to do this, we need to look for the practitioners of these arts where they are, that is, in the various badlands created by industrial modernity.

In short, in recent years, a cluster of ecological, social, and political motivations has tended to place wastelands at the center of attention concerning human and non-human survival in our precarious times. But could wastelands thus move from the margins to the center without losing what makes them worthy of interest?

4 Wastelands at the Center of the Arts of Living in a Precarious World

While wastelands have long been disregarded as pre-modern remains, postmodern thinking has worked to reassess them, until, in recognition of the general threat to living environments in the Anthropocene, they have recently been described as central scenes in environmental and social thinking. From this perspective, wastelands are no longer remnants, they are all that remains in the Anthropocene or in the Capitalocene

(Haraway 2015). In a world in ruins, the ecological and social life forms that emerged from wastelands would be the most appropriate means of inhabiting the Earth. But, if the hypothesis is intellectually challenging, we may wonder whether it is not also risky from an ecological and social point of view. In the remainder of this section, I will examine these two questions.

First, from an ecological point of view, the re-evaluation of the biodiversity of wastelands and the enhancement of the "new nature" that would flourish in the badlands have raised concerns among some environmentalists. Some thinkers and actors feared that the enthusiastic plea for urban wilderness would be paid at the price of a relative disengagement from the battle for nature conservation. Designation of urban badlands as "new biodiversity hotspots" could prevent people from seeing the catastrophic effects of destroying "classic" hotspots such as tropical forests. In summary, the criticism of celebrating the resilience of nature, which would be exemplified by urban biodiversity, is based on the fear that it could weaken the conservation movement. The French philosopher Virginie Maris addresses this point at length in her latest book "The Wild Part of the World" (Maris 2018). Noting the significant development of urban and wastelands ecologies, she points out that:

> in a pendulum swing so frequent in the history of ideas, the legitimate concern for reconnection and ordinary nature is on the verge of marginalizing the pleas for more spectacular, wilder and more distant nature.

And she adds:

> to abandon and discredit the defense of wilderness, to focus attention and resources on ecosystem services, socio-ecosystems, reconnection, urban and peri-urban ecosystems, there is a great risk of allowing the range of possibilities to be progressively impoverished, of admitting without paying attention that, year after year, generation after generation, the reference point that defines the wildest part of the world is approaching the degraded state of intensively exploited ecosystems to the point of disappearance for good (Maris 2018).

In a more severe criticism, the philosopher Frédéric Neyrat had previously stressed the reversibility of the following two "anthropocenic" assertions: "there is nothing wild anymore" or "wild nature is everywhere," even in the most anthropized places in the world, such as the weeds growing on the pavements or the spontaneous nature of the badlands (Neyrat 2018). In both cases, there is no need to worry about wilderness conservation. Proclaiming that nature is dead or very resilient leads to the same conclusion that we do not need to worry about destroying wilderness, since it is either too late or unnecessary.

In summary, these critics fear that descriptions of the rebirth of wilderness in urban wastelands tend to call into question the recognition of the devastation of natural environments in the Anthropocene. They therefore wish to warn against an overly optimistic view of the potential of urban wastelands for wilderness conservation. In their minds, placing urban wastelands at the center of attention could lead to seeing the glass as a hundredth full when it is ninety-nine hundredths empty.

With regard to social issues, Frédéric Neyrat, in his same book, raised a similar concern about the social effects of an unqualified valuation of a mixture of instability and resilience. Challenging what he sees as a controversial attempt to apply "chaos theory" to societies, he argues:

> Everything is unstable, so why should we demand social security or any kind of insurance from the State? Within such a theoretical framework, resilience is nothing more than that which makes humans change in order to better adapt by force to economic, social, and ecological disasters without ever seeking to get at the heart of the primary causes of these disasters (Neyrat 2018).

Do the environmental humanities developed in recent years participate in such a theoretical framework? It is not so clear, but a further look at the work of Anna Tsing might help to clarify this point. In her above-mentioned book, the anthropologist explicitly told her readers that she had written about "[her] travels with mushrooms to explore indeterminacy and the conditions of precarity, that is, life without the promise of stability." She is undoubtedly cautious about the normative conclusions that might be drawn from reading the book, stating that: "To follow matsutake guides us to possibilities of coexistence within environmental disturbance. This is not an excuse for further damage." Yet, a few sentences are more ambivalent in the text, like when Anna Tsing claims that "precarious living is always an adventure" (Tsing 2017a, p. 219). This nuance of romanticism introduces a doubt about the meaning of her interpretation of "the possibility of living in the ruins of capitalism." The ambivalence stems from the fact that life in these ruins appears in the book both as a terrible fate imposed by the global devastation of the planet and as a desirable form of social life in the Anthropocene. This duality is reinforced in the collective book co-edited by Anna Tsing, entitled *Arts of living on a Damaged Planet* (Tsing 2017b), and significantly divided into two parts, respectively, devoted to the "ghosts" that haunt the ruins of capitalism and the "monsters" that are still able to create the arts of living in the Anthropocene.

In his review of Anna Tsing's book, the philosopher Jedediah Purdy sharply criticizes what he considers to be a misdirection. For him, the precarious life in the ruins of capitalism could hardly constitute a new emancipatory narrative. Purdy wonders:

> Tsing has unexpected praise for precarity, the insecurity that increasingly defines the world's economies. A theory-head word for "precariousness," precarity is usually a target for the left, implying as it does gig-based careers, no unions, uncertain pensions, and lifelong work and worry (Purdy 2015).

In other words, Purdy fears that the invitation to honor the semi-spontaneous "assemblages" that allow survival in what Tsing calls a "time of diminished expectations" could be interpreted as a call for capitulation in the face of neo-liberal globalization. He goes on to say:

> An ethics of precarity is too close to taking art photographs of decay in a city we cannot save. [...] It is too soon, and, more important, it surrenders too much, to make ruin our master-metaphor (Purdy 2015).

In the end, Purdy is certainly unfair in his interpretation of Tsing's works, since she has explicitly departed from overly optimistic readings of her book. Yet he rightly points out the drawback of the emergence of a new taste for ruins in the Anthropocene which seems to claim in a Thoreauvian manner that "in the wastelands is the salvation of the World."

5 Conclusion

A growing number of people today live in areas impoverished by capitalist use of the natural world. As we go deeper into the Anthropocene, this number will continue to grow. The transformation of living conditions on Earth concerns not only human beings, but also non-human beings. While the human footprint is everywhere on Earth, the environmental humanities are right when they call for thinking beyond the nature–culture dichotomy to invent new collective ways of inhabiting the world. Our worlds are certainly made up of tangles of humans and non-humans or assemblages of social and ecological processes. Yet there are still many different entanglements or assemblages, which vary according to many social and ecological criteria, including their internal diversity, temporality, and stability. The assumption that we live in a precarious world may tend to hide this plurality. In this sense, the general idea of "life in ruins" does not capture in a descriptive way the diversity of living spaces that still exist in the Anthropocene. Moreover, on a normative level, as Jedediah Purdy might have said, making wasteland our "master metaphor" overestimates the positive results of precariousness and ephemerality. This would bring with it the promise of general instability that could hardly satisfy those who still believe that a just society is based in part on the guarantee of social and environmental protection.

Ultimately, while recent research on wastelands has considerably improved the understanding of contemporary changes, the same question arises when wastelands are defined as a kind of "model," whether it is a biodiversity "hot spot" in brownfield ecologies or a social "ideal type" in environmental humanities. In the Anthropocene, as before, wastelands take on their meaning and value in their capacity to embody the counterpart of socio-ecological orders. They are parts of the world that resist social and ecological norms. We must be careful not to eliminate this critical resource by constituting it into a new order.

References

Benjamin W (1992) Charles Baudelaire: a lyric poet in the era of high capitalism. Verso, London

Bonthoux S, Brun M, Di Pietro F, Greulich S, Bouché-Pillon S (2014) How can wastelands promote biodiversity in cities? A review. Landscape Urban Plan 132:79–88. https://doi.org/10.1016/j.landurbplan.2014.08.010

Bubandt N, Tsing A (2018) Feral dynamics of post-industrial ruin: an introduction. J Ethnobiol 38:1. https://doi.org/10.2993/0278-0771-38.1.001

Callicott JB, Nelson MP (2008) The wilderness debate rages on: continuing the great new wilderness debate. University of Georgia press, Athens

Callicott JB, Nelson MP (1998) The great new wilderness debate. University of Georgia Press

Cannon J (2017) La zone entre classes laborieuses et classes dangereuses: les marges parisiennes de la Belle Époque à la fin des années 1970 [The zone between the working classes and the dangerous classes: the Parisian margins of the Belle Époque at the end of the 1970s]. Espaces Et Sociétés 171:37–54

Cronon W (1998) The trouble with wilderness, or, getting back to the wrong nature. In: Callicott JB, Nelson MP (eds) The great new wilderness debate. University of Georgia Press, Athens, pp 471–499

Foreman D (2008) The Real Wilderness Idea. In: Callicott JB, Nelson MP (eds) The wilderness debate rages on: continuing the great new wilderness debate. University of Georgia Press, Athens, pp 378–397

Gandy M (2013) Marginalia: aesthetics, ecology, and urban wastelands. Ann Assoc Am Geogr 103:1301–1316. https://doi.org/10.1080/00045608.2013.832105

Haber S (2006) Critique de l'antinaturalisme: études sur Foucault, Butler, Habermas. Pratiques théoriques [Critique of antinaturalism: studies on Foucault, Butler, Habermas. Theorical practices]. Presses universitaires de France, Paris

Haraway D (2015) Anthropocene, capitalocene, plantationocene, chthulucene: making kin. Environ Human 6:159–165. https://doi.org/10.1215/22011919-3615934

Haraway DJ (2016) Staying with the trouble: making kin in the Chthulucene, Experimental futures: technological lives, scientific arts, anthropological voices. Duke University Press, Durham

Hugo V (2010) Les Miserables. Vintage Book, London

Latour B (2004) Politics of nature. Harvard University Press

Lefebvre H (1968) Le droit à la ville [The right to the city]. Anthropos, Paris

Lizet B (1989) Naturalistes, herbes folles et terrains vagues [Naturalists, wild grasses, and wastelands]. Ethnologie Française (nouvelle Série) 19:253–272

Lizet B, Wolf AE, Celecia J (1999) Sauvages dans la ville [Wild in the city], Museum Histoire Naturelle ed. Paris

Lorimer J (2015) Wildlife in the anthropocene: conservation after nature. University of Minnesota Press

Lyotard JF (1984) The postmodern condition: a report on knowledge, theory and history of literature. University of Minnesota Press, Minneapolis

Mabey R (2010) The unofficial countryside. Dovecote, Wimborne Minster

Maris V (2018) La part sauvage du monde - Penser la nature dans l'Anthropocène [The wild part of the world. Thinking about nature in the Anthropocene]. Seuil, Paris

Mathis CF, Pépy ÉA (2017) La ville végétale : une histoire de la nature en milieu urbain (France, XVIIe-XXIe siècle) [The vegetal city: a history of nature in an urban environment (France, 17th–21st century)]. Éditions Champ Vallon, Paris

McKibben B (2006) The end of nature. Random House Trade Paperbacks, New York

Merriman JM (1991) The margins of city life: explorations on the French urban frontier, 1815–1851. Oxford University Press, New York

Muratet A (2017) Flore des friches urbaines du nord de la France et des régions voisines [Flora of urban wastelands in northern France and neighboring regions]. Xavier Barral, Paris

Nash R (1967) Wilderness and the American mind. Yale University Press, New Haven

Neyrat F (2018) The unconstructable earth: an ecology of separation. Fordham University Press, New York

Pearce F (2016) The new wild: why invasive species will be nature's salvation. Beacon Press

Plumwood V (1998) Wilderness skepticism and wilderness dualism. In: Callicott JB, Nelson MP (eds) The great new wilderness debate. University of Georgia Press, Athens, pp 652–690

Purdy J (2015) The mushroom that explains the world. The New Republic

Smith J (2017) The new nature writing: rethinking the literature of place. Bloomsbury Academic. https://doi.org/10.5040/9781474275040

Sukopp H, Hejný S (eds) (1990) Urban ecology: plants and plant communities in urban environments; lectures and posters. Presented at the International Botanical Congress, SPB Acad Publ, The Hague, The Netherlands

Tallmadge J (2004) The Cincinnati arch: learning from nature in the city. University of Georgia Press

Tsing AL (2017a) The mushroom at the end of the world: on the possibility of life in capitalist ruins

Tsing AL (ed) (2017b) Arts of living on a damaged planet. University of Minnesota Press, Minneapolis

van Dooren T, Kirksey E, Münster U (2016) Multispecies studies: cultivating arts of attentiveness. Environ Human 8:1–23. https://doi.org/10.1215/22011919-3527695

Wordsworth W (1995) The prelude: the four texts (1798, 1799, 1805, 1850), Penguin classics. Penguin Books, London, New York

Rémi Beau is a Research Fellow in philosophy at the CNRS (National Center for Scientific Research, France; IEES-Paris, Sorbonne University). As a specialist in environmental philosophy, he works on the notions of ecocentrism, moral perfectionism, and the links between environmental ethics and political ecology. In 2017, he published a book entitled *Ethics of ordinary nature. Philosophical research in fields, wastelands and gardens* (*Éthique de la nature ordinaire. Recherches philosophiques dans les champs, les friches et les jardins*, Paris, Éditions de la Sorbonne) and co-edited in 2018, with Catherine Larrère, the collective work *Thinking the Anthropocene* (*Penser l'Anthropocène*, Paris, Presses de Sciences Po). He has also recently co-edited the book *Handbook of the Great Transition. Training to Transform* (Cécile Renouard, Rémi Beau, Christophe Goupil, Christian Koenig: *Manuel de la Grande transition. Former pour transformer*, Paris, Les Liens qui libèrent, 2020).

Afterword: How to Treat Urban Wastelands as a Form of Nature?

Francesca Di Pietro and Amélie Robert

From these studies of urban wastelands, it emerges that, whatever the country considered, the nature of urban wastelands gives rise to contrasting and changing views depending on social groups, but not so much on the cultural areas considered, as urbanization masks regional cultural particularities.

From the ecological point of view, many species use urban wastelands as a refuge from the urban environment, but also as a discontinuous corridor. In contrast with regulated public green spaces, urban wastelands are species-rich and represent a hotspot of biodiversity in cities. They can also be a hotspot for exotic invasive species, especially in industrial brownfields; but exotic species better tolerate high levels of metals in the soil, and even provide ecosystem services with bioremediation processes in this plant community.

From the social point of view, urban wastelands are used by some residents as a green space and/or a refuge or a place to live. It is a deserted space, on the fringes, where people who cannot find their place elsewhere in the city find refuge and are "marginal," like these spaces themselves. There are several reasons for this use: the vegetation, which provides a hideaway from the city; the lack of public (green) space; or because the informal vegetation of urban wastelands, in contrast to the managed vegetation of public green spaces, is appreciated by some (but rejected by others): areas of freedom, and of free expression too. We wished to highlight the informal uses to which these spaces are subject, uses that are often denied; the term urban wasteland itself has sometimes been used to deny these informal uses and to accelerate changes in land use (chapters of Mazy, Verdelli).

Urban wastelands are temporary informal green spaces. From the social and ecological points of view, they partially offset the scarcity of green spaces in cities, especially in working-class neighborhoods (where their transformation into public

F. Di Pietro (✉) · A. Robert
UMR CITERES-CNRS, Université de Tours, Tours, France
e-mail: dipietro@univ-tours.fr

A. Robert
e-mail: amelie.robert@univ-tours.fr

F. Di Pietro and A. Robert (eds.), *Urban Wastelands*, Cities and Nature,
https://doi.org/10.1007/978-3-030-74882-1

green spaces can lead to gentrification). In fact, they are an alternative to the tamed nature of public green spaces.

In order to protect urban wastelands from the developers, should they be included in urban planning and, if so, how? Most of the chapters of this book propose planning suggestions. Four main directions are suggested:

(1) **Include urban wastelands in a global planning project**. Based on a local project, McKinney suggests specific design and planning strategies involving the acquisition of vacant lands. Mathey and Roessler state that "we need ways of preserving these sites or defining them as official green spaces to incorporate such areas into existing networks of urban green infrastructure [...]. However, such green spaces must be well managed in order to answer the needs of the residents for orderliness and security." Mazy suggests planners should expand port-city interfaces as true interstitial breathing spaces within urbanized spaces and exploit the role of these spaces in dense urban environments as a secondary function for ports. Brun and Di Pietro point out that ecological mapping for green networks, in addition to integrating specific urban areas, should consider temporary spaces of nature, such as wastelands. Vanbutsele proposes that urban wastelands be integrated into the park system in order to preserve urban wastelands as informal open spaces, connecting them with metropolitan green continuities could be essential.

(2) **Stimulate public participation**. Kim and Rupprecht suggest participatory management by the residents: management by the local community and user participation. Petit-Berghem and co-authors also suggest giving a voice to the inhabitants. Mathey and Roessler specify that "if unmanaged wastelands are to be included in a development concept, an unusual degree of public communication is required along with a participative process. A vital precondition here is to include citizens in the planning and preparation of the design process and motivate them to get involved. In this way, areas of succession could be better accepted and more closely integrated in future plans for urban wastelands in residential areas."

(3) **Not manage them**. Muratet et al. suggest not intervening, but informing residents. The same idea is defended by Muci and Dorso, who also suggest non-action. They point out that a form of non-action (withdrawal, restraint) has favored the presence and resilience of this urban vegetation, whose positive aspects in biological, spatial, or social terms are taken into consideration, and they envisage this non-action as a possible form of action. "An action that is certainly unorthodox and viable in a different way, and an action that remains ambiguous," evoking the "taboo of non-action."

(4) **Study nature and the uses and perceptions of urban wastelands**. Tritsmans suggests that the use and perception of unofficial space by the citizens is crucial in the future development of research on public spaces within the field of urban history. Moreover, Gallagher observes that urban wastelands call for a change of ecological paradigm.

As we stand at the crossroads of the evolution of thinking in planning and ecology, urban wastelands challenge our perception of nature.

Francesca Di Pietro As Associate Professor of Environmental Geography at the University of Tours (France), in the Planning & Environment Department, **Francesca Di Pietro** works on plant biodiversity in anthropogenic environments, at first in agricultural landscapes, later in urban ones. Her research focuses on plant communities of urban habitats such as public green spaces, urban wastelands, roadsides, and community and individual gardens. Since 2011 she has worked on urban wastelands, coordinating work on their plant communities, informal uses, and how they are perceived.

Amélie Robert is currently a Contract Researcher and Part-Time Lecturer at the University of Tours. Her research focuses on landscape dynamics, the identification of their drivers and ecosystem services, from a geohistorical perspective. More generally, she questions the interrelationships between societies and their environments in different contexts. She pays special attention to novel ecosystems, particularly in urban contexts, having contributed to several research programs in this field.